THIRD EDITION

GODS IN THE
GLOBAL VILLAGE

Titles of Related Interest

THIRD EDITION

GODS IN THE GLOBAL VILLAGE

The World's Religions in Sociological Perspective

LESTER R. KURTZ

George Mason University

Los Angeles | London | New Delhi
Singapore | Washington DC

Los Angeles | London | New Delhi
Singapore | Washington DC

FOR INFORMATION:

SAGE Publications, Inc.
2455 Teller Road
Thousand Oaks, California 91320
E-mail: order@sagepub.com

SAGE Publications Ltd.
1 Oliver's Yard
55 City Road
London EC1Y 1SP
United Kingdom

SAGE Publications India Pvt. Ltd.
B 1/I 1 Mohan Cooperative Industrial Area
Mathura Road, New Delhi 110 044
India

SAGE Publications Asia-Pacific Pte. Ltd.
33 Pekin Street #02-01
Far East Square
Singapore 048763

Acquisitions Editor: David Repetto
Associate Editor: Maggie Stanley
Editorial Assistant: Lydia Balian
Copy Editor: Megan Markanich
Typesetter: C&M Digitals (P) Ltd.
Proofreader: Joyce Li
Indexer: Judy Hunt
Cover Designer: Anupama Krishnan
Marketing Manager: Erica DeLuca
Permissions Editor: Karen Ehrmann

Copyright © 2012 by SAGE Publications, Inc.

Printed in the United States of America

Library of Congress Cataloging-in-Publication Data

Kurtz, Lester R.
Gods in the global village: the world's religions in sociological perspective/Lester R. Kurtz.—3rd ed.

p. cm.—(Sociology for a new century series)
Includes bibliographical references (p.   ) and index.

ISBN 978-1-4129-9125-4 (pbk.)
1. Religion and sociology. 2. Religious pluralism.
3. Religions. I. Title.

BL60.K87 2012
306.6—dc23   2011039706

SUSTAINABLE FORESTRY INITIATIVE   Certified Chain of Custody
Promoting Sustainable Forestry
www.sfiprogram.org
SFI-01268

SFI label applies to text stock

This book is printed on acid-free paper.

11 12 13 14 15 10 9 8 7 6 5 4 3 2 1

Contents

About the Author

Lester R. Kurtz is Professor of Public Sociology at George Mason University, where he teaches the comparative sociology of religion, peace and conflict, social movements, globalization, and social theory. He lectures regularly at the European Peace University and was previously director of religious studies at the University of Texas at Austin. He holds a master's in religion from Yale Divinity School and a PhD in sociology from the University of Chicago. He is editor of the three-volume *Encyclopedia of Violence, Peace, & Conflict* (Elsevier), coeditor of *Nonviolent Social Movements: A Geographical Perspective* (Blackwell), and coeditor of *The Web of Violence: From Interpersonal to Global* (University of Illinois Press). He is the author of numerous books and articles on religion and conflict, including *The Nuclear Cage: A Sociology of the Arms Race* (Prentice Hall) and *The Politics of Heresy: The Modernist Crisis in Roman Catholicism* (University of California Press), which received the Society for the Scientific Study of Religion's Distinguished Book Award. He is currently working on books titled *Gandhi's Paradox* and *Gods and Bombs*.

Dr. Kurtz is the past chair of the Peace and Justice Studies Association as well as the Peace, War, and Social Conflict Section of the American Sociological Association, which awarded him its Robin Williams Distinguished Career Award in 2005. He has lectured in Europe, Asia, Africa, and North America and has taught at the University of Chicago, Northwestern University, Delhi University in India, and Tunghai University in Taiwan.

*Dedicated to Jeannie Kurtz and the
memory of Merwin Kurtz, my first teachers*

Preface

T he Buddha is said to have argued that if a house is on fire we should not sit around debating how to put it out but should set to work immediately. A scholar's inclination, however, is to think carefully about a problem before writing about it, let alone acting on it.

Humanity's common house is on fire in a very real sense: Even after the end of the Cold War and its reign of nuclear terror, we live in a time of acute crisis, especially in the age of terrorism and the widespread war against it as well as numerous civil conflicts around the world. Despite our technological advances and abundant natural resources, millions of people—especially children—die of starvation every day and millions more are malnourished. Wars and various armed conflicts between religious and ethnic communities plague the planet, which is still booby-trapped for self-destruction with thousands of nuclear weapons. Peoples around the world find their familiar societies being torn apart by a rapid globalization of the economy and culture of the human community that is producing a dizzying pace of change. Earth itself has been seriously abused by civilization, and it is not clear that we have the will to stop further damage, let alone repair what we have already done.

This book focuses on a central aspect of that common crisis—the relationship among the major faith traditions that inform the thinking and ethical standards of most people in the emerging global social order. It would be a better book if I spent another 25 years revising it, but the topic's urgency justifies this preliminary attempt to use sociological tools to assess the state of religious life in a globalizing world. Trying to synthesize the material in this book has been a humbling experience, and I hope that my readers will accept it in the spirit in which it is offered—as a tentative analysis, a first step in a process that will require more research and more action. This new edition provides a more accurate picture of the human communities' religious beliefs and practices than the previous attempts. One major addition is a chapter on indigenous religious written by Bonnie Mitchell-Green, who is well versed in Native American traditions.

Since Max Weber's plea in 1918 for a value-neutral sociology and the professionalization of academia and the social sciences after World War II, sociologists have been divided about the manner in which they may properly address social issues—or whether these issues should be addressed at all. Mainstream sociology, especially in the United States, has tended to favor the objective, dispassionate approach avoiding both personal biases and political positions according to the canons of science. Alongside this trend, however, a more critical sociology has persisted since the founding of the discipline, gaining added momentum in the 1960s, when questions were raised about not only the morality but even the possibility of objectivity in the study of pressing social problems. *Gods in the Global Village* falls clearly in the latter camp; it employs the scientific model to investigate religious life but does not pretend to be entirely value free. The pursuit of objectivity by bracketing one's own biases is important—it is the truth we are after, rather than evidence to bolster our own prejudices—but we always start the quest with our own presuppositions, and they determine the trajectory of our inquiries. It is necessary, therefore, to begin our argument by outlining some underlying assumptions.

A major assumption of this book is that all knowledge is shaped by the social context of the knower; therefore, both religious traditions and our studies of them are shaped by the context in which we construct them. That is not to say that the pursuit of objectivity is not valuable—we must always try to overcome our own biases in our pursuit of the truth—but that we will never fully attain it. We cannot know the world in and of itself but only as it is filtered through the categories of the mind. Like other scholars, sociologists cannot pretend to be objective because their biases, explicit or implicit, will shape their work; nonetheless, all scholars should attempt to transcend their social and disciplinary contexts. This apparent relativism does not mean that there is no absolute but only that our knowledge of it is always imperfect.

A second assumption is that religious pluralism will be a necessary precondition of the global village for the foreseeable future. The question that faces us as a human community is not "Which religious tradition is true?" or even "Is any religious tradition true?" but rather "How can we enable the various religious and secular traditions to coexist peacefully on the planet?" Mahatma Gandhi believed that even tolerance was inadequate—we must learn from and respect others, even our enemies.

A third assumption is the belief that the sociology of religion—itself a pluralistic discipline—can provide invaluable insight into the most pressing problems of our time. That is not to say that sociologists should therefore lead the way in solving these problems (as Auguste Comte thought) nor that they should simply loosen the soil for contemplative thought (as Max Weber contended). Rather, sociologists should use their analytical tools to assess the role of religion in the global community and become involved in the lively debates about the future of humanity that will ensue.

Finally, in the classroom, I believe it is important to inform my students of my biases from the beginning so that they do not have to play guessing games. I will do the same here, because I think my background shapes what I see and how I interpret it. I am the son of a United Methodist preacher in Kansas, from a long line of clergy ancestors. I became politically conscious in the 1960s under the instruction of my parents, who took me on a civil rights march in Wichita, Kansas, when I was in junior high school. I continue to be a practicing Christian although I have reinterpreted some of the church's doctrines in my own way and leave the details of theology to some anticipated future revelation in which, as the Qur'an puts it, God will inform us on all that wherein we have differed. My comparative study of the world's religions, along with time spent living in India and Taiwan, has further shaped my personal beliefs, as did 2 years working on a master of arts in religion at Yale Divinity School. My wife Mariam (named after the mother of Jesus) is Muslim, and my family regularly attends the United Methodist Church, as well as participating in religious ceremonies from various faiths. At our Muslim wedding at the Abraham mosque on the eve of an interfaith conference, I declared "I believe in one God and Muhammad is God's messenger." The story about our wedding in an Istanbul newspaper reported my comment about claiming dual citizenship, Christian and Muslim (although not without controversy). I am convinced from my own experience and intellectual inquiries that cultural innovation comes from combining opposites at crucial moments in history and that one's personal and intellectual lives can complement and enrich each other. Humanity's current quagmire requires cultural innovation and spiritual quests that—like Joseph Campbell's *Hero With a Thousand Faces*—move beyond the ordinary routines of conventional life in search of new insights and benefits for humanity. The astute reader will detect all of these shaping influences in the pages that follow.

Acknowledgments

A s the author of this book, I am responsible for its errors; its correct insights are only partially mine. The list of contributors to the work's form and content would fill several hundred pages and would include everyone from my family to authors whose work informed me to teachers both past and present. Others to single out are the editors of the *Sociology for a New Century* series, starting with the first round's Larry Griffin, Charles Ragin, and especially Wendy Griswold, whose guidance was invaluable. Of the SAGE team, Steve Rutter became as much a valued friend and colleague as a publisher, as did Rebecca Holland. I am grateful for Ben Penner then David Repetto, who picked up the banner and encouraged me, and for Annie Louden followed by Maggie Stanley, who patiently called in all the chapters and details. Editor Victoria Nelson reshaped every page of the first edition with care as Taryn Bigelow did for the second and Megan Markanich on the third. Finally, hundreds of students—especially Katharine Teleki—energized this project in many ways, as did friends, colleagues, teachers, and strangers from around the global village, notably S. Jeyapragasam, Yuan Horng Chu, Steven Dubin, Christopher Ellison, Robert Herrick, Fred Kniss, Juan Linz, Edgar Polome, Edward Shils, Gideon Sjoberg, Teresa Sullivan, David Tracy, Stephen Warner, Andrew Weigert, Robert Wuthnow, and now my new colleagues at George Mason University.

I give special thanks to those who reviewed the book:

Valerie J. Gunter, Indiana University of Pennsylvania

Clark D. Hudspeth, Jacksonville State University

Ahmad Khalili, Slippery Rock University

Riad Nasser, Fairleigh Dickinson University

Claude A. Perrottet, University of Bridgeport

Kiran Sagoo, University of Hawaii at Manoa

Mary Sawyer, Iowa State University

Gerhard Schutte, University of Wisconsin–Parkside

Darren Sherkat, Vanderbilt University

Mark Shibley, Loyola University

Anson Shupe, Indiana University–Purdue University Fort Wayne

Kristen Wenzel, Sacred Heart University (2nd edition)

—Lester R. Kurtz

1

Religious Life
in the Global Village

Terrorism and the war on terrorism, as well as waves of nonviolent civil resistance to dictatorship, are accelerating many of the changes taking place on our planet and revealing the continued role of **religion** in our daily lives and in global politics. The whirlwind of change in our economic, social, political, and personal lives that we are experiencing may change the course of human history, but at the same time we are rooted in ancient faith traditions that both resist and promote social change. Never has the study of religion been more important or an understanding of the various traditions more crucial. Can we call ourselves educated or understand what is going on in today's world without having a basic knowledge of our shared faith traditions?

Fortune-tellers in China now provide computer-generated astrological charts. Telecommunication satellites link isolated religious communities at separate ends of the earth; American television offers its viewers Christian preachers and Buddhist teachers. In the summer of 1993, representatives of religious communities met at a Parliament of the World's Religions in Chicago to establish a process for ongoing dialogue and to initiate a debate on a declaration of a global ethic. The parliament had not met since 1893 but met again in South Africa in 1999 and in Spain in 2004. At lectures given by a Hindu teacher in Texas, a large color portrait of the Indian guru Sai Baba is framed by a vase of fresh flowers and a candle painted with an image of the Virgin of Guadalupe. In the middle of Colombo, Sri Lanka, sits St. Anthony's Shrine, a pilgrimage center for hundreds of thousands each week, 90% of whom are not Christians but Buddhists and Hindus. Capital, goods, and people move about the planet with increasing frequency, changing the spiritual as well as economic and political landscapes of our world.

The pleasant coexistence of religious traditions is only one side of the story, however, as anyone who follows the news already knows. The first decade of the 21st century—designated the decade of peace and nonviolence by the United Nations—started off with a bang, literally, apparently continuing a pattern of intercommunal violence well established in the previous century. In early 1994, for example, an Israeli doctor entered a mosque in Hebron at the Cave of the Patriarchs, where Abraham is supposedly buried, and murdered more than 30 Muslims at prayer. He was beaten to death on the spot by the worshippers, and violence broke out again between Jews and Muslims throughout the region. Meanwhile in India, Hindus and Muslims were killing one another in a flare-up of a centuries-old conflict, with current issues including the political status of Kashmir, the destruction of a Muslim mosque by Hindu nationalists at a disputed site in Ayodhya, and the development of nuclear weapons by India and Pakistan. Catholics and Protestants were fighting one another viciously in Northern Ireland. Militants were killing and injuring thousands in the name of Islam, while the U.S. administration called on the name of God to justify its war against terrorism and invasions of Iraq. The Ku Klux Klan still marches in the United States, using religious arguments to denounce African Americans, Jews, and others. In the former Yugoslavia, Serbian Orthodox Christians engaged in a campaign of "ethnic cleansing" of Muslims that involved wholesale slaughter.

The **global village** is becoming a reality economically and socially, if not politically, as every isolated corner of the planet is being knit together into a world system. This global order, emerging for several centuries, has become a reality in the 20th century, as all humans increasingly participate in a "shared fate" (Joseph, 1993; Wallerstein, 1984; cf. Durkheim, 1915/1965). Our economic and social institutions; our culture, art, music; and many of our aspirations are now tied together around the world. The human race, however, is constructing a multicultural global village full of conflict and violence as well as promise.

Just as the Cold War between the United States and the Soviet Union ended in the 1990s and we made astounding progress in solving old conflicts, ethnic and religious nationalism exploded in violence around the world. Mark Juergensmeyer (1993) contended that rather than witnessing the "end of history" (see Fukuyama, 1992) and the emergence of a worldwide consensus in favor of secular liberal democracy, we may see the coming of a new Cold War, one between the secular West and numerous new religious nationalisms. "Like the old Cold War," said Juergensmeyer (1993, p. 2), "the confrontation between these new forms of culture-based politics and the secular state is global in its scope, binary in its opposition, occasionally violent, and essentially a difference of ideologies." The new millennium ushered in not an end to the bloodshed of the last century but the terror of a new violence.

Social life may be fundamentally different in the coming century, although many features of today's life will persist, just as there was much continuity between preagricultural and agricultural eras, premodern and modern times. A major task of the coming millennium will be to order our lives together and to create an **ethos**, or style of life, with a moral basis. The ethos must include sufficient agreement about common norms to facilitate cross-cultural interactions, international commerce, and conflict resolution while permitting considerable cultural diversity on the planet. The process of coming together, however, will not be an easy one. Religious traditions are central to that process because of their role in defining norms, values, and meaning; in providing the ethical underpinning for collective life; and in forging the cultural tools for cooperation and conflict.

Much of the best and worst of human history is in the name of its Gods, Goddesses, and religious traditions that continue to provide both an ethical critique of, as well as a justification for, much bloodletting. The central thesis of this book is that the sociological study of religion has important insights into the central issues of how we can live together in our multicultural global village as well as helpful tools for investigating the problems created by our newly created common life with its diverse norms and values. The task here is to review those insights, assess the tools, raise questions, look at what does work, and develop some tentative conclusions about the role of religions in promoting chaos or community as humanity moves into the 21st century. Whether or not we can discover a means for sustaining a diversity of religious traditions and a wide range of ethical values and still live together remains an unanswered question.

The world's religions will be an integral part of the process, for better or for worse. Faith traditions "work" because they seek to answer fundamental questions in a comprehensive way. That strength, however, sometimes results in exclusivist claims to a monopoly on the "Truth," which, in a multicultural global village, often precipitates fatal conflicts among competing religious claims and the people who make them. The very things that hold a community together can also tear it apart.[1] The next few generations may decide, in this new global context, whether we continue to escalate the violence until the last shred of human civilization lies under the rubble or, alternatively, develop what Martin Luther King Jr. called a "beloved community" that ushers in an era of peace and justice.

Religion and the Globalization of Social Life

Our ancient ancestors sat around the fire and heard stories about their forebears—about the time when life first emerged in the universe, about lessons for living their lives. When people gather today, the flickering light usually comes from a television rather than a fire, but we still hear stories about the nature of reality as we perceive it in our own cultures. Many of Earth's previous inhabitants heard only one story about creation during their lifetime, but today most people hear more than one as the various religious traditions of the world—as well as newer scientific ones—diffuse widely through modern means of mass communication. We are surrounded by not only our own cultures but those of countless other peoples. Encountering these different perspectives on life is stimulating and enticing, but the overall process of cross-cultural contact is highly complicated because meaningful differences do exist among religions and sometimes provide the basis or excuse for confrontation.

Historically, religious ideas have provided the major organizing principles for explaining the world and defining ethical life for elites and masses alike, and they continue to do so, but modern critiques of religion have shaken them to the root. The globalization of our "lifeworlds" (Habermas, 1987) will have as great an impact on religious life as industrialization did. Just when humanity most needs an ethical system that enables diverse peoples to coexist peacefully and justly, the traditional sources of such guidelines are being daily undermined by the challenge of modern science and the increased cross-cultural contact.

Many conflicts occurred throughout the history of Christianity, of course, but none so radical as those precipitated by the crisis of modernism in the last two centuries. Scientific arguments called into question not just specific dogmas but the very notion of dogma. As the Roman Catholic pope put it in 1907, modernism lays "the axe not to the branches and shoots, but to the very root" of the faith (Pius X, 1907). Cross-culturally, meanwhile, competing religious traditions were offering alternative religious explanations to fundamental questions about life and how it should be lived.

Even before the changes in society and culture associated with industrialization had time to become fully absorbed, however, the world changed again—just as profoundly—when the various human communities were thrust into intimate contact by late-20th-century communications and transportation technologies and the globalization of an advanced capitalist economy that relies on far-flung networks of production and consumption. Most scholars in the 19th century predicted a new era of peace and prosperity; yet the 20th century brought bloodshed on a scale never before experienced and prosperity for a privileged few, accompanied by mass starvation and misery for many more.

The communications and transportation revolutions of the 20th century took off in the post–World War II era. By the time Marshall McLuhan (1960) introduced the term *global village* into our vocabulary, a new awareness of the interconnectedness of our lives was emerging. In the 1960s and 1970s, a massive increase in international trade transformed the nature of economic processes. Capital from the industrialized countries, in search of cheap labor, shifted to so-called "less developed" nations so that much of the actual production process moved outside the United States and Western Europe and into Third World countries. By the early 1970s, the 500 major U.S. corporations were making 40% of their profits abroad.

These economic changes were intertwined with dramatic transformations in the civil society and political spheres as well. In 1900, there were about 200 nongovernmental organizations in the world—that is, noneconomic institutions organized to take care of some aspect of human life. By 1990, the number had risen to 6,000 and rose another 50% to 9,000 by 1993 (Smith, 1999), creating a web of structures ranging from religious organizations to humanitarian, activist, and other civic organizations. Cultural diffusion, driven in part by economic developments, has resulted in a global greed for consumer goods among those people who can afford to participate in the system (and often a hope for participating among those who cannot afford to do so). In addition to nation-states, regional and international political alliances and institutions are playing an increasingly important role, right up to the United Nations, which functions as something of a quasi-state at the global level.

At the close of the 19th century, the sociologist Emile Durkheim (1893/1933) observed that the emerging world system of his day showed two separate and contradictory trends: (1) increasing unity and (2) increasing diversity. This insight proved to be an enduring one. Even as our lives become ever more intertwined, the people who exist in our everyday world are increasingly diverse. Most people live not in isolated homogeneous villages but in heterogeneous cities. International trade, global social networks, and telecommunications locate us all in the same shared space. Even rural villagers are linked in an unprecedented way to the world economy as they send and receive goods around the globe.

Most people are probably ambivalent about the new world order. Many enjoy the material benefits, but they have come at a high price—including the destruction of many of the world's indigenous cultures and radical

transformations of other societies as well as widespread ecological devastation. The past two centuries have seen violence and misery on an unprecedented scale, but a large portion of the world is healthier, eats better, and
lives longer than the royalty of past civilizations. In the 19th century, the
people with the most advanced technology, Western Europeans, subjugated
most of the rest of the world. In the 20th century, they began slaughtering
one another at an unprecedented rate as militarized conflict was industrialized and the technology of war created "total war," in which—for the first
time in history—all humanity is involved and all are potential victims.

The academic study of religion provides one valuable approach to a serious study of the dilemmas plaguing modern culture. In the 19th century,
science seemed to be replacing religion in the cultural centers of Europe; the
Christian church cast its lot with the monarchy and appeared to be dying
along with the old order. More than a century later, however, religion persists as a vital force in the world. Because of its persistent importance, the
study of religion remains central to any adequate understanding of the
nature of human life. The discussion that follows introduces the history of
the scholarly (mostly sociological) analysis of religion, some of the analytical
tools that can be used to explore current trends in religious life, and the
series of themes that will inform this book.

Religion and the Sociological Tradition

Nineteenth-century sociologists, even as they mistakenly anticipated the
imminent demise of religion, created a new approach to the study of
religion that is rich with insights relevant to our lives. By identifying the
very issues that define our present struggle, the intellectual quest initiated
by Durkheim, Weber, Nietzsche, Marx, and Freud can help us as we move
into a 21st century that teeters between destructive conflict and harmony.
Why study the sociology of religion? Because it offers theories and methods
for understanding a central fact of human life that grapples with core issues
of what it means to be human and also a part of a natural order (see Glock
& Hammond, 1973).

The creation of the discipline of sociology in the 19th century was largely
an effort to come to grips with the crisis of faith and the revolutionary turmoil of the post–Enlightenment West. With the modern sensibility came a
new level of self-consciousness about fundamental questions ordinarily
taken for granted or explained by religious tradition. This intense reflexivity
of the 18th and 19th centuries gave birth to the modern social sciences and
explains why the earliest social scientists attempted to use scientific methods

not only to explain social life but to create a new basis for morality as well. From Immanuel Kant, Adam Smith, Auguste Comte, and G. W. F. Hegel to Karl Marx, Sigmund Freud, and Emile Durkheim, most of the major European intellectuals of this period sought to formulate a scientific moral basis for collective human life that would replace the religious foundations of European culture.

It is no accident that three of those intellectual giants—Marx, Freud, and Durkheim—were Jews living in a culture built on a Christian tradition that was being widely challenged by science and competing religious perspectives. Cross-cultural encounters and social and cultural revolutions usually precipitate innovation, and the post–Enlightenment West was no exception. Moreover, the personal torments of these men—and of others who built the social sciences—were representative, in many ways, of the experience of millions buffeted by the storms of modernism.

The founder of sociology, the French philosopher Auguste Comte (1798–1857), was trapped between his traditional family—his father was a fervent Catholic and royalist who supported the monarchy and opposed democratic reform—and the rebellious democratic, anticlerical milieu he encountered when he left home. Comte became a champion of scientific inquiry, contending that antiquated theological thinking gave way first to metaphysics and then to science, or what he called "positive philosophy." He insisted on applying the methods of physics and the natural sciences to social life to construct a rational social order, solve the profound problems of human life, and elevate the intellectual over the rest of humanity with its coarser affective faculties.

Comte initiated a series of lectures outlining his master plan of human knowledge—the "Course of Positive Philosophy"—in which he proposed that a scientific sociology could solve the burning social problems of the day that the monarchy, the church, and the Revolution alike had failed to address. He insisted that religion was simply a residue from an earlier era and that science could replace it to everyone's benefit. Just as his lectures began, however, Comte suffered a mental collapse and attempted suicide. The vociferous opponent of religion and champion of science eventually became a practitioner of "cerebral hygiene" and refused to read anything but the medieval devotional classic The Imitation of Christ. In his final years, isolated from his peers, Comte founded a "Religion of Humanity" in which he championed affect over intellect.

Karl Marx (1818–1883) came from a long line of Jewish **rabbis**, but his father converted to Christianity as a compromise to advance his position in the predominantly Lutheran community in Germany in which he lived. Marx's own disenchantment with religion was fueled by what he saw as the co-opting of religion by elites to control the dependent classes. He attacked religious ideologies of repression with a moral passion, called on the oppressed to turn from theology to politics, and was marginalized and exiled for his views, as well as lauded by many seeking change.

Similarly, Sigmund Freud (1856–1939)—himself a victim of anti-Semitism in Vienna—saw religion as a psychological defense mechanism to compensate people for the deprivations they suffered as a consequence of social organization, such as the repression of sexuality and the channeling of energies into building civilization rather than meeting personal needs. Freud advocated that rational mastery of one's environment replace the "illusions" of religion.

Emile Durkheim (1858–1917), as the son of a French rabbi, made a sharp break with the Ashkenazi Jewish community that had nurtured him from infancy when he took up Auguste Comte's sociological banner. Durkheim's lifelong intellectual struggle to find a scientific substitute for the civic morality French culture lost when it rejected its Catholic past resonated deeply with his own personal trauma as he converted to the anticlericalism of the Parisian intellectual scene of the late 19th century.

German sociologist Max Weber (1864–1920) spent his life struggling with the contradictions between the religious faith of his pious Protestant mother and the secular bureaucratic world of his politician father. He became increasingly estranged from his father, who died shortly after a visit to the younger Weber's home, where they fought and Weber asked his father to leave. After this event, Weber fell into a deep depression for 5 years. When he recovered, he grappled with the issue that became the central agenda of his intellectual life: the tension between religious faith and modern Western rationality.

The groundwork for contemporary sociology of religion was thus laid by five men—(1) Comte, (2) Marx, (3) Freud, (4) Durkheim, and (5) Weber—who were personally as well as intellectually caught up in the broad historical currents of change sweeping through Western civilization (see Table 1.1 for more information). Religion was one of the first phenomena to occupy sociologists because it lay at the center of the intellectual, religious, and political controversies of their time just as it does today. When Comte coined the term *sociology* in the mid-19th century and advocated replacing the "arbitrary" authority of the church with a new authority based on science, Christian clerics branded sociologists as the devil's workers. Sociologists and practitioners of religion have enjoyed a love–hate relationship ever since. Although its foundation is European, the contemporary field is dominated by American scholars; fortunately, the field—like all scholarly endeavors—has started to expand beyond the Euro-American context to incorporate

Table 1.1 Major Early Scholars

Scholar	Dates	Nationality	Religious Heritage
Auguste Comte	1798–1857	French	Catholic
Karl Marx	1818–1883	German	Jewish
Sigmund Freud	1856–1939	Austrian	Jewish
Emile Durkheim	1858–1917	French	Jewish
Max Weber	1864–1920	German	Protestant

content and theories from other parts of the world as well, which is essential in order to be more accurate about how religion works in the global village. After all, most of the world's population is neither American nor European.

Tools of the Trade

Methods

The methodological tools sociologists use in studying religion are part of the intellectual heritage of Western European social thought from the 17th through the 20th centuries. Like the other social sciences, sociology draws from the post–Enlightenment concept of the scientific method, which requires the investigator to disengage as much as possible from personal biases to gather and interpret information about the world.

The first task of sociologists, like other social scientists, is to gather data and attempt to define their subject matter with clarity. Sociologists of religion utilize standard data-gathering techniques such as social surveys and interviews, ethnographies (direct observations of settings), and textual analysis (of religious writings, speeches, etc.). All scientists simply look for indicators of the reality they are exploring and instruments to measure the phenomena in question. Social science methodologies have become more sophisticated in recent years, and many scholars of religion make skillful use of traditional statistical methods. Because of the nature of their subject matter, however, sociologists who study religion tend to be eclectic in their approach. They often draw upon techniques from the humanities as well, and use such conceptual tools as **metaphors** and what Herbert Blumer (1954) calls a **"sensitizing concept"**—that is, one that "gives the user a general sense of reference and guidance in approaching empirical instances . . . [and] rest on a general sense of what is relevant" (p. 7). Some of the best work in the sociology of religion relies on metaphors, sensitizing concepts, and statistics.

Ethnography involves close-up observation, description, and analysis of a community or social setting over an extended period. It has long been a favorite method of anthropologists and qualitative sociologists and often involves narrative and discourse analysis—that is, looking for patterns in the writing and speech of a community. By examining the words and frames people use to explain the world and interpret events, ethnographers can learn much about how people think. Ethnography also involves observations of actions as well as words—what kinds of body language do people use and how do they organize themselves spatially? Do people stand close to one another when they talk or keep a distance from each other? Are there differences in how people relate to others who are of another gender, race, or class? If so, are those actions explained by the rhetoric of the actors? Do believers explain their behavior in religious terms or link it to religious beliefs and **sacred** texts? These are the kinds of questions easily explored with close-up ethnographic methods.

Textual and historical criticism allows scholars to examine historical or sacred texts of a religious community from a scientific or disinterested point of view. It is a favorite method of historians of religion, who sometimes do not have communities available to study with ethnographic methods, and has the advantage of allowing a scholar to see what people in religious institutions actually wrote or said. Much of it is now available on the Internet, including ancient primary documents in original languages and in translations. Comparative analysts can compare a wide range of religious traditions on a particular theme, like Joseph Campbell's (1968; cf. Campbell, 1988) famous study of *The Hero With a Thousand Faces*, which looks at the hero motif across the globe and notes similarities and differences. It would take many lifetimes to do such a study with ethnography, but it is possible to analyze the actual texts of many traditions.

Criticism of sacred texts is a standard method of religious studies scholars as well and is highly developed. Such scholars will look, for example, at the authorship, composition, and sources of a particular text, often comparing parallels in other texts, both within and outside of the tradition. A number of texts include flood stories, for example—not only the story of Noah, but also the ancient *Epic of Gilgamesh* and in Hindu, Aztec, and other traditions as well.

These methods not only are valuable in discovering the nature and history of faith traditions but also create considerable controversy when scholars find problems in comparing the texts. The lineage of Jesus, for example, is strikingly different in two books of the New Testament, Matthew and Luke (see Table 1.2). On the one hand, people who adhere to

Table 1.2 The Lineage of Jesus in Matthew and Luke

Matthew 1:1–16	Luke 3:23–38
Jesus	Jesus
Joseph	Joseph
Jacob	Heli
Matthan	Matthat
Elea'zar	Levi
Eli'ud	Melchi
Achim	Jan'na-I
Zadok	Joseph
Azor	Mattathi'as

biblical literalism—that is, those who believe that every word in the **Bible** comes from God and is literally true—may find such discoveries problematic and may discredit the method. On the other hand, some opponents of Christianity may use such comparison to attack the faith on religious and political grounds, missing the other spiritual meanings that believers find in the text. This issue, as we shall see later in the discussion of **modernism**, was the subject of great controversy in the 19th and early 20th centuries.

Different academic disciplines have affinities for particular methods of study. Anthropologists traditionally use ethnography, whereas historians often employ textual analysis. Other social scientists prefer statistical analysis. Sociologists use all of these different methods, depending on the particular subject they are studying and the customs of their training and personal preferences. In recent years, a "triangulation" of methods has become popular—that is, looking at a particular phenomenon from several methodological angles, such as statistical survey analysis, ethnography, and textual or discourse analysis. With the development of personal computers and sophisticated statistical methods, statistics have become particularly popular because they allow scholars to study broad patterns efficiently. The statistical manipulation of data can sometimes enable a scholar to discern patterns and trends in attitudes and behavior and to see broad relationships among different kinds of attitudes and behavior. One very interesting and successful area of inquiry in the sociology of religion in recent years is the study of the relationship between religious practice and health, which will be discussed in Chapter 4 (see, e.g., Ellison, 1999; Flannelly, Ellison, & Strock, 2004; Hummer, Ellison, Rogers, Moulton, & Romero, 2004).

Sociological Definitions of Religion

One of the first conceptual tools necessary to begin any intellectual inquiry is a good definition, but religion poses an obstacle. Weber admits that the term is impossible to define, at least at the beginning of a study (Weber, 1968; cf. Plock, 1987); William James (1902/1960, p. 46) similarly advises us not to look for a single essence but rather to explore the many characters of the phenomenon. Nevertheless, as Barbara Hargrove (1989, p. 21) correctly observed, we cannot beg the question of definition, because it will shape "the questions we ask, the behavior we observe, and the type of analyses we make." A website devoted to religious tolerance has an interesting compilation of various definitions of religion and a discussion of problems with them, for people interested in pursuing this issue further (see http://www.religioustolerance .org/rel_defn.htm).

The classic sociological definition comes from Durkheim (1915/1965, p. 62), who says that religion is a "unified system of beliefs and practices relative to sacred things, that is to say, things set apart and forbidden—beliefs and practices which unite into one single moral community called a Church, all those who adhere to them." Deleting the ethnocentric term *church* gives us a still serviceable definition that points to the three major sociological components or pillars of the academic study of religion. Religion, then, consists of the following three things: (1) the beliefs about the sacred, (2) practices (rituals), and (3) the community or social organization of people drawn together by a religious tradition.

A religious tradition's **worldview** is outlined in a set of interrelated beliefs that explain the world and guide people in living their lives. These ideas are expressed through narratives (myths and legends) that incorporate the oppositions and contradictions of life. They are reenacted and reinforced through rituals that are sustained by, and in turn provide legitimacy for, the institutions of each religious movement. Every religion has a system of beliefs about the world and what should be considered sacred or held in awe (Durkheim, 1915/1965) or what is of "ultimate concern" (Tillich, 1967) or "unrestricted value" (Hall, Pilgrim, & Cavanagh, 1986), which are expressed in narratives that encompass a wide range of possibilities. Viable religious traditions usually incorporate some answers to the fundamental questions of the meaning of life and how the world was created; they offer some comfort, perhaps even joy, in the face of suffering and death. They provide standards of Truth and Beauty and ways of seeing and interpreting the world, its seen and unseen forces. Ancestors and other significant figures usually weave such ideas together into a series of narratives that include stories about unusual encounters with the sacred experience. The beliefs of a religious tradition never stand in isolation, either from one another or from the life of the community. Any given religion is also part of a people's culture; in societies with little institutional differentiation, religion and culture are often essentially the same. Heterogeneous societies with multiple religious communities will develop a culture in which religion (even all religions represented in the society) constitutes only a part of the culture, especially when there is a secular state. The fundamental "truths" they contain are persistently recalled and reinforced in **ritual practices** that also sustain the social order. Rituals include religious festivals, rites of passage (including births, marriages, and funerals), and the like, which hold the spiritual and material world together.

The elements of religion we explore will no doubt continue to play a role in human societies. The form and content of these inherited traditions will persist, although they will also be transformed. When the world changes, so does tradition—even when presented as immutable truth.

Sociological Metaphors and Sensitizing Concepts

Sociologists have used a number of metaphors to describe elements of religious life—that is, analogies that sensitize observers of religion to aspects that might not be immediately apparent. Among the most common metaphors are four that will be employed throughout this book: (1) the **sacred canopy**, (2) **secularization** and the **religious marketplace**, (3) **elective affinities**, and (4) the Durkheimian "trinity" of beliefs, rituals, and institutions.

Constructing a Sacred Canopy

Sociologists of religion often see religious and cultural traditions as outcomes of the construction of what Peter Berger (1969) calls a sacred canopy over the life of a people, answering questions about meaning and "why" the world is the way it is. That is, they provide a sheltering fabric of security and answers for both the profound and the mundane questions of human life: What is the meaning of our existence? Why do people suffer and die? How can we get food for our family today? A particular social group's answers to these questions usually provide an overarching vision of the universe as well as a perception of how best to organize individual and collective life. That canopy may cover only a small subculture (such as a religious commune or an isolated tribe), or it may cover an entire national culture (such as Iran's). People may use it to legitimate either a resistance movement or a national elite.

Meant to function in a small, homogeneous society, these unilateral belief systems are difficult to construct and maintain in a multicultural society. In pluralistic societies such as the United States or India, the belief systems of social groups differ so sharply that even the most vague consensus is difficult to reach. Here, the canopy metaphor is not adequate for describing the rich religious life of the entire society unless we think of it either as sewn from numerous differing threads or as a patchwork quilt. Perhaps a "force field" would be a more appropriate metaphor than a canopy, because it implies a dynamic system that has a reality of its own but one that is constantly changing. I will, however, cautiously use the sacred canopy metaphor in this analysis because it has been widely used in the past 25 years, and it points to a key aspect of religious phenomena: Believers do try to construct a sacred canopy that shields them from the vicissitudes of life, and they often think they have succeeded. The problem with the metaphor is that it presents an image that is too static for this dynamic phenomenon: The canopy is never a finished product; its construction is a process that is constantly under way.

Whether a society is small and homogeneous or large and diverse, its religious symbols grow out of, and in turn act back upon, social life. Religion is a matter of what Berger (1969) calls **world construction**— that is, it is an attempt to make sense out of the universe. Although our natural and social worlds are given to us when we are born into them, humans are also cocreators of their world. Certain fundamental parameters (e.g., the law of gravity and the inevitability of death) impinge on us. Yet we continue to form our own interpretations of the ecosystem, creating perceptional models that significantly affect the reality outside us. Because we are all involved in world constructions, this active creation of ours is a dynamic process that continuously acts back upon us, its producers.

Religion is at the core of the world-constructing process because it involves the highest level of the process: what a people holds sacred. Berger (1969) suggested that the construction of the sacred canopy involves three basic elements: (1) **externalization**, (2) objectivation, and (3) internalization. The first element, externalization, is simply the ongoing outpouring of human beings into the world around them, both physically and mentally. In our daily lives, our thoughts and actions affect and shape the world in which we exist. Through our activity, we create **material objects** (e.g., buildings, machinery, toys, paintings, books) and **cultural objects** (e.g., theology, money, institutions, social networks; see Griswold, 1987) that change the world in some small or great way.

In the second stage of this process, our creations become objects external to us. This objectivation means that after we project our creations onto the world, they confront us, their original creators, as facts external to and separate from us. Sometimes authors who create fictional characters find that they lose some of their control over those "people" as the figures develop personalities of their own. In the same way, once we create an institution—a university, a corporation, or a church—it seems to take on a life of its own, functioning independently and sometimes even in opposition to its designers.

Finally, in the third stage of world construction, we reappropriate the reality that has become objective and transform it from structures of the external world back into structures of our subjective consciousness through internalization. In other words, we internalize the outside world through the process of socialization. In relating to other people, individuals learn to accept their culture's sacred canopy as a given and natural reality. Each society, according to Berger (1969), thus creates a *nomos*, a meaningful order that people impose upon the experiences and meanings of individuals and provides norms, or rules, for every situation and every social role. Traditionally, this process of constructing a worldview and nomos has been a religious quest, although it has become more self-conscious and dispersed since the arrival of the modern era. Broad theories of the universe and the ethical systems and rituals that grow out of them are now created by a variety of institutions, some traditionally religious, like churches, synagogues, and religious orders, and also by people who are deliberately independent of religious institutions, such as mythmakers and other shapers of culture, like writers, artists, scientists, journalists, entertainers, and intellectuals. Although the struggle to control the production of culture has been widespread throughout human history, clearly no one has a monopoly on it in the postmodern world (see Griswold, 1994). Religious leaders find themselves competing in a cultural marketplace even when they try to make exclusive claims to the truth.

Secularization and Religious Marketplaces

Since cultural and social diversity are the distinguishing characteristics of modern life, individuals or groups in the global village can choose their religious orientations from a variety of options rather than simply accepting the specific sacred canopy transmitted to them by their family and friends in early childhood. Thus, a second central metaphor in the study of religion is that of the religious marketplace and its related concept of secularization. Dissatisfied with the sacred canopy metaphor, scholars coined this term to emphasize the fact that in a multicultural society religious institutions and traditions compete for adherents, and worshippers shop for a religion in much the same way that consumers choose among goods and services in the marketplace (Warner, 1993); it is sometimes referred to as a supply-side theory of religion.

The secularization of society generally refers to the removal of responsibility and authority from religious institutions resulting in **privatization** of religion, although many sociologists, especially in the early phases of the field, believed that personal piety would disappear as well. These two concepts refer to basic responses to religion in the modern era, which we will discuss in more detail in Chapter 5, although it is helpful first to understand the role they played in the formation of the field as they have generated much research and debate.

For some scholars, especially early in the development of the social scientific study of religion, the concept of secularization was central to its understanding. As Rodney Stark (1999, p. 249) observed, "For nearly three centuries, social scientists and assorted western intellectuals have been promising the end of religion. Each generation has been confident that within another few decades or possibly a bit longer, humans will 'outgrow' belief in the supernatural." Unfortunately, the intellectuals who founded the field seem to have gotten it wrong—Stark contended that as "astounding as it may seem the secularization thesis has been inconsistent with plain facts from the very start" (p. 254). Not only has American church members increased over the last 150 years, he claimed, but there been no long-term decline in religious participation in Europe.

The secularization debate has two major issues: (1) Has the role of religions institutions become less important? (2) Have people become less pious, believing and participating less in spiritual practices? On the first question, it is not so difficult to see that religious institutions have become less significant in some spheres of life such as politics and economics, but even that is complicated. As Jonathan Fox (2008) noted in *A World Survey of Religion and the State*, although modernity is causing a decline in some parts of the religious economy, as has been predicted by many in the social sciences, "other parts of the religious economy are reacting to compensate" (p. 10). What Fox called the "modernization-secularization" theory, although noting that it includes two distinct ideas, will be discussed in more detail in Chapter 6. The second question, whether belief itself has declined, is more problematic.

As Warner (1993) correctly noted, the reality addressed by the religious marketplace metaphor in the American case is not so much economic viability as the disestablishment of religion. It turns out, in fact, that the secularization contested in the post–Enlightenment European sociology of religion is actually the exception worldwide rather than the rule. Europe's established Christian churches of the Middle Ages were something of an anomaly, though one often taken as a universal norm by European and American sociologists of religion and the extent of the Catholic monopoly may have been overdrawn (see Finke & Stark, 1992). Moreover, it is unlikely that medieval or early modern Europe was really the religious utopia people often think, as Stark (1999) showed in his review of detailed historical studies of the period. Far from revealing a pious culture, historians of the period have uncovered evidence of widespread lack of participation in church and remarkable ignorance of the basics of Christianity, even among the clergy.

Visitations by Anglican bishops and archbishops in 1738 led to reports of meager participation—in 30 Oxfordshire parishes, less than 5% of the population took communion during a given year. Moreover, church membership in specific congregations in Britain was only 12% of the population in 1800, rising to 17% in 1850, which is about the same percentage found in 1990 (Stark, 1999, p. 259).

Although religious perspectives may be relatively uniform in small, homogeneous societies, they are never so in heterogeneous ones and attempts to impose a single sacred canopy over such societies are never fully successful. The more pluralistic a society is, the more likely it is that people can choose their religious preferences. Students of American religion have thus found the marketplace metaphor helpful for examining various developments in the United States (Finke & Stark, 1992; Iannaccone, 1991; Lee, 1992; Stark & Bainbridge, 1985) in which people seeking religious experience make a rational choice among various "spiritual entrepreneurs" (Greeley, 1989). Indeed, the religious markets or economic interpretation of religion has taken on the character of a new paradigm especially in the study of American religion and other diverse societies (Turner, 2010). It is not easily applied to all situations, however, and takes different forms in some cultures where people are encouraged to practice multiple traditions simultaneously rather than make exclusive choices, as in many Asian cultures where people will be simultaneously Buddhist, Taoist, and perhaps Confucian or **Shinto** without seeing any contradiction in doing so. Even in Christianity, which tends to be relatively exclusivist, we find some practicing Buddhist of Hindu rituals such as meditation.

Sherkat and Ellison (1999) identified the following:

Two schools of rational choice thought about religion. Supply-side theorists emphasize the importance of constraining and facilitating factors on the collective production of religious value—and assume that underlying preferences for religious goods remain stable. Demand-side theorists highlight shifting preferences and the influence of social constraints on individuals' choices. At the heart of all rational choice perspectives is the market analogy applied to religion, and the following axioms are common to studies in this genre: that religious markets involve exchanges for general supernatural compensators—promises of future rewards and supernatural explanations for life events and meaning (Stark & Bainbridge, 1985). Like other commodities, religious goods are produced, chosen, and consumed. (p. 378)

A number of forces shape the religious market: individual preferences of consumers (see Iannaccone, 1990; Sherkat & Wilson, 1995); the process of cultural production and the creators of the narratives embodied in a tradition (Stark & Bainbridge, 1985); and the social world in which cultural constructions are found, including both the religious community with its norms that shape individual preferences (Ellison & Sherkat, 1999; Sherkat & Wilson, 1995) and the broader world in which that group exists. The marketplace metaphor has its shortcomings too, of course. Ethical systems and beliefs of ultimate concern are not bought and discarded as easily as shoes or houses, and ancient religious practices persist in the most advanced technological societies. Often seen as a rational choice model that views individual as free agents make rational decisions, this approach carries with it some of the intellectual baggage of that approach but also benefits from its insights.

As Iannaccone (1988, 1992) observed, subgroups gain from their distinctiveness in the religious marketplace, so that such issues as sacrifice and stigma—which are usually seen as costs when an individual is making choices—actually become benefits to a religious group that deliberately seeks tension with the dominant culture to provide participants with a distinctive identity (see Finke & Stark, 1992; Iannaccone, 1994; Stark & Bainbridge, 1985). Religious worldviews usually acknowledge that believers might incur costs or be labeled negatively for their beliefs but claim that future rewards will compensate them for any current sufferings. This insight suggests that religious particularism will thrive even as the globalization process intensifies; membership in a religious community labeled deviant by the mainstream becomes, for many believers, a way of protesting the trends of modernism and postmodernism, which they abhor. Whether they are located within the Islamic, Christian, Jewish, Hindu, Buddhist, or some other tradition, religious traditionalists, as I shall call them (see Chapter 7), cling to localized versions of a religious tradition in defiance of broader global developments.

Elective Affinities

Since, as we have seen, a sacred canopy usually does not span the life of an entire society in the global village, it still may serve to protect a particular social stratum or group. Because religious beliefs and expressions are always closely linked to social life, individual social groups are drawn toward their own cultural styles and definitions of the sacred. Certain ideas seem particularly suited to some status groups and lack any sensible fit with others.

Weber (1947, p. 83) uses the metaphor of elective affinities to describe this relationship between ideas and interests (cf. Howe, 1979). He takes the metaphor from Goethe's famous novel of the same name in which two people are inexorably drawn to each other despite being already married to other people. The concept of elective affinities is an extremely useful one for examining the relationship between culture and social structure, in part because it shows the connection between the two phenomena in a dynamic and nondeterministic way. Farmers and businesspeople are drawn to pragmatic theologies, for example, whereas university professors might prefer more abstract religious ideas. Affinities between the interests of some status groups and particular ideas or belief systems do emerge, but that does not mean that people have no free choice in selecting their own beliefs and practices.

The extent to which a religion is attached to particular social strata or ethnic groups varies over time and across traditions. Judaism has always been closely linked with a particular ethnic group and continues to be tied to the phenomenon of being Jewish by blood. Similarly, Hinduism is closely linked to the South Asian subcontinent; few people outside that geographical region are practicing Hindus unless their ancestral roots are there. The other major world religions are less clearly linked to ethnic groups, although Buddhists are most likely to be Asian (but do not have to be) and Arabs are predominantly Muslim (although most Muslims are not Arab). Some subgroups link their identity to religious traditions: Latin people are more likely to be Roman Catholic than Muslim or Quaker, for example.

These tendencies are partly a result of historic circumstances (such as who conquered whom) but may also be related to Weber's (1947, p. 83) notion of elective affinities between the ideas and the interests of particular social strata. Specific groups of people sometimes use religion as a way of promoting their own interests, but they also may find a given religious orientation more helpful in explaining the world as they experience it. Ethnic variations in religious expression become socially significant only when ethnic status is meaningful in a society—that is, when lifestyles and social status are based at least in part on ethnic criteria.

Religion has traditionally been linked to specific geographic locations in the social world. Because faith traditions can either sustain or subvert the social systems in those places, some traditions—or versions of them—attract a system's elites, whereas the rebels in a society have a natural affinity for other religious beliefs or interpretations of the same tradition. Elites in virtually every culture use religious legitimations to explain why they are in control and others are not. Similarly, the most effective dissident movements often employ religious arguments to legitimate their own positions. Some of the most successful movements for social change are religiously motivated and religiously framed, which gives the struggle an intensity and legitimation otherwise unavailable, and makes it easier for reformers and revolutionaries to mobilize popular support. Religious traditions have provided both ideological support and institutional resources for a number of significant social change movements in the 20th century, from the Indian freedom movement to the U.S. civil rights movement, from the People Power Revolution in the Philippines to antiapartheid forces in South Africa and pro-democracy forces in Eastern Europe and the Soviet Union (see Chapter 6).

In the analysis that follows, I shall use the elective affinity metaphor to identify connections between religious traditions and various social groups in the global village. Of special interest is the common tendency for intense ethnic, class, and even gender conflicts to emerge along religious cleavages and to be framed in religious rhetoric.

Contemporary Approaches to the Sociology of Religion

Like the discipline of sociology in general, the sociology of religion has become more empirical and quantitative in its methodologies since Weber and Durkheim, largely because of the development of computer technology, statistics, and social surveys and in part because of the theoretical orientation of modern researchers. Consequently, even though contemporary sociologists of religion usually examine the same issues as the discipline's founders and in much the same way, they are more precise and therefore often more narrow (see Wuthnow, Davison Hunter, Bergesen, & Kurzweil, 1984).[2] I will now examine four current theoretical frameworks that form yet another set of tools for the contemporary sociologist of religion.

The classical sociological tradition of Marx, Weber, and Durkheim has produced a "neoclassical" perspective—represented by Peter Berger, Clifford Geertz, Robert Bellah, and Thomas Luckmann—that tends to be a **subjective approach**, emphasizing individual beliefs and attitudes, opinions, and values. This perspective, which dominates much current work in the sociology of religion, adapts well to survey methodologies and often uses the sacred canopy metaphor to frame its questions. The focus of these social psychological studies is usually on the problem of meaning and an individual's interpretations of reality.

In recent years, a second, **structural approach** has explored patterns and relations among cultural elements. Its central task is the identification of structures (orderly relations and rules) that give culture coherence and identity. Structural studies by such scholars as Mary Douglas examine such phenomena as boundaries, categories, and elements of behavior (as opposed to attitudes, beliefs, and values), a category that includes discourse, gestures, objects, acts, and events that are amenable to observation. Thus the structural approach examines not so much the content of the tradition's beliefs as the relationship between the religious system and the structures of social life.

A third, **dramaturgical approach** examines expressive or communicative properties of culture and its interaction with social structure. This approach, as Wuthnow and colleagues (1984) suggested, explores the expressive dimensions of social relations over either individual feelings (the subjective approach) or structural categories (the structural approach). Much of this analysis focuses on **ritual** and its symbolic expressions of a moral order as a prototype of other symbol systems. It can be traced historically to Durkheim's (1915/1965) and Malinowski's (1954) studies of ritual, embellished by Kai Erikson's (1966) exploration of witchcraft trials in colonial New England and more recent work inspired by Erving Goffman and others.

A final contemporary school—and one especially relevant for examining culture in the global village—is the **institutional approach**, represented in work by Guy Swanson and others. From this perspective, "actors who have special competencies" produce culture and sustain it through institutions that ritualize, codify, and transmit cultural products (Wuthnow, 1987, p. 15). Proponents of this approach maintain that these social institutions of religion are often more securely understood than the more abstract notions of myth and ritual and thus provide the firm empirical ground on which to address the larger issues.

In this volume, I will weave together elements of the four approaches—(1) subjective, (2) structural, (3) dramaturgical, and (4) institutional—as I analyze the world's religions, because each identifies a significant element of religious life. The subjective element is important in identifying how individuals are linked with the broader human community through the worldview and ethos options available to them in a diverse social setting. The various beliefs of religious systems are interconnected structurally and tied in patterned ways to each other and to the social system in which they are developed. Sociologists tend to be especially interested in religious institutions as key players in the social world and the ways in which religious ideas are performed on the world stage. An individual Hindu, for example, interprets the world and acts in it according to the cultural patterns provided by his or her social networks and institutions as they collide with the institutions, beliefs, and practices of people from other religious traditions by means of the globalization process.

Anyone trying to understand religion in the global village will find in sociology a fruitful analytical framework. It is only fair, however, to warn the reader of some problems with the way sociologists look at religion. Religious traditions are, from a sociological perspective, comprehensible constructions of the human mind, yet they transcend comprehension. They constitute a collective effort to make sense out of life and death but (like life itself) are riddled with contradictions and paradoxes. Religion tries to bridge the gap between temporal reality—what sociologists are rather good at describing—and the mysterious aspects of reality that cannot be easily examined, if they can be studied at all, by empirical methods. Consequently, sociologists often focus on those elements of religious life that are immediately observable: religious institutions, written texts and patterns of behavior, opinions about religious matters studied through surveys. Much of this subject does not fit neatly into our narrow conceptual boxes of survey instruments, however. We cannot discern by data gathering if the Gods exist, let alone interview them face-to-face, so our conclusions seem always inadequate. Like astronomers looking for faint evidence from distant galaxies or archaeologists examining potsherds from the bottom of a 3,000-year-old well, we often have to choose between focusing on inconsequential details and constructing explanations that go far beyond what our data allow us to say.

The sociologist of religion tries to discern patterns, but because religious expression is so varied, the enterprise is fraught with danger. The sociologist of religion may also offend a person's religious sensibilities by subjecting his or her beliefs to rational scrutiny. The historical prejudices of the sociological tradition, developed primarily by white Western men, also stands in the way of objective observation. Even the language we use to discuss religion is riddled with prejudice. In talking about the deities of the world's religions, for example, it is difficult to generalize. If we talk about a "God," we imply a monotheism common in Western—but not Eastern—religions. If we refer instead, to "the Gods," monotheists may object. Some, like Buddhists, are uncomfortable with the idea of a transcendental deity, because they believe that all creation is ultimately a unity.

Do the Gods—or God—exist? The norms of science require us to be as objective as possible, yet science cannot answer this question because it involves a faith stance, not a strictly empirical one. Scientists examine phenomena indirectly by looking at indicators. If we could agree on what or whom God is and what indicators might prove his/her/their existence, then we could test its reality. We could not, however, agree on the most basic indicators, and if we did, measuring them would be difficult as well.

Each of the sets of sociological tools I have examined here—definitions, metaphors, and theoretical frameworks—aids the task undertaken in this book. The sociology of religion, growing as it did out of the social turmoil of 19th-century Europe, identifies the struggles of late-20th-century multiculturalism and points us in a direction that will assist us in understanding the current state of religion in the global village. The tumultuous history of the field betrays its assets and liabilities: The sociology of religion is relevant and valuable because it was born out of the early stages of battles we continue to fight. Yet those who forged it were partisans in the fight, and we must remain conscious of their limitations as well as of their insights.

Three Pillars of Analysis: Beliefs, Rituals, and Institutions

The focus of the following discussion is the interplay among beliefs, rituals, and institutions, which Durkheim (1915/1965) identified as the central components of religious life in his classic study *The Elementary Forms of the Religious Life*. Social scientists studying comparative religions have often focused on this aspect of the faith traditions because so many texts are readily available to explore, analyze, and compare across space and over time. I will undertake a brief overview of many contemporary religious patterns not only in terms of how they function (Durkheim's primary focus) but also in terms of how they have been interpreted and changed over time. The change process especially has profound implications for the future of religious life and collective life in the 21st century.

Anatomy of a Belief System

Each religious tradition has a set of interdependent beliefs that are woven together in such a way that the integrity of the entire fabric is dependent on each strand. The structure of these cultural systems involves the identification of what is considered sacred and meaningful, a set of theories about how and why the world was created, and an explanation for suffering and death.

Belief systems express a worldview—that is, a culture's "picture of the way things in sheer actuality are, their concept of nature, of self, of society. It contains their most comprehensive ideas of order" (Geertz, 1973, p. 127). In contrast, Geertz continued, the ethos of a people encompasses the culture's "tone, character, quality of life, its moral and aesthetic style and mood; it is the underlying attitude that people have about themselves and the world that life reflects" (p. 173).

Religious myths both reflect and inform the world or, as Geertz (1973) put it, provide both *models of* and *models for* reality. First, they are models of reality in the sense that they offer information and explanations about the world. As models *for* reality, sacred stories also show how the world "is," in spite of appearances to the contrary. They often highligh between appearance and reality or between the sacred and the **profane**. I will now examine each of these elements of a belief system.

The Sacred and the Profane

Emile Durkheim (1915/1965) believed that the entire world of human experience could be divided into two categories: (1) the sacred, what is of ultimate concern, and (2) the profane, what is considered ordinary and mundane. The fact that Durkheim was the son of a Jewish rabbi may have influenced his theoretical model because the division of the world into these conceptual categories is central to Judaic thought and ritual. Durkheim's distinction became central to the academic study of religion even though some religious traditions insist that all of life, not simply one sector, is sacred. To understand most of the world's religions, however, one must grasp this fundamental distinction, which has been elaborated by comparative religions master Mircea Eliade in his 1959 classic *The Sacred and the Profane*.

The two categories of sacred and profane may actually lie on a continuum. Some things are considered more sacred than others and are ranked according to their sacrality. What constitutes the sacred varies from culture to culture and changes over time. The sacred may be recognized (or "manifest itself") in the form of a stone or tree, a flag or mountain, or even an idea. Scholars use the term *hierophany* to identify the process in which people encounter and experience the sacred. Sometimes two phenomena defined as sacred may come into conflict, and people must choose which is the most sacred. A theophany is a type of hierophany referring to the appearance of the sacred in a visible form to a human being, either a human or humanlike figure (Jesus, Radha, Krishna, angels) or a natural object such as the "burning bush" that many believe Moses saw in the wilderness. The problem of representing the sacred in art or imagery led to prohibitions on doing so in Judaism and Islam.

The idea is that it is impossible to represent God accurately because human efforts to do so are always inadequate and therefore misleading. All one has to do is to go to a major art gallery in the West to see the problem— the famous classical European paintings of biblical characters (including God) usually represent them as fair-skinned Europeans. Even ichelangelo may present a very misleading idea of what God g God as an aging bearded white man reaching down from angels could be represented was an interesting controversy tium culture (see Peers, 2001). , specific times and places are identified as sacred. Most holidays ("holy days") and sacred sites (Eliade, 1959).

A certain location—a temple, mosque, or cathedral; a war memorial or cemetery; the birthplace or grave of a famous person—often affords an encounter with the sacred. Crossing from profane to sacred space often requires certain actions, dress, or attitudes: taking off shoes, genuflecting in front of the cross, bowing in respect, refraining from loud talk, and so on. In the modern world, some of these spaces become sources of great political controversy (such as Jerusalem) or tourist sites (see MacCannell, 1976).

Religions also traditionally divide time into the sacred and the profane. Religious festivals and rituals involve the **reactualization** of sacred events that took place "in the beginning" or during some significant hierophany; the sacred elements of time are thus transformed and shifted to the present moment. Belief systems of the various religious traditions are almost always systematized and disseminated by a select group of religious elites but also in the speech and rhythms of everyday life; for most people, their faith is encountered not in the subtle theologies or massive writings of their tradition, but in its rituals, that is, the regularly repeated behaviors that symbolize the values of their belief system. Ritual behavior, as Durkheim (1915/1965) observed, provides the occasion for an encounter with the sacred, and it is socially organized in such a way as to reinforce the values and authority of the community.

Religious pilgrimages, such as the **Hajj**, the Islamic duty to go to Mecca if possible during one's lifetime, sometimes combine sacred space and time, so that it is optimal to visit a sacred location during a particular season or time of the year. Muslims may try to go to Mecca during the month of Dhu Al-Hijjah. Similarly, a Catholic might try to visit the shrine of the Virgin of Guadalupe on December 12. That is the day she was believed to have appeared to the Native American peasant Juan Diego in 1531 and an official national holiday in Mexico complete with pilgrimages, masses, fiestas, and processions honoring the event.

Faith communities preserve their religious beliefs in often contradictory narratives united at an abstract level by the structure of the stories themselves (see Kurtz, 1979). Life and death, good and evil, the divine and the human all exist side by side within the narrative. The world gets created twice in every culture: First, in the material sense, the world comes into being. Then it is re-created through sacred stories or mythologies (or by means of scientific theories) in a **cosmogony** that links the present with the past. Some sacred events are more significant than their frequency would suggest (like religious visions, death, and sexual intercourse).[3] They may even be statistically insignificant but still have a profound impact on our lives—what is significant statistically may not be the best indicator of what is important to people, especially in the realm of the sacred.

Cosmogonies

Most religious systems have a cosmogony, a story about how and why the world was created. These stories tell us a great deal about a religion's most significant ideas. All the world's creation stories fall into a few basic patterns: A God creates the world out of nothing (ex nihilo). Life emerges from a cosmic egg. A prechaotic animal pulls mud up out of the water. A mountain rises up. A giant sustains the sky. The world takes shape as a spiral. A God delegates power to a demiurge (minor deity)—and so forth.

These cosmogonies link the people who tell the story with the creation process, because their Gods and/or ancestors were involved in giving birth to the world. Most creation stories also have an **anthropogeny**—that is, a theory about the creation of humans and how they should think of themselves. Sometimes the Gods are bored and want to play; other creation stories provide no clear reason why humans come into existence. Sometimes people are created out of clay, mud, water, or blood; other times, they are chiseled out of stone or a primeval tree trunk or are brought out of a plant. Not everything always goes right in the process of creation, nor are the Gods uniformly good or evil. Sometimes the Creators are good, other times they are amoral jokers or tricksters. In most cases, however, the events at the beginning of time have special significance for the faith community's present situation. The stories provide a paradigm for all creation, blueprints of necessary accomplishments to prevent evil or the reversion to chaos. A cosmogony is one of several basic building blocks of the world-constructing process—the means by which a system of beliefs is developed that explains the world and its implications for daily life.

Not all religious traditions posit a clear-cut beginning point for creation. The Vedas, for example, present a vision of vast time and space that involves endless cycles of birth and rebirth. The cosmos has no beginning or end, although the known ages of the world are born and die just as individuals do. Consequently, the Hindu belief system suggests there are strict limitations on what one individual can change about this vast world and even on the given cycle of time in which one now lives. Buddhist cosmogonies present a similar philosophical view of the universe, in which the five cosmic elements (**skandhas**)—(1) form, (2) sensation, (3) name, (4) conformation, and (5) consciousness—participate.

Because a religion's worldview and ethos are closely related, its creation story provides clues into the nature of the culture to which that religion is linked. In the Babylonian creation story, a great God made minor Gods out of stone, brick, and other materials; because these lesser Gods had to do menial tasks, they in turn created humans to do their work. Such a cosmogony implies that the meaning of human life is to be found in serving God, or perhaps God's representatives. In other religions' creation stories, God created humans in God's own image, thus setting homo sapiens apart from the rest of creation. Such a creation story may empower the species to take control of the rest of nature in an effort to force it to serve humans.

Theodicies

Theodicies are the explanations a religion offers for the presence of evil, suffering, and death in the world, a perennial concern of religious traditions. What causes these events, how can they be alleviated or transcended, and why should the righteous suffer? Each tradition has a theory about suffering and a wide range of explanations occur, including punishment for an individual's sinful behavior, a battle between good and evil forces, the result of natural processes, and so forth. Just as nature has cycles of death and regeneration, so an individual may experience death and rebirth or patterns of illness followed by health. Because suffering and death are such obviously universal and mysterious elements of the human experience, a tradition's theodicy plays an important role not only in religious traditions but in social relations as well. Though we are not usually conscious of it on a daily basis, the ever-present fact of death affects the way in which we organize our lives (see Dunne, 1965). In addition, a theodicy that explains suffering primarily as punishment for wrongdoing, for example, can lead to a justification of social oppression of those who suffer. Different traditions and subtraditions have varying themes and emphases in their theodicies (see Table 1.3).

One significant element of most theodicies is the social construction of evil—that is, identifying the sources of evil in the world. This framing of evil often has a profound impact on collective life because it affects the nature and intensity of social boundaries; as a social construct, it encourages certain kinds of conflict and discourages others. The most potent construction of evil occurs when a social group's enemies are defined as a deity's enemies so that the latter's destruction or subordination is seen as divine retribution. Sometimes, the same God is claimed by both sides, as in the Iran–Iraq War of the 1980s or in World War II, when Allies and Germans alike claimed a Christian God on their side.

Evil may be of divine or human origin, although the two are often intertwined, especially in the heat of conflict. **Dualistic theodicies** connote a struggle between the powers of light (good) and those of darkness (evil). Suffering is thus caused by the evil forces as part of the ongoing battle for control of the universe. Dualistic theories, most highly developed in **Zoroastrianism**, Mandaeism, Gnosticism, and Manichaeism but also found

Table 1.3   Comparative Theodicies

"Primitive Religion"

- No sharp boundaries: between individual and society, between society and nature
- Significant boundaries: between social groups
- Suffering: result of crossing social boundaries, violating taboos
- Death: part of the natural rhythms of nature

Mysticism

- Seeks union with sacred forces or beings
- Individual death and suffering are trivial and insignificant

Messianic/Millenarian Eschatologies

- Millenarian: an ideal society will come (often through revolutionary action)
- Eschatology: concern with final events of history
- This-worldly: transformation of society; a hero or messiah will bring justice; suffering may be rewarded
- Otherworldly: change will come in the afterlife or next world

Dualism

- Mandaeism, Gnosticism, Manichaeism
- Struggle of powers of darkness (evil) versus light (good)
- Zoroastrianism: suffering caused by evil forces as part of an ongoing, often cosmic, battle

Karma and Transmigration

- No bifurcation of the world as in dualism, although there is guilt and merit
- Karma: action or action energy; has effects that later become causes as part of an inexorable law of cause and effect ruling all actions
- The world is completely connected
- Cosmos of ethical retribution
- Transmigration: a wheel of rebirths (samsara) in which guilt and merit are compensated by fate in successive lives of the soul

Sources: Adapted from Weber (1968) and Berger (1969).

in other religious traditions including Christianity, often lead to contentious political views. Enemies are identified with evil cosmic forces, making peaceful coexistence difficult and affecting the style and tone of interpersonal and intergroup relations as well as cross-cultural and international conflicts.

Eastern religions do not delineate good and evil as clearly as Western religions because they tend to claim that the cosmos is unified and that such distinctions are mere illusions. On another level, however, Eastern sacred texts (especially Hindu) are richly populated with demonic figures that battle with heroes. Even the worst characters have some element of good in them and the Gods themselves can cause pain and suffering, so that evil can still be seen as something independent of persons or even the Gods themselves. Mahatma Gandhi, as I will discuss later, emphasized this characterization of evil as an independent entity in his effort to promote the idea of struggling against ideas and systems rather than other people.

One of the most troubling realities for a religion to explain is why the righteous suffer. Because most theodicies offer a way out of suffering through ethical behavior, the fact that the righteous suffer presents a difficult anomaly. The major religions prefer to suggest that suffering is simply embedded in the nature of the universe and leaves no one, including the righteous, untouched.

This tension between admonitions to live an ethical life and the apparent lack of immediate reward for doing so has inspired theodicies that promote the idea of reward or release from suffering in another world. Those who fulfill their religious duties can sometimes escape suffering by leaving this existence altogether, either in the Hindu–Buddhist effort to liberate themselves from the cycle of rebirths or in some sort of heaven or otherworldly plane, as in Christianity or Islam. A second major solution to the problem of the suffering of the righteous is to hope for some positive result to come out of the struggle itself—the "no pain, no gain" belief that pervades many theodicies. A third solution is to recognize that one's current woes are not as bad as they could be; suffering is thus reframed and placed in another context.

Religious Rituals

Ritual, a regularly repeated, traditional, and carefully prescribed set of behaviors that symbolizes a value or belief, plays a central role in all the world's religions and is usually studied ethnographically in the field. Much can now be learned, however, by getting on the Internet, where we can observe not only documentary films the represent rituals but weddings, funerals, processions, festivals, and other rituals posted by participants or institutions wanting to share their rituals with other members of their community or the general public. Rituals come in a wide variety of forms; some help people to show devotion to the Gods, as in corporate worship and certain modes of communicating with the Gods, such as praying, chanting, singing, and dancing. Others, such as meditations and mantras, facilitate the process of life organization, on both personal and collective levels. Rituals help to frame daily life by regulating such matters as hygiene, diet, and sex; rites of passage surround major transitions such as birth, puberty, marriage, and death; and still other rituals, such as seasonal festivals, processions, and holidays, help people cope with the cycles of nature.

Rituals are not only limited to religious practice, however, and the idea is a sensitizing concept that helps us understand more about what is going on in much human behavior; once you get it in your mind, you see it everywhere. All social institutions rely on ritual behavior to sustain their values and participants' consciousness of their authority as well as to build solidarity among its members. Rituals solve problems of the collective life in a time-tested way by identifying evil, marking social and ideological boundaries, and reinforcing the institutions that sponsor them. Religious rituals in particular link the experience of ordinary life with the sacred and place both trauma and joy within the context of a worldview that orders a people's life and provides them with meaning.

Although rituals often preserve a social order and sustain old habits, they are crucial to cultural innovation and change as well. Victor Turner (1967) observed that religious rituals often signify a special or **liminal period** set apart from ordinary reality. During the liminal period, which involves a separation from, or marginalization of, ordinary reality, participants leave ordinary time and space and enter a sacred region in which the problematic aspects of everyday life—filled with suffering and injustice—are solved, or perhaps denied. Thus, rituals fall "betwixt and between" different worlds: the world of everyday life and the sacred time set apart from mundane reality. During the ritual, normal rules of interaction and social structure no longer apply or are inverted.

Some religious rituals assure people that death is not final because it is followed by rebirth and that social change is not ultimately destructive; the person who stays in the fold of the community is never alone in times of crisis. Thus, the world is renewed annually through the cycles of the seasons and the rituals that mark them, such as Mardi Gras, Divali, and the Chinese New Year. These rituals signal the paradigmatic nature of time; life is regenerated through a return to the time of the origins. This liminal period is filled with inversions and transformations (see Babcock, 1978; Turner, 1967). During those moments of ritual time (e.g., when the crucifixion of

Jesus is relived at Easter), the mundane characteristics of life are reshaped: The children and the Princess of Spring run Old King Winter off of his throne at Mardi Gras; the God is taken out of his or her temple and paraded through the streets. The births, deaths, and sometimes resurrections of the Gods are celebrated with a renewed emphasis on the things that, contrary to the appearances of earthly, everyday life, "really matter": family, justice, love, and order.

Rituals as a social form share a number of common structural characteristics, no matter what their content: (1) they provide solutions to problems; (2) they are rooted in experience; (3) they involve the demarcation of boundaries and the identification of evil; (4) they include nonrational as well as rational aspects of behavior; and (5) they reinforce, or reify, social processes. I will examine these characteristics in turn.

First, how do rituals solve a variety of human problems? This is a key to the important role of religion in human life: On the abstract level, religious rituals bind a social order together, linking it to the culture's worldview and ethos. On a practical level, rituals provide a proven repertoire for social action, especially at times of crisis. When people are confronted with suffering or death, going through major passages in their life cycle, or experiencing rapid change, religious rituals may guide them through the crisis. Rituals provide socially approved responses, preestablished scripts, and social support for those who have suffered a loss.

Perhaps the best example of religious ritual as problem solver is the funeral rite: A death creates a crisis among the living that must be acknowledged; something must be done and something said, and ritual packages provide a repertoire of words and actions appropriate to the occasion. The funeral ceremony provides a sense of closure on that stage of the grieving process and places the death within the broader worldview of the tradition. As long as the sacred canopy is in place and retains its legitimacy, the scripts provide a modicum of relief for sufferers. People surrounding the bereaved are familiar with the rituals, and specialized institutions and ritual experts usually guide the victims of the crisis. A ritual package for the ceremony includes a repertoire of appropriate comments for the bereaved and their social networks—some general in character ("It was God's will that he go at this time"), others tailored to a specific situation ("At least she's out of pain now"). A major feature of ritual problem solving is the provision that allows the bereaved to focus their energies on arranging details of the rituals. By accomplishing the detail work surrounding ritual, the individuals involved feel as though they have accomplished something in the midst of a crisis that would otherwise make them feel impotent; taking care of the deceased's needs after death can address lingering unresolved tensions.

Second, how are ritual practices rooted in experience? Because rituals were constructed in the past, they have the authority of time-tested formulas: As Weber (1968, p. 226) observed of traditional authority, rituals are "believed in by virtue of the sanctity of age-old rules and powers." People wiser than we were confronted in the past with a similar situation and created this solution. Evidence confirming that ritual brings about the desired effect is essentially anecdotal, passed on in mythical stories that contain the underlying principle: "We've always done it that way, and it's always worked." The elders recall a time, long ago, when a drought brought disaster to the village. A rain dance was performed, the rain Gods were pleased, the clouds rolled in, rain fell like a monsoon, and the village was saved. The efficacy of rituals is in the eye of the performers—they appear verifiable to those who believe in them, if not always to outsiders. Similarly, evidence calling a ritual into question can be discounted by pointing to flaws in the *performance* rather than the ritual itself. If any ritual performance is scrutinized closely enough, mistakes can be identified that could have been responsible for the ritual's failure. The reason for failure may be attributed to sabotage by an outsider, such as an enemy who performed a more powerful counterritual.

The fact that rituals ensure continuity and reliability is both a blessing and a curse. Because the world is constantly changing, the conditions for which a specific ritual was created may alter to such an extent that the ritual is no longer appropriate. Yet the change may be unnoticed or may be considered unimportant by the ritual's advocates, who continue to perform it anyway. This is the phenomenon known as **cultural lag** (Ogburn, 1922). Rituals are valuable because people can rely upon a known procedure to maintain the social order, but they resist changing procedures when conditions change, thereby rendering the ritual behavior counterproductive. A ritual designed to protect people from drought may not be useful to a tribe moving from a desert to a rainforest.

Third, we usually need to identify a problem's cause in order to solve it, and most rituals contain a theory about the origin of the problems they are designed to alleviate; in short, they contain a theory of evil. Rituals mark boundaries between good and bad ideas, between "us" and "them." An "evil" force, situation, or group is identified as the source of the difficulty, opening the way to a solution. Ritual behavior is thus an integral part of the social construction of evil by which an image of the enemy is created and then spread throughout the culture by media, folktales, and jokes. Evil is often attached to enemies outside the belief system or to those within the system who can be labeled as heretics (see Kurtz, 1986). Many religious traditions personify evil in a particular figure, such as a monster or humanlike creature, which needs to be defeated in battle, making the evil appear more manageable.

Whether the figure deemed responsible for evil is personified in a devil or mythical figure or found in a human enemy, rituals identify it and give people something concrete to do about mastering it. Sometimes the simple act of naming an evil and denouncing it is useful; at other times, the evil is physically punished or ridiculed or exiled from the social order. Some will engage in exorcisms to drive away evil spirits; others will repeatedly denounce a group of people for a misdeed. Psychologically, this process of ritual identification of evil is sometimes helpful in the short run but destructive in the end if we simply project evil onto another person or group (like the Nazis did to the Jews during the Holocaust). Such naive projections distort our view of reality:

> The enemy appears as the embodiment of all evil because all evil that I feel in myself is projected onto him. Logically, after this has happened, I consider myself as the embodiment of all good since the evil has been transferred to the other side. (Fromm, 1961, p. 22)

Fourth, rituals include nonrational as well as rational aspects of human experience. Although scholars often tend to emphasize the cognitive aspects of human attitudes and behavior, we are also emotional, creative, and affective beings whose lives are influenced by many different factors. The idea of ritual goes beyond the rational, although that makes it difficult to examine empirically, since our scientific methods generally limit us to observable, rational behavior and ideas. We can still observe expressions of love, for example, as empirical facts that may not appear rational to outside observers. From an objective point of view, kissing someone or risking one's life for another may not be simply rational or it may even appear to defy all rationality. In the liminality of a ritual, the injustices of the existing world are replaced by the justice of another world or an age to come, giving people hope to carry on in this one. The idea that "the last shall be first" may be enacted in the ritual, as when the young Princess Spring dethrones Old Man Winter in the Carnival ritual preceding Lent; the male authority figure is driven off by the people, replaced by the youthful feminine.

Finally, rituals have social as well as psychological consequences, one of which is the **reification** of social processes. Reification means the treatme of an abstraction as a concrete material object. Social institutions that sor rituals gain authority and legitimation from doing so in the ey participants. People who never "darken the door" of a temple or church on a regular basis may still turn to their religious institutions and its authorized agents (rabbis, priests, etc.) in times of crisis or celebration.

Because rituals establish the link between a people and their worldview, social transformations and ritual changes are mutually interactive. The move toward universalism is common in religious traditions that shift from being locally oriented to being more cosmopolitan. When religions diffuse across different cultures—or when the territory in which they are based is invaded by traders or conquering armies—religious practices are almost inevitably modified. When Buddhism moved into China and the rest of East Asia, the indigenous festivals and Gods of the various cultures became part of the overall Buddhist practice or lived side by side in the same ceremonies and temples.

Ritual "Packages"

Religious belief systems integrate different kinds of rituals into sets of interdependent practices through which people can engage in showing devotion to the Gods while organizing their individual and collective lives. These sets of "packaged" rituals, though containing diverse practices, fit together neatly as a whole, as when one goes to a place of worship and undertakes a series of rituals designed to fit together. Going to a Buddhist temple, for example, involves a combination of a number of acts, which may all be performed or isolated individually. One might wish to gain guidance about a particular question, so a believer may consult the yarrow sticks, or increase one's chances of a good grade on an exam or of success in romance by giving offerings of fruit, or demonstrate respect by bowing in front of an image with incense, and so forth.

Two "packages" of rituals, **Yoga** in the Hindu tradition and the **Five Pillars of Islam**, deserve a closer look, and I will examine them in the sections in subsequent chapters on Hinduism and Islam. Both of these ritual packages are highly developed intellectually but also contain a rich repertoire of symbolic acts constructed over many years, with many active participants.

It is helpful—to understand how these ritual packages evolve—to examine ... ls found across religious traditions, such as acts of ... with the Gods, and collective rites of worship.

... often designed to draw a people closer to the ideals ... hem of their Gods, their values, and the principles ... nportant events and figures of their shared history.

Religious rituals thus involve many forms of communication with the Gods, or with saints, **bodhisattvas,** and other heroic figures from the tradition, as well as occasions for collective worship in public gatherings of the believers. Although they vary widely, some kinds of devotional acts are virtually universal in the world's religious traditions.

Communication With the Gods

A central aspect of religious practice is the set of rituals designed to facilitate communication with the Gods, such as prayer, chanting, singing, dancing, reading from the scriptures, and offering sacrifices. From a sociological point of view, messages designed for communication with the Gods are also ways of communicating with oneself and with others, especially those within the religious community.

The devotee's relationship with the deity is sometimes highly personal, especially in popular traditions. Protestant Christianity often emphasizes the believer's personal relationship with Jesus, who is perceived as a friend or a brother, and with Mary who, although not formally a Goddess, is considered endowed with great power and yet empathetic, especially with the plight of poor women. When the image of God is transcendent, sometimes intermediary figures like Mary, the saints, the prophets, or living members and officials of the community may intercede between the individual and a God who is not easily approachable.

The notion of sacrifice to the deity also lies at the heart of many religious rituals, sometimes including acts of considerable violence to humans and animals. In the earliest rituals, the totem animal, usually protected from harm, was sometimes killed and devoured in an orgiastic frenzy (see Girard, 1977). In ancient Judaism, sacrifices to the deity were performed in the same way as in other religions of the region: Burned offerings were made, including animal sacrifices, along with other offerings representing the first fruits of the harvest, tithes, and taxes. Human sacrifices may have been part of the original set of ritual practices, but if so, the tradition was rejected in early Judaism, as shown in the story of Abraham's sacrificing a ram in place of his son (Genesis 22). The relative merit of animal and vegetable offerings, as in the story of the conflict between the brothers Cain and Abel, probably reflects an early conflict between people growing crops and those raising animals.

Many religions shifted from human to animal to purely symbolic sacrifices over time; ritual practices became increasingly abstract as personalistic ferociousness declined in public behavior (see Collins, 1974). Perhaps a large majority of today's sacrifices are either symbolic acts (e.g., bowing, kneeling, or prostrating) or the offering of money, fruit, vegetables, and so forth.

Collective Rites of Worship

Many religious rituals are performed in identical fashion by believers at home and in temples by priests and laity alike. The temple is usually the "home" of a deity, represented by an image that takes on the power of the God who resides in it. In Hinduism, the *puraris*, or temple priests, follow a prescribed daily pattern in which an image is awakened, bathed, fed, visited and honored, anointed, decorated, and retired for the night. Sometimes worship is collective, especially in **Bhakti Yoga**, with chanting or hymn singing. The combination of the congregation members' mutually reinforcing attitudes; the presence of the Gods; and the stimulation of all senses through music, incense, touching of the God or offerings given to God sometimes produces trances in zealous devotees.

In the Abrahamic religions, collective worship centers on the Sabbath, a ritual that re-creates the primal creation myth in which God rested on the seventh day after bringing the world into being. As Abraham Heschel (1962) noted of the historically oriented, geographically mobile Jewish tradition,

> Judaism teaches us to be attached to holiness in time, to be attached to sacred events. . . . The Sabbaths are our great cathedrals; and our Holy of Holies is a shrine that neither the Romans nor the Germans were able to burn. (p. 119)

In this tradition, place is not as important as time: Work ceases on the Sabbath, which is observed in both home and synagogue, and the sacred interval is marked by the lighting of candles, the reciting of prayers, the drinking of wine, and the eating of bread.

Ritual Organization of Life

Religious rituals link daily life routines with the broader order, allowing individuals and groups alike to place life's crises within a broader religious frame. A variety of rituals organize personal and collective lives: (1) rituals that regulate daily life, from hygiene and diet to sexual practices, link that daily life to the broader worldview; (2) rites of passage routinize life cycle changes at birth, puberty, marriage, and death; and (3) seasonal festivals and processions bring self, society, and nature into harmony by linking human activity to the seasonal cycles of the natural environment. I will look briefly at each of these categories of rites.

Daily Life

Most traditions advocate a practice of regular meditating, praying, or chanting that provides moments for introspection. Daily devotional rituals sacralize the profane existence of a worshipper's life, giving it larger meaning and purpose and framing it within the ongoing processes of the cosmos. Hindu devotees—especially **Brahmans**—are to engage in ritual washings and daily **poojas**, usually a ritual with chanting and the offering of food or flowers to a God. Buddhists can frame their day by meditating or chanting *om*, the fundamental sound of the one undivided universe. St. Paul told the early Christians that they should "pray without ceasing." Devout Muslims traditionally stop whatever they are doing five times daily, bow toward Mecca, and pray to Allah.

Many daily activities such as eating become acts of faith that reflect a worldview and/or link the individual in a special way with the religious community. The act of sharing a meal together frequently functions as a ritual for the creation of group solidarity, as in the Jewish Passover meal (*seder*) or potluck dinners in American Protestant churches. Membership in a religious community is sometimes expressed through dietary restrictions; in India, for example, most Muslims eat meat and most Hindus do not. The vegetarian diet of many Hindus and Buddhists reflects a worldview in which all creatures are of equal importance and should therefore not be harmed. Jewish **kosher regulations** and Islamic **halal regulations** about diet make the simple act of eating a sacred symbol of belonging and worshipping.

Ritual sometimes transforms the profane into the sacred, thus inverting the normal order of things. In this framework, eating becomes not just a way of gaining nourishment for the body but an act of worship that makes a material action spiritual. One of the most striking examples of the effort to sacralize bodily activity is the use of sexual acts and imageries as part of worship rituals. One of the best developed of the sexual rituals is *Tantric Yoga*, a spiritual discipline that opens a permissible avenue of sensuality in a sexually repressed society. Sinha and Sinha (1978) explained it as follows:

> Rites of Tantra affirm the need for intelligent and organized fulfillment of natural instinctual desires. . . . The essential element underlying the Tantric practices is the belief that these rites cleanse or purify mundane or profane acts and in the process sacralize the acts themselves by superimposing certain constraints on them. Hence Maithuna [participation in a sexual act] for a Tantric Yogi is a rite and not a profane act; since the partners are no longer human beings, but "detached" like gods, sexual union is elevated to the cosmic plane. (pp. 142–143)

Paradoxically, then, the transformation of the sexual act into a sacred one results not in condemnation but salvation. The Tantric texts themselves are conscious of this paradox and comment frequently: "By the same acts that cause some men to burn in hell for thousands of years, the Yogi gains his eternal salvation" (Eliade, 1958/1970, p. 263, quoted in Sinha & Sinha, 1978, p. 143).

Rites of Passage

Religions provide rituals for all major life passages; these rites sanctify each transition of an individual's life (see Weightman, 1984, p. 216ff.). Within the Hindu tradition, at the occasion of a birth, horoscopes are drawn up, a name-giving ceremony is performed on the 6th or 12th day, the house is purified, and restrictions on the new mother are relaxed. On the first birthday, the baby's head will often be shaved (sometimes at a temple or festival) as a sign of thanks for his or her health.

The ritual of **baptism** in Christianity serves, in some congregations, as a birth ritual as well as a public expression of belief for converts. Baptism, as I have noted—borrowed from ancient Judaism and used by Jesus—was extremely significant in the early church as a rite of passage into the religious community because of the absence of social boundaries around the church. The literalness of the earlier practice gave way to a more symbolic interpretation in which the death and resurrection of Jesus are celebrated by immersion in water followed by a rebirth, as if the believer first died and then emerged from the womb a second time.

Most religious traditions have a rite for the passage from childhood to adulthood, such as the bar mitzvah for boys and bat mitzvah for girls in the Jewish faith. In some Protestant Christian congregations, baptism comes at puberty rather than at birth. In orthodox Brahman families, an initiation ceremony occurs when the child enters the **Brahmachari**, or student stage of life, which I will revisit when I discuss Hinduism.

One of the most significant rituals in most religious traditions is the funeral, and the process of dying is often full of religious significance as a passage from one world to another. The problem-solving character of rituals can be seen in the way that they frame the end of a human life. Typically the lifestyles, values, and social structures of an entire culture are reproduced in funeral ceremonies. Filial piety and honoring the dead, for example, are important aspects of traditional Chinese funerals as they are in daily life.

Cycles of Nature

The more spectacular ritual events are often pilgrimages and festivals, usually associated with seasonal cycles of nature or milestones in a tradition's sacred history. Virtually every religious tradition acknowledges the importance of special times and places during which, and at which, the sacred is experienced. Pilgrimages to places where the Gods were born or engaged in heroic encounters with each other or with humans hold special promise to believers for their own experience with God. Often a pilgrimage site offers a specialized benefit, such as healing or the prospect of a son.

Similarly, the annual cycle of festivals provides regular occasions for lavish ritual practice. Many are related to the seasons and have ancient roots in the cycles of nature; others are associated with specific Gods, such as Krishna, **Shiva, Ram, Lakshmi,** Durga, Ganesha, and Hanuman in popular Hinduism; various bodhisattvas in Buddhism; and saints in the Christian calendar. Often, modern believers who participate in ancient festivals rooted in natural cycles, such as the winter and spring rituals now associated with Christmas and Easter, have lost touch with their original meaning.

In many religions, on significant festival days a God is brought out of his or her resting place in the temple, anointed and dressed, and paraded through the streets with much ceremony. People crowd around to greet the God and pay homage, and—depending on the powers of the specific God— special benefits are received by those who participate. This process serves to remind people of the ideas and moral implications of their faith and reinforces the status of the temple and its personnel. Through regular cycles of poojas, prayers, pilgrimages, and festivals, the sacred canopy remains intact over the millions of Hindus on the Asian subcontinent.

Religious Institutions

Religious traditions do not exist in isolation but are institutionalized and often highly bureaucratized. The institutional nature of religious practice is one of its most sociologically significant aspects and has changed substantially over time despite its seemingly immutable nature. The most significant development is the shift from local to cosmopolitan religious institutions, a transformation that each of the major religions has undergone and intensified in the 20th century as institutions in every social sphere become linked to global processes.

Religious institutions create a base from which religious beliefs and practices can have regularity over time. Indeed, the provision of continuity is the very essence of institutionalization. As social organization changes over time, however, so too does the organization of religious life. This perception of permanence in the midst of change is one of the most interesting aspects of institutional life. What we now know as established religious traditions were once small religious movements, often sustained only by the charismatic authority of a religious figure—Moses, Jesus, Muhammad, the Buddha—and a small group of devoted followers. If a movement is successful, the initial period is followed by a process Max Weber (1968) called the "routinization of charisma," in which the mobilizing energy of founder and followers becomes routinized and "crystallized" in a social organization that sustains the beliefs and sponsors the founder's practices.

Religious institutions tend to reflect the more general types of social organization in a given society. Wallace (1966, pp. 84–88) suggested the following categories of religious organization:

1. Individualistic cult institutions, which tend to be "magical"—that is, they sponsor ritual acts to implore or coerce forces into meeting specific needs, such as general "luck" or guardian spirits or success in economic or traveling ventures

2. Shamanic cult institutions, which are an early and persistent form of specialization in religious practice because the **shaman** (either full or part time) is a specialist, a private entrepreneur who aids clients in ritual matters

3. Communal cults, which are not led by specialists but meet the needs of a particular community, such as that found in a family, kinship, or locality group or other social groupings that have a common membership characteristic such as age or sex

4. **Ecclesiastical institutions,** which have religious professionals organized into a bureaucracy along the same lines as other nonreligious organizations in the culture as well as a clear-cut division of labor between the professionals (or clergy) and the laity

The first three types are found in the primal religions of preagricultural (i.e., primarily hunting and gathering) cultures and contemporary folk religion, including what Redfield (1957) calls the "little tradition" variants of the more well-known "great traditions." In primal cultures, as will be discussed in Chapter 2, religious practices and beliefs are generally more integrated into the broader social organization of the environment than in modern cultures, and their institutions are not as highly specialized.

That is not to say that these primal institutions—or elements of them—have not persisted into the 21st century. Shamanism still represents a large proportion of the world's religious practice, although it is usually associated with primal cultures. The shaman is an individual thought to have special powers for performing beneficial religious rites and may even be "possessed" by one or more spirits, thus enabling others to encounter these forces or at least to use their services (see Wach, 1944). Even postmodern societies have shamanic roles, such as that of medium or faith healer. Hargrove (1989, p. 94) suggested that televangelists and media celebrities function much like shamans do. Ironically, television has also made it possible for this ancient cultural artifact to become highly developed and institutionalized, sometimes in the form of multimillion-dollar corporations.

As religious traditions become part of complex societies, their institutional forms become highly embellished and specialized—as seen in the case of the shaman–evangelist, who represents something of a mixed type, with the personal charisma of the leader supported by an elaborate institutional framework. Every religious movement, especially in the global village, experiences the pressure to become increasingly institutionalized. The structural form ultimately adopted is related both to the social context in which the movement is born and takes shape and to the organizational preferences of the group that is the new religion's carrier.

Today, all the major religious traditions are what Wallace (1966) called ecclesiastical, because they have a highly developed institutional structure run at least in part by professionals. Each tradition also has elements of the other three types of institutional structure as well, sometimes as a residue of earlier organizational forms that persist and sometimes as a revival responding to particular needs of the time or abilities of particular leaders or groups. Examples of revivalism include the shaman–evangelist already mentioned, communal cults of protest movements, or individualistic cult institutions like the Reverend Gene Ewing's "Church by Mail," which provide people with specific "religious services" for tasks that promise spiritual and financial blessings to those who participate in the program.

Considerable variation has emerged even within the ecclesiastical institutions of the major traditions, which I shall now examine briefly. Of particular interest sociologically—and because of the democratic ethos associated with the global village—is the elective affinity between particular traditions and certain types of institutional forms. Although this is not an inevitable outcome, religious institutions that encompass a diverse population and a pluralistic religious worldview may tend to be decentralized in their authority structures; monotheistic religious traditions and those with more homogeneous populations and belief systems, in contrast, may tend toward more centralized structures. There is no strict relationship, it should be noted, between degree of centralization and the extent to which an institution is formally democratized, although a decentralized system may be more lenient toward local deviations than a centralized one. Moreover, a system that is nonhierarchical across a broad geographical area (such as the Southern Baptist Church in the United States) can have local branches that are extremely hierarchical.

The broad variation within each religious tradition makes it difficult to generalize about differences among them, but some further general statements may be useful. Whereas Islam, Hinduism, and Buddhism tend to be less centralized and hierarchical, Christian churches have often become more formalized as a response to the universalism of their membership. Judaism, despite its highly rational tone, has until very recently retained a more tribal, less differentiated social organization, in part as an attempt to retain the sacred canopy over a specific ethnic identity that has been reinforced by persecution and adversity as well as sharp exclusionary boundaries.

Major Themes in the Sociology of Religion

Sociologists cannot answer questions about the ultimate meaning of life or the relative truth or fallacy of religious traditions, but they can examine systematically how these issues are dealt with in human societies. The chapters that follow will introduce the major religious traditions inherited from the premodern world and apply the conceptual and methodological tools of sociology to the task of understanding religious beliefs and practices. A short tour, in Chapters 2 through 4, of the central beliefs, rituals, and institutions of each major religion and the parallel indigenous religious will focus on these traditions not as static belief systems but as dynamic processes that have changed dramatically over time as various civilizations have risen, fallen, and come into contact with one another. The historical and sociological nature of all of these traditions is an essential starting point for this analysis.

Chapter 5 focuses on the religious ethos—that is, the relationship between religious and social life. Three themes emerge from the sociological literature on the nature of religious life, especially as it has been practiced up to the modern period:

1. Religion is a social phenomenon. Each religious tradition grows out of, and in turn acts back upon, the social life of the people who participate in it.

2. Religious traditions contain a systematic set of beliefs that are acted upon and sustained by *rituals* and *institutions*.

3. Each tradition constructs a religious ethos that defines the taboo lines between acceptable and inappropriate behavior, defines identities, legitimates social orders, and provides guidelines for everyday life.

Once I have described the historical religious context of the emerging global village, I will examine the twin crises of modernism and multiculturalism, identifying three themes that are usually implicit rather than explicit in sociological literature:

1. The advent of the modern world created a crisis for religious communities, challenging traditional beliefs, rituals, and institutions with scientific critiques and competing views of the world.

2. The multicultural context of the global village precipitated contradictory responses: a revitalization of ancient traditions (e.g., "fundamentalism"), **civil religion** and nationalism, and religious **syncretism**.

3. Religious traditions in the global village can promote chaos or community, either facilitating the construction of a peaceful world or intensifying and justifying violence between conflicting social groups.

The focus of Chapter 6 is the crisis of modernism and multiculturalism. As religious communities absorbed scientific teachings while concurrently confronting other faiths on a daily basis, people within each tradition struggled to find a way to solve the crisis posed by these assaults on the absoluteness of their belief system. People within each community often develop protest theologies that revitalize their own tradition as a way of resisting elements of the modern world or look at new ways to think about their faith. Chapter 7 examines the impact of social change movements on the faith traditions, including nationalism, alternative religious movements, and women's and environmental movements. Chapter 8 focuses on the relationship between religion and social conflict, first in the frequent link between religion and violence and second on the ways in which religious traditions develop nonviolent means of conflict.

The book concludes by exploring the ways in which religious communities can either promote violent conflict and warfare or cultivate a global culture that facilitates the creation of a community in which diverse groups coexist peacefully. Religious life, in its various forms, involves people striving to transcend their current existence—either in positive, heroic ways like the heroes and saints of human history and their everyday counterparts or to gain power and wealth by manipulating the reservoir of power deep within the human psyche for their own benefit.

I now embark on this inventory of the major religious narratives, practices, and communities inherited from the past to assess the current state of religion in the global village and consider the possibilities of its future direction.

2

A Sociological Tour: Turning East

I f a group of 10 people were taken to represent the world's religious communities, 3 would identify themselves as Christians, 2 as Muslims, 2 would be unaffiliated or atheists, 1 would be Hindu, and 1 would be Buddhist or a related East Asian religion (like **Taoism**). Another would have to represent every other religious group, including various folk religions, tribal and shamanist traditions, and Judaism (see Figure 2.1).

This model is complicated by the enormous number of people actually involved. On closer examination, for example, you would discover that each of these persons represents not only a religious tradition but particular elements of the world's social organization as well. Some religious traditions are surprisingly underrepresented; for example, Jews do not get a full representative because their numbers are relatively small. Because of the broad dispersion of Jews around the world (in more than 100 countries) and their impact on religious and political affairs, however, their influence is much larger than the numbers would suggest. That groups such as shamanists would not get their own representative would not surprise most of the readers of this book. More people, however, identify themselves as shamanists around the world (almost 10,000,000) than as United Methodists in the United States.

In this chapter, I will briefly examine some of the major beliefs and practices of the major religious communities in South and East Asia—especially Hinduism and Buddhism but also **Confucianism**, Taoism, and Shintoism. In Chapter 3, I will explore Judaism, Christianity, and Islam. Finally, we will look at indigenous religions in Chapter 4, which lie outside of the world's

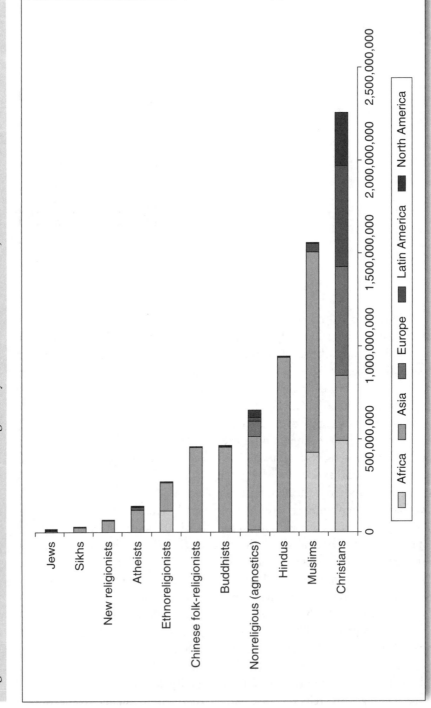

Figure 2.1 Worldwide Adherents of All Religions by Five Continental Areas, Mid-2010

Legend: Africa | Asia | Europe | Latin America | North America

Categories (top to bottom): Jews, Sikhs, New religionists, Atheists, Ethnoreligionists, Chinese folk-religionists, Buddhists, Nonreligious (agnostics), Hindus, Muslims, Christians

Axis: 0, 500,000,000, 1,000,000,000, 1,500,000,000, 2,000,000,000, 2,500,000,000

Source: Data from the *Encyclopædia Britannica Online.* Retrieved from http://www.britannica.com/EBchecked/topic/1731588/Religion-Year-In-Review-2010/298437/Worldwide-Adherents-of-All-Religions

major religions in more local forms but which also helped to shape the large **cosmopolitan faiths** and still interact with them. Because of the diversity within each tradition, the discussion will be highly oversimplified. It will, however, introduce the basics of the most prevalent traditions. Of special interest are the ways in which each of these traditions relates organically to a specific local social group and then changes over time as it diffuses and interacts with other religious perspectives and cultures. Although sometimes presented as immutable, each tradition proves in fact to be the product of centuries of transformation. Before beginning this tour, I will provide a brief sociological orientation regarding the various kinds of religions that will be encountered along the way.

Types of Religious Traditions

Most efforts to classify religions are, out of convenience, theological: The texts for most of the world's religions are readily available, although not always in good translation, so scholars can at least construct theological categories within which to place them. Do people believe in one God, many male or female Gods, or both? Are they more concerned with individual or collective religious issues, with the actions of Gods or humans? Is history presented as linear or cyclical? Was the world created by one God or many? Do people suffer because of their own actions or because of the actions of the Gods?

The very questions we ask about religious traditions reflect our own culture-bound interests, however, and it is often impossible to define "correct" or "orthodox" beliefs and practices from within any given tradition, let alone to identify them from the outside. How are scholars to judge which sacred texts are significant, and how should they be interpreted? To what extent have time and translation altered the original sacred texts? What was the sociohistorical climate when it was produced, and how does it compare with the current context? How wide is the gap between what Redfield (1957) called the "big tradition," the "official" beliefs as defined by elites, and the "little tradition"—that is, the popular version of a religion to which everyday people adhere? How much of each of these traditions should be used to characterize a religion's theology (if it even has one)? Which is more authoritative: the ancient or the recent versions of a tradition?

One theological distinction widely used by sociologists is that between **this-worldly** and **otherworldly religious orientations**—that is, does a religion emphasize ethical activity in this lifetime, or does it focus on what happens to people after they die or after a major transformation in the world, such as the coming of a **messiah**? Relying solely on religious texts to determine whether a tradition is this-worldly or otherworldly may be misleading, however; a text emphasizing rewards in the next life, for example, may serve to focus a believer's attention on how he or she lives in this one. Sometimes otherworldly language may refer to both an afterlife and to hope within this world, sometimes in disguised rhetoric as in most African American spirituals. Moreover, each tradition has such internal diversity that it is hard to characterize a whole tradition. Christianity may appear this-worldly in one congregation and otherworldly in another congregation just down the street.

Other issues must be addressed outside the textual sources: Is the tradition more concerned with doctrine or with practice? What are the socioeconomic and demographic characteristics of the religion's practitioners? Have these characteristics changed over time? Are they similar around the world? To what extent do beliefs and practices vary between cultures and across social strata within cultures? Are certain kinds of people or social groups more likely to practice the tradition than others are?

Those few broad, cross-cultural characterizations of religious traditions that brave scholars have attempted always fall short of the mark, but it is still helpful to get some sense of the varieties of religious life from a sociological perspective. Despite the impossibility of finding fully satisfactory answers, I will keep asking these questions throughout the book.

Evolutionary Schemes of Classification

A number of scholars of religion, especially during the 19th and early 20th centuries, developed evolutionary schemes of classification as they attempted to make sense of the broad range of religious behavior observed around the world during the colonial period. Using the metaphor of physical evolution, folklorist Sir James Frazer (1854–1941) argued in his classic *The Golden Bough* (1950/1992) that religion grew out of magic practices, an argument now widely discredited because of its ethnocentric implications about a model of progressive development leading from "primitive" to more "advanced" (the theorists' traditions are, of course, always advanced). Frazer's theory was similar to earlier contentions by Auguste Comte that humans have progressed from a theological to a metaphysical and finally to a scientific stage, in which arbitrary superstition gives way to rational scientific knowledge.

E. B. Tylor (1832–1917), sometimes called the father of modern anthropology, contended in *Primitive Culture* (1871) that the earliest and most basic religious forms were animistic, giving way first to fetishism, then a belief in demons, then polytheism, and finally monotheism. Marx and Engels (1844/1975) developed a theory of religious evolution as well, linked to their dialectical materialism. Because religion is an epiphenomenon of economic processes, they said, as the modes of production changed so did religion.

A more sophisticated evolutionary approach is Robert Bellah's (1964, 1970) typology of evolutionary stages. Not an entirely satisfactory model because it describes more the experience of Western culture than of all human culture, Bellah's typology is still helpful in examining some of the broad changes that have taken place over the course of human history. Bellah claimed to describe not a unilinear development from lower to higher forms but the ways in which religions and societies develop increasing differentiation and complexity of organization, thus becoming more autonomous relative to their environment. Such historical perspectives are important to understand continuities and discontinuities across historical epochs that shed light on the nature of contemporary religious beliefs, practices, and organization.

Sociological Classifications

Because the task is so difficult and the field lacks sufficient comparative research, sociologists have not developed a satisfactory scheme for classifying the world's religions. Ideally, a sociological approach would classify religions according to variations in the relationship between religious orientations and a variety of social criteria, such as (1) social strata, (2) different modes of production, (3) institutional and social diversity and complexity, and (4) various functions of religion, especially social functions.

1. *Social strata*, as examined by Weber (1922–1923/1946). To what extent do different strata have elective affinities with, and serve as carriers of, particular religious orientations, as suggested in Table 2.1?[1] One provocative treatment in the same spirit is the exploration of the differences between Goddess worship in "partnership" societies, on the one hand, and the warlike and hierarchical religious systems introduced by patriarchal invaders at the end of the Neolithic Age on the other (Eisler, 1988; Gimbutas 1982, 1989).

2. *Different modes of production,* suggested by Marx (1932/1972b), such as the relationship between the possible simultaneous emergence of monotheism and kingship during the agricultural revolution. Particular types of economic systems may have an affinity with specific kinds of belief structures.

3. *Degrees of institutional and social diversity and complexity*, as developed by Bellah (1964, 1970). This approach might also include different institutional strategies for "marketing" religious beliefs and rituals.

4. *Emphasis on various social functions fulfilled by religious life*, exemplified in Durkheim's (1915/1965) study of aboriginal religious life in Australia and elaborated by a number of contemporary scholars. O'Dea and O'Dea Aviad (1983) suggested that religion (a) provides support, consolation, and reconciliation; (b) offers a transcendental relationship that promotes security; (c) sacralizes norms and values of established society; (d) provides standards for critically examining established norms; (e) performs important identity functions; and (f) aids passage through the life cycle. Functional analysis also explores ways in which religion is dysfunctional in a social order.

I will now take a closer look at Max Weber's (1922–1923/1946) schema (see Table 2.1), which explores the religious determination of life-conduct in each tradition, beginning with the "directive elements in the life-conduct of those social strata which have most strongly influenced the practical ethic of their respective religions" (p. 268). Weber argued that "these elements have stamped the most characteristic features upon practical ethics" (p. 268) in each tradition. The Confucians who received prebends—that is, a stipend from the state for their religious duties—were drawn to a secular **rationalism** that shaped the direction of the Confucian ethos. The high-status Brahmans of India promoted status stratification as a religious ideal, whereas the mendicant Buddhist monks who lived by begging and owned no personal property favored an ethic of world rejection. Islamic warriors

Table 2.1 Weber: Elective Affinities Between Religious Ethics and Social Strata

Tradition	Stratum	Characteristics
Confucianism	Literary prebendaries	Secular rationalism
Hinduism	Cultured literati (Brahmans)	Status stratification
Buddhism	Contemplative mendicant monks	World rejection
Islamism	World-conquering warriors	Disciplined crusading
Judaism	Pariah people, intellectuals trained in literature and ritual	Rationalist
Christianity	Itinerant artisan journeymen	Urban and civic

promoted a theology of disciplined crusading; the intellectually oriented Jews coped with their pariah, or marginal, status through the cultivation of rationalist religious ethics; and the itinerant artisans of Christianity developed a civic ethic. Weber failed to note, of course, that Christianity also has its own crusading ethic much like Islam's.

Weber did not contend that the tradition is a "function" of the social situation of the stratum that appears as its "characteristic bearer," or a mere reflection of the stratum's interest. On the contrary, he claimed the following:

> However incisive the social influence, economically and politically determined, may have been upon a religious ethic in a particular case, it receives its stamp primarily from religious sources, and, first of all, from the content of its annunciation and its promise. (Weber, 1922–1923/1946, p. 270)

Weber's theory centers on his notion of the elective affinities between ideas and interests, as discussed in Chapter 1. Thus, the religion of intellectuals tends to be relatively theoretical and that of business classes more practical, because of the affinity between their respective material interests and the different orientations. Nonetheless, once worldviews are established, they have their own force: "Very frequently the 'world images' that have been created by 'ideas' have, like switchmen, determined the tracks along which action has been pushed by the dynamics of interest" (Weber, 1922–1923/1946, p. 280).

Of particular interest to Weber was the development of an emphasis on ethical systems, especially religions that call for action in the world. He finds this call more frequently in religious traditions carried by civic strata, especially when "they have been torn from bonds of taboo and from divisions into sibs [kinship groups] and castes" (Weber, 1922–1923/1946, p. 285). Out of these strata comes the tendency for an ethical prophecy to emerge, which articulates the gap between the world as it is and as it should be. It calls on believers and elites to close the gap by remaking the world.

Weber undertook a long-term historical–comparative study of the world's religions from a sociological point of view, thereby founding the discipline of modern comparative religion; however, he relied upon secondary sources that, though in most cases the best scholarship of his time, are now considered inconsistent and, in some cases, highly unreliable. Because of the increasing specialization of academia and the complexity of the task, no one has attempted since Weber to undertake a similarly ambitious task. Because no one has undertaken Weber's comparative project, the sociology of religion tends to be unfortunately parochial and often more focused on North

American, and especially Christian, religious life than I think it should be, although it has become less so in recent years as sociologists have become more integrated into the global community.

Classification by Local Versus Cosmopolitan Religions

The world's religious traditions have thus been classified in a number of ways by social scientists, such as by the relationship between religious orientation and various social characteristics (e.g., social strata and modes of production) and by stages of evolution. For my purposes, it is helpful to look at the ways in which all the existing religious traditions have transformed from local to cosmopolitan orientations. This will become more apparent as I describe the five dominant religions currently practiced around the world.

Whatever the reasons, it is clear that changes in society occur when religious and social life become more complex and differentiated. When social circles expand and social institutions encompass increasingly heterogeneous populations, changes in both cultural and social organization proliferate. Given the dramatic changes in today's world, one can expect many religious changes as well, interacting dialectically with social, political, and economic transformations.

German sociologist Georg Simmel (1858–1918) proposed in 1908 that "individuality in being and action generally increases to the degree that the social circle encompassing the individual expands" (1908/1971, p. 252). He expected that when homogeneous social circles originally isolated from one another come into contact, social differentiation would increase at first, but gradually an increasing likeness would emerge between the groups. Processes of "differentiation and individualization loosen the bond of the individual with those who are most near in order to weave in its place a new one—both real and ideal—with those who are more distant" (Simmel, 1908/1971, p. 256). What Simmel did not know is that a localizing backlash would emerge against the globalization tendencies occurring in his era at the turn of the 20th century as people rebel against being overwhelmed by large-scale global structures. Many empires and large nation-states have split apart, for example, as various ethnic groups demand their own nation, as happened in the disintegration of the Soviet Union. Still, ironically, this localizing process is part of a global trend—claims by ethnic groups for their own country in one part of the world are inspired by movements for self-determination elsewhere.

The primal, **local religions** in relatively isolated locations focus on direct connections between life and existence as it is experienced locally, thus highlighting the continuities between human and natural life; the cycles of life and death; and the significance of particular animals, plants, and elements of the local ecology. A particular mountain or animal species may be considered sacred and be featured in religious ceremonies because of its importance in the local environment and the lives of humans living there. As social circles and religious traditions expand across diverse cultures, a transformation occurs. At first, differentiation emerges within the circles and the traditions: encounters with other groups and religions with different characteristics accentuate the previously unnoticed differences within each group as well as commonalities between them.

When a religious tradition crosses cultural boundaries with a conquering force that imposes it upon indigenous worldviews and lifestyles not only is the preexisting culture transformed but the conquering tradition is as well. In the spread of Buddhism across Asia and the diffusion of Christianity along with the European colonialists, each tradition became more diverse as it adapted to a wide variety of local religious traditions and institutions. All of the dominant religions in the world today are, by definition, cosmopolitan rather than local, because they have diffused around the world over time and thus have been transformed. Even the traditions that remain more tribal—Judaism and Hinduism and to a lesser extent Islam—have been so altered by encounters with countless local traditions that they have lost their geographic particularity as their beliefs and practices have become more varied. Primal elements run deep within each cosmopolitan tradition, however, and the continuity of symbols and stories across the centuries persists to some extent even among the most cosmopolitan sectors.

Of course, for many people religion is simply a tool to achieve a sense of security or meaning or even the basics of life such as food, clothing, and shelter. If you spend a few hours surfing the net you will find thousands of "varieties of religious experience," as William James (1902/1960) put it more than a century ago. You will find everything from sophisticated theories about the universe and profound ideas about spiritual growth to "secrets" about how to pray for large sums of instant cash. In the meantime, I will commence this tour of some of the major traditions as viewed from a sociologist's eye. I begin with the oldest identifiable traditions, which came from ancient South Asia.

Hinduism, or Sanatana Dharma

> *Spiritual knowledge is the only thing that can destroy our miseries forever; any other knowledge removes wants only for a time.*
>
> —Swami Vivekananda

From one part of South Asia to another, the rich mixture of cultures appears in the panoply of religious customs varying from region to region. India itself is much like Europe in that it has an overall culture but many different cultures within it. We can identify an Indian culture, certainly, but also a Tamil and Maharashtran culture, a Bengali and a Rajastani, a Karnatakan, Keralan, Telegu, and Gujarati culture (just as Europe has French, German, Spanish, British, etc., cultures). Within each location, there are also extreme differences. A wide range of Gods and rituals reflect and inform the diversity of roles and situations in which humans find themselves as they move through their life cycles. To some extent, each temple, household, and individual chooses Gods and rituals because of situational requirements, personal disposition, gender, caste, occupation, and stage in the life cycle. The underlying theme of this elaborate religious system is that each individual must discover his or her own path—with the aid of religious experts—to cope with and transcend life's suffering and eventually to escape it.

Hinduism—or Sanatana Dharma, as some believers prefer to call it—is a religious tradition that encompasses layers of complex deposits from many different cultures over the centuries. Its remarkable diversity and doctrinal tolerance combine with a highly elaborated worldview and ethos that provide hundreds of millions of people with compelling answers to the basic questions of human life. Because of this religious anchor, they can live out their daily lives with a sense of dignity, despite widespread poverty and deprivation, and feel connected to a larger community.

As in many other societies, however, the stratification system is sacralized— that is, made a part of the religious belief system. The exploitation of the masses by a small number of extremely wealthy and powerful families is legitimated by the traditional Hindu worldview. In recent decades, as these problems have been addressed by Hindu reformers, this ancient tradition is incorporating the egalitarian norms of modern culture under its sacred canopy alongside ideas and symbols that have endured for thousands of years.

The wide-ranging collection of beliefs and practices known as Hinduism dates back about 3,500 years, linking contemporary and ancient India. Scholars dispute the origins of Hinduism; some believers claim that it had no beginning. Contemporary Hinduism probably emerged out of the encounter between the indigenous Dravidians and the Indo-Aryans[2] who migrated into India from Central Asia or Iran about 1500 BCE.[3] Some contemporary scholars believe that practitioners of the **Vedic religion**—that is, the faith based on the ancient Vedas now called Hinduism—encountered Jainism and Buddhism between the 5th and 2nd centuries BCE, laying the foundations of modern Hinduism. Hindu ideas and practices evolved over the centuries across the subcontinent in a myriad of forms as various cultures of the region became intertwined, prefiguring the process of intercultural contact now occurring in the larger global village.

The strength of Hinduism has been its rich combination of highly rational with nonrational symbolism on the one hand and its adaptability and theological tolerance on the other. Perhaps the most consistent theme in Hinduism is variety, and a central tenet of the perspective from the religion's earliest periods has been the belief that there are many paths to the "Truth," with the result that Hinduism has taken root and grown in a wide diversity of cultures on the Asian subcontinent. One of its most essential ideas is that people can reach their ultimate goal—to break the chain of rebirths in a process of liberation called **moksha**—by discovering one's own **dharma** (duties or responsibilities) and performing them well, thus enabling one to advance in a new lifetime or eventually to escape the cycle altogether. The idea that different paths lead to the same summit opens the door to remarkable flexibility in overall orientation and specific practices under a single broad sacred canopy.

At the center of the tradition is an ancient and diverse collection of hymns, ritual chants, and stories about encounters with the sacred in its myriad forms. The religion's earliest beliefs and rituals are expressed in the hymns called the Vedas, which were originally transmitted orally, like most ancient religious literature, and compiled about 1500 BCE, at the time of the Dravidian–Aryan encounters. The Vedas themselves, according to tradition, were not authored— not even by the Gods—but have always existed along with the very sounds of the universe itself, and were "discovered" by the sages who then transmitted them. Gods and humans interact regularly in these texts, creating stories that have been the lifeblood of social life in Indian villages over many centuries.

The *Rig Veda*, which consists of 10 books with liturgical chants and sacrificial formulas, is seen as the source of all Truth and provides the basis for much of the material in other Vedas, each of which has a huge body of interpretive literature. "Ekam sat, vipraha bahudha vadanti," the *Rig Veda* declares the following: "Truth is One, though the sages call it by various names" (O'Flaherty, 1981). This verse and others lead some to insist somewhat ironically that monotheism began not with the ancient Hebrews but at least 3,500 years ago in the Vedas.

Six or seven centuries before the Common Era, a collection of philosophical speculations was compiled as the **Upanishads**, first as part of a secret cult threatening the religious establishment but later as a central component of Hindu literature. Finally, the massive epics of the *Mahabharata* and the Ramayana are the classics of Hindu literature, read and told by village storytellers, priests, and elders for centuries. The former consists of about 100,000 verses (twice the size of the *Iliad* and the *Odyssey*) and includes the famous **Bhagavad Gita**, the "Song of the Lord." The *Mahabharata* has been revived in the late 20th century in a somewhat sensationalized form, in a television series that has riveted Indian audiences.

The Gods serve as models both for everyday life and for the structure of authority in Indian life. Just as the Gods' authority flows from their heroic deeds, so individuals with authority must prove their worthiness and receive considerable deference once they establish their legitimacy. The Gods of Hinduism are highly specialized in character and function. They number, according to some estimates, about 33 million, and at least one God addresses each social and psychological need, whether it be safety (**Vishnu**), wealth (Lakshmi), or liberation from the pursuit of safety and wealth. The elephant God Ganesha is the overcomer of obstacles who aids people in need of courage or assistance in times of trouble; consequently, most temples contain his image even when devoted primarily to another God. Most modern Hindus, however, believe that only one God stands behind the multitude of these manifestations (just as most Christians believe that the Father, Son, and Holy Spirit are different manifestations of the same God).

This complex belief structure serves to unite the ethnic and economic diversity of a large majority of India's 1 billion people. These social groups and subcultures interacting with one another across the vast Asian subcontinent are loosely bound in a shared worldview while maintaining their autonomy. Each temple, and most households, has its own style of worship, choosing those Gods and rituals that have an affinity with its interests. Nonetheless, the Vedic texts and Hindu epics, the stories of Gods and humans interacting, stretch a loosely woven sacred canopy over the heterogeneous social order of the Indian subcontinent.

A central task of Hindu philosophers and theologians has been making sense of these diverse cultures within a single geographical territory, and an effort to address that social problem lies at the root of major doctrinal development within Hinduism. The famous teacher **Shankara** (ca. 788–829 CE) attempted to systematize Hinduism by claiming that all reality is one (monism) and that we live in the illusion (**maya**) that individuals have a separate existence from the universe. The Supreme Reality, the Brahman, is unqualified and absolute in contrast to the world in which we live with its cycles of suffering, death, and rebirth. In the same way that actors perform a play, our lives have a provisional or dependent reality, dependent on the Brahman, or ultimate reality. Thus the unity and diversity of the world are simultaneously affirmed in Shankara's school and the problem of social diversity is thereby solved theologically: There appear to be many Gods, but there is really only one. There appear to be many social orders, but there is only one—many individuals but each soul contains and is contained in the Infinite. Shankara's ideas prevail in Indian philosophy today, reinforced by the elaborate institution of his monastic order, which controls many of India's most important Hindu temples.

A second major school of Hinduism was started by **Ramanuja** (ca. 11th century CE), who claimed that a single unified reality exists but it is qualifiable. Matter, souls, and God all exist in the world, but the first two are simply qualities of God. The popular appeal of this perspective is that it does not involve the rejection of the world implied by the monism of the Shankara. According to Ramanuja, the material universe, including one's family and all the various Gods, are all aspects of the one ultimate God. A third major Hindu school emerged in **Madhua**'s 13th-century dualism—called *Duaita Vedanta* that views souls as distinct from God and matter. Madhua believed that at least some souls could become enlightened and that one can achieve moksha through devotion to Vishnu. The two major branches of Hinduism—the **Shaivites** (embracing the teachings of Shankara and Ramanuja) and the **Vishnaivites** (embracing the teachings of Madhua)—continue the dialogue between the monism of Shankara and the dualism of Madhua, respectively. Such philosophical debates are not central to Hinduism, however, which is rooted in ritual rather than doctrine.

In the many manifestations of Hinduism on the Indian subcontinent, the boundaries among the natural, human, and divine worlds are often unclear, reflecting the reality of a life that is much more integrated across species than that most Westerners experience. Even in the modern cities of India, cows, pigs, water buffalo, goats, and sometimes monkeys and other animals live side by side with humans. People who exhibit great spiritual powers and are revered because of their accomplishments become deified in popular culture.

Hindu religion expresses this fluidity of boundaries among the diverse parts of a unified world in a number of ways: Hanuman is sometimes a monkey, sometimes a God; Ganesha is sometimes an elephant, other times divine. Ram, a name widely used to refer to God, is a legendary human hero and is also an incarnation of the God Vishnu. Major figures in human history become godlike, and people treat them as such. In addition to small shrines to Vishnu and Shiva, for example, one also finds shrines to Mahatma ("Great Soul") Gandhi. People take some Gods' names for their own, and even businesses may be named after a God.

The focus of Hinduism for most believers lies not so much in theological arguments but in the actual practice of worship, primarily in acts demon-strating respect for the divine in one's home or temple. Humans and Gods are mutually dependent: The Gods require sacrifices or offerings, especially in the form of food, to stave off chaos and sustain nature; if they thrive, they offer benefits to humans. Moreover, Hindu theology reveals its social origins in the fact that each God represents an important aspect of human life.

Many Hindus believe in a single "Godhead," made up of three elements: (1) **Brahma** and **Saraswati**, the Creator; (2) Vishnu and Lakshmi, the Preserver; and (3) Shiva and **Parvati**, the Destroyer (see Table 2.2). Each of these deities has a stylized form of representation in both female and male forms.

Each of the mysteries of human life is thus personified in a particular God. Issues of creation and death are embodied in Brahma and Shiva, respectively. The tension between the two for preservation and survival is addressed in the person of Vishnu, who intervenes with heroic action at times of crisis to give people hope (see Table 2.3). When Mother Earth sank to the bottom of the ocean, for example, Vishnu took the form of a boar, dived down into the water, and rescued her.

Table 2.2   The Hindu Trinity

Creator	$\longrightarrow$	Brahma (male) Saraswati (female)
Preserver	$\longrightarrow$	Vishnu (male) Lakshmi (female)
Destroyer	$\longrightarrow$	Shiva (male) Parvati (female)

Table 2.3   Avatars or Incarnations of Vishnu

Avatar	Form	Heroic Action or Purpose
Matsya	Fish	Help re-create the world after a flood
Kurma	Tortoise or crocodile	Preserve immortality for the gods
Varaha	Boar	Save sinking Mother Earth
Narasimha	Half man, half lion	Destroy demon Hiranyakashipu
Vamana	Dwarf	Destroy Bali, king of demons
Parasurama	Angry man	Destroy autocratic princes
Rama	Perfect man	Overcome injustice, establish just rule
Krishna	Divine statesman	Teach action without desire
Buddha	Compassionate man	Purify Hinduism of ritualism
Kalkin	Incarnation to come	Save humans at the end of this age

The male aspects of God represent the rhythms of natural life, whereas the female aspects are more closely linked with human nature and therefore more accessible to ordinary people. A Mother **Goddess movement** emerged around Durga-Parvati, a body embodying the immanent active energy (shakti) of the distant and terrifying Shiva, who is transcendent and generally inaccessible to common mortals.

Part of the key to Hinduism's durability is its ability to incorporate other religious symbols. The God Vishnu plays a particularly important role in this process because he becomes identified in various **avatars**, or incarnations (including the ever-popular Krishna), with local deities as Hinduism becomes more cosmopolitan over time. Some Hindus reincorporated Buddhism by viewing the **Buddha** as an avatar of Vishnu; others have gone so far as to include Jesus in the list. When asked about that idea, many Hindus respond "Yes, of course," whereas others scoff at the idea. This is an interesting example of the flexibility of Hinduism to adapt to new situations, especially when dealing with foreign invasions. During the late 19th century, with the arrival of British colonialists, several important Hindu authorities began to explore the relationship between Christianity and their native faith and found a number of parallels. Some became exponents of Christianity as well as Hinduism: Sri Ramakrishna had a vision in which he was embraced by Christ, and Swami Vivekananda called on his fellow monks to become "Christs" and "to pledge themselves to aid in the redemption of the world" (Prabhavananda, 1963, p. xiv). Swâmi Abhedânanda, in his 1902 essay "How to Be a Yogi," expressed Hinduism's affinity with Christianity in a particularly Hindu way:

> A genuine seeker after Truth does not limit his study to one particular example, but looks for similar events in the lives of all the great ones, and does not draw any conclusion until he has discovered the universal law which governs them all. For instance, Jesus the Christ said, "I and my Father are one." Did He alone say it, or did many others who lived before and after Him and who knew nothing of His sayings, utter similar expressions? Krishna declared, "I am the Lord of the universe." Buddha said, "I am the Absolute Truth." A Mahometan Sufi says, "I am He" while every true Yogi declares, "I am Brahman."

Hinduism enables people to cope with life as it is. Some characteristics of one's life change over time; others, such as socioeconomic status, nationality, ethnicity, and gender, remain relatively stable. Hinduism teaches people that their happiness and fulfillment lies in discerning their own dharma (duty) and carrying it out to the best of their ability. In this analytical framework, given stages in the life cycle, occupational tasks, and other roles show affinities with particular Gods and rituals under a loosely woven sacred canopy that integrates the diversity of individuals, roles, groups, and ethnic communities living on the subcontinent. At its best, Hinduism enables people to identify, yet in some ways transcend, the givens of their life situations; at its worst, it legitimates a system of social stratification in which privileged and poor alike become convinced that the inequalities of society are both fair and immutable.

The ancient tradition of Hinduism provides some interesting models for the cultivation of religious tolerance within a broader social tradition. Deities and their representatives alike are highly revered, yet they are not exclusivistic. They must prove their worthiness by providing perceived benefits, whether spiritual or material, psychological or social. The pluralism of Hinduism within this broad unity is accomplished by means of an extraordinarily large perspective from which it is viewed, as shown in Hindu cosmogonies. Just as the details of life on a planet seem unimportant when viewed from a distance in space, so the teeming particularity of subcultures, castes, families, and individuals becomes less urgent from the universalistic perspective of the Hindu cosmogony.

Hindu Theodicies

The explanation for suffering and death in Hinduism is related to the ultimate goal of existence—that is, to become liberated from the world by uniting the individual soul or spirit known as the **atman,** with the Brahman, or Universal Soul, that encompasses the entire universe. That goal is not easily reached, however, and is accomplished only by struggling through successive lifetimes during which one must fulfill one's dharma. This worldview results in a highly rational theodicy, which explains suffering by referring to one's **karma,** or actions from this and previous lifetimes. Each individual soul goes through a cycle of rebirths, known as *samsara,* or the wheel of life, the endless round of deaths and rebirths: When a person dies, the soul leaves its body and transmigrates to another. The nature of the next reincarnation is determined by the person's karma—that is, the collective consequences of all individual actions. The Gods do not punish or reward; negative actions bring their own dire consequences, and positive actions bring their own rewards. The law of karma, or action, is a basic notion of cause and effect: "As we sow, so shall we reap" is the saying one Hindu author uses to explain the law (Jagannathan, 1984, p. 54), which has a status in Asian thought similar to the law of gravity in Western science.

The consequences of every thought, word, and deed grow exponentially. As Lama Sopa, a Tibetan authority, explains, when one sows seeds, a tree grows that in turn bears fruit with new seeds that become new trees. The impact of karmic action thus has a ripple effect or is exponential in its consequences. All elements of the cosmos are interdependent, constituting a self-contained cosmos of ethical retribution, which does not require the intervention of Gods to punish or reward. Encounters with the Gods are primarily for educational purposes so that people can learn how to be rewarded, rather than punished, for their actions. You are responsible for all of your actions—touching fire will cause a burn whether knowingly or unknowingly. One's individual life at any point is the summation of all previous actions so that together we create the kind of social environment in which we live.

In this system, one is not to fear or mourn death; it is simply another passage. The soul leaves one body and enters another in the same way that a person changes clothes. The broad cosmology of the Hindu tradition allows the individual to look somewhat philosophically upon the transitory pain of present existence. Even if life seems intolerable, a person can work diligently to do the best with his or her current lot and thereby look forward to a better life in the future.

The karma–samsara concept contains a deterministic element that sometimes convinces people to accept their fate and not try to change their immediate life circumstances, which amount to rewards and punishments for actions in a previous life. Some karma (praradbha karma), such as family or environment, is beyond our control. The karma–samsara theory is supposed to facilitate the individual's transcendence of the profane life by endowing it with religious duty. The system leads to remarkable abuse when exploited, however, by providing a powerful rationale for the ruling classes and legitimating a false consciousness among the poor, who are taught that their poverty is punishment for their deeds in a previous lifetime. This is an ideal rationale for what sociologists call "blaming the victim" (Piven & Cloward, 1971; Ryan, 1976).

The idea of karma is not entirely deterministic, however. First, two of the three stages of karma are amenable to change. The accumulated karma of all previous births (samchita karma) and actions in the present life determine a person's future (agami karma). The impact of the habits of previous lifetimes can be altered by cultivating new habits and ridding oneself of evil thoughts and desires. Finally, a better life in the future can be constructed through attention to present life duties. Moreover, we have the freedom to choose whether or not to live according to our dharma—that is, the duty appropriate to the state produced by individual karma—just as we can decide to ignore gravity if we are willing to face the consequences. Finally, even though we should not strive to change our own life situation, we should attempt to improve others', an idea promulgated both by Hindu activist Mahatma Gandhi and the current Tibetan Buddhist leader, the 14th Dalai Lama (Gyatso, personal communication, 1990; Kurtz, 2005a).

In the Asian religions, then, suffering is seen as integral to the fabric of existence, as we experience it. It enters into human consciousness when people develop desires for worldly or material objects and become excessively attached. The solution to the problem of suffering is moksha (liberation) in

Hinduism, or the attainment of enlightenment or *nirvana* (that is, supreme bliss) in Buddhism, which allows one to escape the "wheel of karma–samsara." In Hinduism, the paths of **Yoga** cultivate detachment of the self from dependence on this world, allowing one to escape it. Similarly, in Buddhism, following the Eightfold Path prescribed by the Buddha allows us to reach Enlightenment, or nirvana.

Death in Hinduism is sometimes a means of liberation, and a believer does not even have to wait for its natural occurrence. A widow may jump onto her husband's burning funeral pyre, or a devout Hindu can become a *sanyasi*, who renounces the world and lives a life of isolation and self-denial. The sanyasi "dies" to his own life; funeral rites are celebrated and the person's inheritance is passed on as if he were dead. The Hindu theodicy is most fully developed in the legends of the God Shiva, who is simultaneously destructive and life giving, representing death but also the recreation of life, as in the cycles of death and rebirth in nature. This close connection with the natural world reveals the primal origins of Hindu beliefs. Although many Indians live highly urbanized lifestyles, the country is still based on its villages, and popular religion is intimately tied to the struggle to survive in a relatively hostile natural environment in which resources are scarce and months of drought are followed by violent monsoons in which millions are sometimes left homeless or killed.

The Hindu concepts of dharma and **transmigration** explain the deprivations of individual existence in a highly rational manner, at the same time giving hope for a better lifetime in the future. They provide a source of comfort at times of death as well, because the end of this life is merely a passage into the next one. One's fate in each successive lifetime is self-determined, within the boundaries of the law of karma, so that one can earn a better life in the next incarnation and hope for ultimate release from the struggle. Although this theodicy solves many problems on both psychological and social levels, it has also been exploited to maintain a strict social hierarchy that has come under attack in contemporary Hinduism.

Hindu Rituals

Yoga and Three Paths to Enlightenment

Yoga (literally, to yoke or unite) is the Hindu ritual for uniting the soul with God through meditation and certain ethical practices. Hindu rituals are extremely well developed, spelled out in rich complexity and elaborate detail in ancient ritual manuals. At the most abstract level, Hindu rituals solve the threat of chaos. Discord is held at bay by the Gods, who in turn require gifts (especially signs of deference and food) to do their work. Thus, it is the obligation of each Hindu to carry out the rituals of the faith as an individual contribution to maintaining the cosmic order.

Because of the syncretic and flexible nature of Hinduism, a number of paths to enlightenment are possible, all leading to the same summit. All of them involve disciplined self-transcendence to overcome excessive attachments to this world. Hinduism identifies three main paths, with a different yogic discipline to facilitate progress along each route:

1. The Path of Wisdom or Knowledge, **Jnana Yoga** (for reflective persons)

2. The Path of Action, **Karma Yoga** (for active persons)

3. The Path of Devotion, Bhakti Yoga (the most popular path)

Jnana Yoga, the quickest but therefore steepest and most difficult path, is rarely chosen. It requires not only cognitive but also intuitive knowledge gained from a disciplined reading of the Vedas, the Upanishads, the Bhagavad Gita, and other works along with the guidance of a guru who helps the inquirer to reflect upon the readings and engage in deep meditation on the Absolute.

Karma Yoga, also relatively rare, focuses on selfless acts or service, so that action—even in one's work—becomes a form of worshipping God. The key to liberation through Karma Yoga is doing a task for its own sake rather than for any reward. Consequently, the process contains a paradoxical element: If practiced only for one's own liberation, it will not work because it will not be selfless action at all.

Finally, Bhakti Yoga, the broadest and most popular path, consists of acts of devotional worship and involves daily poojas—that is, expressions of respect toward a representation of God in one's home, a temple, or one of the small shrines that dot the Indian landscape. The aim of these rituals of personal worship in Bhakti Yoga is to become consumed with love for God and therefore to be intoxicated by a divine vision.

Choosing the path most suitable to personal inclinations and abilities, the believer strives to transcend selfish desires and the routine of profane life through disciplined practice of the appropriate rituals. Such practices may be as elaborate as a sustained pilgrimage to a holy place or as simple as chanting the Lord's name in the privacy of one's home. For every devout Hindu, power accumulates in the performance of rituals so that the use of ritual packages sustains them in daily life and at times of crisis.

Rites of Passage

The rich ritualism of Hinduism extends beyond both cradle and grave, beginning with the fervent prayer for a child called Garbhadana (conception) and continuing to Punsavana (fetus protection) during the third or fourth month of pregnancy; Simantonnya, satisfying the craving of the pregnant mother in the seventh month; and Jatakarma that welcomes the baby into the family at birth. Mantras are recited for a healthy long life; Namakarna is a naming ceremony, and Nishkarma is performed to take the child outdoors for the first time.

In the initiation ceremony that occurs when the child enters the Brahmachari, or student, stage of life, a sacred thread is put around the neck of the child, who is taught a mantra for worship and begins school or training with a guru, or religious teacher. Usually only male children receive the thread, which is to be worn for life, but ancient temple sculptures showing women wearing the thread may suggest that the practice was formerly universal. Some reformers advocate introducing the ceremony for both boys and girls of all classes and sects (Jagannathan, 1984, p. 63).

The ritual climax of the Hindu life cycle is the marriage ceremony, in which the couple is treated as Gods in the sacred interval when the bride, the groom, and (just as importantly) their two families are united. A priest chants appropriate scriptural passages for several hours while the wedding party engages in an elaborate procession through the streets and then feasts. In this way, the union is sacralized and made a part of the universal order; the bride and groom enter another stage of their life cycle and thus obtain a new dharma, which they must identify and fulfill.

The final ritual ceremony of a Hindu's life is the antyeshti samskara, in which the body is cremated on a pyre lit by the deceased's eldest son, if possible; the body is burned since it is no longer needed, and the soul is released. The ashes are often scattered on a river, especially the Ganges. The funeral is followed by a period of restrictions placed on the relatives, and then offerings are made to the departed soul. Traditionally, a wife sometimes joined her husband in passing from one lifetime to the next by throwing herself on the funeral pyre, a practice called setee, which still occurs but is widely denounced.

Thus a sacred canopy is drawn over the individual's and the community's life. The 1,028 hymns of the *Rig Veda* are recited at initiations, weddings, and funerals not only because their very sound is holy but also because they guide everyone through each lifetime, past death, and into the next life.

Decentralized Structures in Hinduism and Buddhism

Hindu institutions and their priests do not have the same authority as priests in the traditional **Roman Catholic Church**, for example. In traditional Catholicism (especially before the Second Vatican Council substantially democratized the church in the 1960s), the bishop appointed priests, and people were expected to defer to them no matter how competent they were. This was an "authority of the office" rather than of the person. The Hindu system rather resembles American Protestant churches, in which the authority of a religious leader or institution is based primarily on charismatic qualities such as the ability to perform and deliver in a way that satisfies people in a congregation. The authority of the ancient texts persists in popular Indian culture, as does, therefore, the importance of their interpreters, the **pandits**. Hindus believe that answers to any question can be found in these writings, and the pandits can interpret their advice on everything from the origins and purpose of the cosmos to the meaning of an individual's own life, from ethical dilemmas to important decisions such as a choice of occupation or marriage partner.

Eastern religious traditions have never had strong, centralized institutions as in the West, and there is no central authority like a pope or patriarch. This both reflects and shapes the nature of the beliefs and practices of the faith, which are thus more decentralized and flexible. Moreover, Hindu worship takes place in households as well as in temples, so that a wide range of authority structures sponsors religious practices and the family remains a religious institution as well (as in the "primal religions") despite specialized institutions and a priesthood. Since people can choose their Gods and gurus, the religious pluralism of India cultivates a diffusion of authority. This might give us a clue about the nature of religion in the global village, in which a large variety of religious institutions exist side by side and people have considerable latitude in choosing from the sacred marketplace (see Warner, 1993; cf. Iannaccone, 1990; Lee, 1992; Robertson, 1992b).

Two broad branches of Hinduism center around two of the most important Gods, the Vishnaivites (or Vaisnavas) who worship Vishnu, and the Shaivites (Saivas), or followers of Shiva, and each branch has its own temples, although people can choose freely to go to more than one. A third significant group, the **Saktas**, is usually treated as a subdivision of the Shaivites. Temple priests, called *pujaris*, serve the deity in each temple, treating the God's representation as an honored guest. Being responsible for the God's maintenance, the priest has considerable authority that is linked to the God's reputation. Hinduism is, of course, replete with famous teachers who claim to have become enlightened, gather a body of disciples, and become known as gurus, whose authority is essentially of the charismatic variety, confirmed by the testimony of his or her followers but sometimes routinized into a more formal movement.

Some of the most sacred sites in Hinduism, such as the locations of encounters with God, or **dharshans**, are given special authority because of the traditions and stories of miracles, serving as places of pilgrimage for believers. Consequently, those in charge of the famous temples built at those spots have a particular authority. Of particular importance are the five **Shankaracharyas**—that is, *acharyas* (teachers), who succeeded the famous 8th-century Shankara (who systematized Hinduism), traveled by foot throughout India and established a monastic order with *matams* or *ashrams* (dwelling places for the monks) in each part of the country. A Shankaracharya governs the matams of the order in each of five locations around the country and often manages the major temples in his respective region. These leaders are not appointed or self-appointed but are "found" through a series of signs and clues given by the Gods and have great charismatic authority among the faithful. When I went to interview the Shankaracharya of Kanchipuram in South India, there was a huge line of pilgrims waiting to see him. I was told that he would meet each one briefly and give them a dharshans and a blessing. Fortunately for me, I was escorted by his brother who took me in a back door; I felt a bit guilty that the long line was kept waiting while we talked.

The Internet now plays a role in transmitting centuries-old instructions about rituals. The Hindu Temple of Atlanta, for example, has a website that provides not only a virtual tour of the temple and a way to sign up for e-mail notices about events but also information about various Gods and Goddesses, complete with prayers and chants that can be downloaded (http://www.hindutempleofatlanta.org/). The website even contains detailed instructions for members of the community to help them with their rituals and prayers, for example, what supplies to bring to a funeral ceremony.

The Hindu solution to questions of ordering the collective life of a highly diverse population is reflected in this religion's theology and its organization, which in turn help shape the nature of Hindu society. Respect for authority in general is highly regarded and internalized; Hindus are expected to be submissive to legitimated authorities. Any claim to authority must be proven, though, and a variety of options are available, some specific to particular social functions, allowing believers to choose Gods and teachers they find most appropriate for their needs. The fact that charismatic authority is highly regarded prevents the excessive rigidity of institutional forms. Moreover, the criteria for authority are somewhat functional—that is, any prospective God or guru must bring benefits to his or her followers or suffer from neglect.

Jain and Buddhist Protest Movements

Virtually every dominant religious system eventually faces opposition that either reforms a tradition internally or even breaks off as a **new religious**

movement (NRM). Hinduism is no exception—by the 6th century BCE, Hinduism was well entrenched in South Asian society and had serious critics. The most striking movements to react against it were Buddhism and Jainism, both of which not only became religious traditions in their own right, but also changed Hinduism forever. Both Buddhists and Jains, who appeared in the Ganges basin region of Eastern India, rebelled against the caste system and rigid ritualism of Hinduism and the alleged abuses of the Brahmanic authorities. (These issues were taken up again in 20th-century movements as well.)

The most well-known story is that of Prince Siddhartha, born in the 6th century BCE to the wealthy King Suddhodana and Queen Maya. Dissatisfied first with his privilege and then with ascetic renunciation, he became the Buddha and founded one of the most significant religious movements on the planet that persists and flourishes today. I will discuss it in a moment as a separate tradition.

The other revolt against the Hindu establishment was the tradition of the Tirthankaras, the Jainist teachers culminating in Lord Mahavir, with their hallmark teachings of *ahimsa*, or nonviolence toward all living creatures. Born around the same time as Prince Siddhartha, Mahavir offers a parallel story. Unhappy with his status as a prince, Mahavir became an ascetic in the solitude of the forest for 12 years, from the age of 30 to 42. According to the Jain tradition, he reached perfection, or enlightenment, through concentrated meditation, intense purification of the self through spiritual practice and discipline. He combined the scientific knowledge of his time, and its emphasis on reason, with a quest for spiritual knowledge, and recruited 11 disciples (*ganadharas*) from among the Brahmans. Through years of fasting and sleep deprivation, he developed an intense meditative life that allowed him to become oblivious to physical pain and discomfort, detached from attachment to physical appearance and material goods.

A forerunner of egalitarianism and contemporary nonviolence, Mahavir reportedly spoke to people in their own language and addressed all castes, creeds, and genders without discrimination. He urged people to find happiness within the self through disciplined spiritual advances toward the ultimate goal of freedom from the cycles of rebirths, to fully realize the self and achieve a permanent blissful state. He taught that continual rebirths are a product of defiled conditions of the soul, characterized by desire that in turn leads to actions that draw karmic matter leading to further defilement and embodiment. One is weighted down, as it were, into the cycle of rebirths; the way out is to quench the power of the mind's cravings by meditation and physical deprivation to empower the soul.

Jain philosophy, following Mahavir, emphasized Right Knowledge (regarding self-understanding and self-control), Right Faith, and Right Conduct, or the Five Vows:

1. Truthfulness

2. Nontaking or nongreed

3. Celibacy, renunciation of sensual pleasure

4. Nonattachment, nonpossessiveness

5. Ahimsa, nonharmfulness

In carrying out these vows, which are stricter for monks than for nuns, Jains try to live as nonviolent a lifestyle as possible. The lifestyle of monks and nuns is so filled with scripture study and meditation that there is little time for eating and sleeping. Great care is taken not to harm or kill any life forms. Violence in thought precedes violent behavior, so all forms of injury must be avoided—whether by body, mind, or speech. A strict vegetarianism prevents Jains from eating not only meat but also vegetables grown below the ground, as that disturbs the lifeworlds of various creatures. Members of religious orders will thus have only a few possessions—a robe, an alms bowl, a whisk broom, and a *mukhavastrika*, a piece of cloth worn over the mouth to protect against accidental ingestion of small insects. They do not eat after dark or drink water that is not boiled; to do so might involve harming insects.

Because the current age is so corrupt, it is not possible to seek enlightenment in this lifetime; rather, Jains hope to move through disciplined and meritorious nonviolence toward a rebirth that will bring them closer to the desired state. Although it is not possible for most of us to imagine such an austere lifestyle, the presence of the monks in our violence-ridden, environmentally destructive time is a reminder of the violence of our lifestyles. Whenever I drive my car at night and find the windshield splattered with insects, I am reminded of the Jain monks and the harm that the convenience of my car causes to the environment. Moreover, we have all been influenced by the Jains through Raychandrabhai Mehta, a friend of Mahatma Gandhi. Gandhi said Mehta had an important impact on his development of his philosophy of nonviolence. The Jain effort to avoid harming all creatures is echoed in the other great tradition born in South Asia, Buddhism.

Buddhism

India is a land rich in religious pageantry and history, and it is in that fertile soil that the seeds of Buddhism were planted. The enormous gap between wealth and poverty at the time (which persists to the present) helped pre-cipitate the founding of this religious tradition, which asserts that life is full of suffering that can be escaped by showing compassion to all creatures in the world. In Buddhism, a person gains merit by serving others and rejoicing in their good fortune. From this starting premise, we can see that Buddhist thought is extraordinarily rational yet also deliberately nonrational, straight-forward, and thus often paradoxical. The Buddha's Middle Path, in fact, attempts to unite a series of opposites that people encounter (as the Buddha himself did) in their life experiences: materialism versus rationalism, asceti-cism versus indulgence, skepticism versus belief, reality versus illusion, and so on. In short, Buddhists accept the contradictions of the world but also strive to overcome them.

No religious tradition has flourished for so long in such disparate cultures as Buddhism, which has adapted to a broad range of indigenous cultures and changes over time while retaining considerable continuity over the centuries (see Cousins, 1984, p. 278ff.). The complexities of these Buddhisms are compounded by their flexibility and diffusion over many cultures during a period of 2,500 years. More than half of the world's population lives in areas in which Buddhism has at some time been the dominant religion. Even in places where Buddhism has been politically sup-pressed (notably the People's Republic of China), it continues to exert a substantial influence both at the level of popular culture, where religious rituals often persist, and in the general culture, which bears the stamp of centuries of Buddhist influence.

As noted earlier, religious traditions emerge or change during times of social change, and Buddhism is a product of just such an era. The 6th cen-tury BCE—the period when both the founder of Jainism and Gautama the Buddha emerged as vital religious forces in India—was a time of consider-able social and cultural ferment throughout Asia. Cyrus the Great was extending the Persian Empire into Central Asia, and the Indus Valley was rapidly becoming the richest province of the empire. At about the same time, **K'ung-Fu-tzu (Confucius)** was responding to a period of war and chaos in China with his teachings about respect and order in heaven and on earth. The Buddha's religious teachings flourished in an environment in which diverse cultures were encountering one another, and Buddhism itself (like the other major religious traditions) was constructed as a response to multicul-tural contact.

Gautama Buddha, also known as Sakyamuni and Prince Siddhartha, was born about 563 BCE as the son of King Suddhodana. More interested in fathering a ruler than a monk, the king tried to keep his son bound to this world by surrounding him with luxury and shielding him from ugliness, sickness, decrepitude, and death. According to Buddhist tradition, however,

Siddhartha slipped out of the palace and saw a decrepit, broken-toothed, gray-haired, bent old man, thereby learning about old age. On a second ride, he saw a body racked with illness and learned about disease. On a third ride, he encountered a corpse and learned about death. Finally, he saw a monk with a shaven head, robe, and bowl and learned about the option of withdrawing from the world of wealth and power into which he was born. The young Siddhartha escaped the palace at night and joined the wandering holy men of the forest, studying with various teachers and following a life of strict asceticism. He found that path no more satisfying than the life of indulgence in the palace, however, so he experimented with more moderate means of seeking fulfillment. In the midst of intense meditation under the famous *Bodhi* tree, he obtained enlightenment, or **Bodhi:**

> I thus knew and thus perceived, my mind was emancipated from the asava [canker] of sensual desire, from the asava of desire for existence, and from the *asava* of ignorance. . . . Ignorance was dispelled, knowledge arose. Darkness was dispelled, light arose. (Buddha, 1954, p. 249)

Although he probably did not intend to found a new religion, a large number of followers gathered around Siddhartha, and their ideas became popular throughout Asia. The complex network of Buddhist traditions is difficult to characterize briefly, but scholars identify various branches emerging from the adaptation of Buddhism to local cultures during its diffusion in Asia (see Buddha, 1954, p. 80). Different elements of the Buddha's teachings were emphasized in different regions of the continent, according to their affinities with indigenous cultures, interests of the local ruling classes, and so forth.

The basic teachings of the Buddha are summarized in famous images such as those of the lotus and the river. The lotus is a ubiquitous symbol in Buddhist art and literature. This water lily is a beautiful pale blossom that emerges in its fragile splendor from the mud of shallow ponds, suggesting that purity can spring up from the anguish of the world; it thus becomes symbolic of the Buddha's teachings that humans can rise above the suffering of the world. Widely used in both Buddhism and Hinduism, the lotus image pervades Eastern thought. The river occupies a similarly archetypical spot in Buddhist imagery. Gautama Buddha himself was much inspired by the symbolism of rivers, which helped him to understand the unity in the diversity of life. A ferryman taught him to listen to the river, in which the water continually flowed and flowed, yet was always there; it was always the same yet at every moment was new. In just this way, the role of the Buddha figure would also vary widely with numerous reincarnations or manifestations that have made this religion, like Hinduism with its many avatars, so adaptable to indigenous cultures.

The Four Noble Truths are a series of four propositions about the nature of life that also encapsulate the Buddha's teachings:

1. Life is *dukkha* (usually translated as suffering, pain, or anguish, although these are merely subjective attributes of the larger phenomenon).

2. This suffering is rooted in *tanha* (craving, desire, attachment).

3. One can overcome tanha and be released into Ultimate Freedom in Perfect Existence (nirvana).

4. Overcoming desire can be accomplished through the Way, or the Eightfold Path to nirvana.

The suffering of which the First Truth speaks is the pain that seeps into all finite existence. Like an axle dislodged from the center of a wheel or a bone slipped out of its socket, life becomes dislocated, especially on six occasions: (1) in the trauma of birth, (2) in the pathology of sickness, (3) in the morbidity of decrepitude, (4) in the phobia of death, (5) in the entrapment in what one abhors, and (6) in the separation from what one loves. The cause of this dislocation is complex and may come from a former, present, or future life (see Gard, 1962, p. 113ff.). Once the cause of suffering has been identified, it can also be overcome, by means of the fourth principle. As the Buddha put it, "I lay down simply anguish [*dukkha*] and the stopping of anguish [*nirodha*]."[4] This last principle asks that adherents follow the Eightfold Path, which requires the following:

1. Right knowledge: This is to be sought not by itself but in conjunction with the other seven attributes.

2. Right aspiration: Seek liberation with single-mindedness.

3. Right speech: Master one's use of language, so that it moves toward charity. Because it is an indicator of motives, one should avoid false witness, idle chatter, abuse, slander, and the like.

4. Right behavior: This lies at the center of the moral code of the Five Precepts, which will be examined in a moment.

5. Right livelihood: Avoid occupations incompatible with spiritual advance.

6. Right effort: Will and exertion are important.

7. Right mindfulness: Note the importance of the mind and its influence on life ("All we are is the result of what we have thought," Buddha, 1954, p. 180).

8. Right absorption: Use techniques like those of *Raja Yoga*.

The early Buddhists had a pantheon that began with the five cosmic elements (*skandhas*): (1) form, (2) sensation, (3) name, (4) conformation, and (5) consciousness. Known as the five *Dhyani* Buddhas, these elements were eventually conceived as the five primordial Gods responsible for creation. A sixth Dhyani Buddha, Vajrasattva, is often added to the list, although he is of a rather different order because he serves as a priest to the others. This system, though somewhat polytheistic, still posited ultimate reality as indivisible, a problem complicated by the fact that wherever Buddhism traveled, it incorporated indigenous religious practices—and along with them, indigenous Gods.

Perhaps as a response to the millions of Gods in the Hindu tradition against which Gautama Buddha and his followers were reacting, contemporary Buddhism relies more on the teachings of the Buddha than on the action of any particular Gods. Moreover, the objects of worship in Buddhism tend to be humans who become divinized because of their acts during a lifetime or local Gods who are incorporated into the Buddhist tradition and are then perceived to be Buddhas themselves. The question of whether there are any Gods in Buddhism is a difficult one. Although many Buddhist scholars and sages will insist that there is no deity because the forms are mere illusion, Buddhism as practiced often appears to have Gods. As in Hinduism, the line between humanity and divinity is not a clear one, however, and Buddhists tend to believe that every individual has the Buddha nature within himself or herself.

The story of the Queen Mother of the West is interesting in that regard. According to the legend (*100 Celebrated Chinese Women*, n.d.),

> During the reign of Han Emperor Wu (140–86 BCE), the Queen Mother of the West boarded her chariot of purple clouds and traveled to the imperial palace with an entourage of celestial maidens, her green birds flying in front to clear the path.
>
> Emperor Wu watched in awe and then rushed outside to greet his guests. During the feast held in her honour, she commanded her celestial maidens to sing and dance. . . . Emperor Wu saw that she had the power to give orders to the celestial maidens and consequently believed that she must also be the dowager mother of all celestial beings.

In **Mahayana Buddhism,** the school dominant in Northern Asia, including China, emphasis is placed on a Buddha's decision to remain a bodhisattva—that is, one who qualifies for entrance into nirvana but chooses to work for the salvation of all beings before doing so. The importance of a bodhisattva in popular religion is quite understandable; as such a figure plays a vital intermediary role between the human and the divine worlds, just as Jesus, Mary, and the saints in Christianity, Muhammad in Islam, and the prophets in Judaism help to bridge the chasm between heaven and earth in their respective religions.

In practice, Buddhist religion places emphasis on the "Three Jewels"—the Buddha, the dharma, and the **Sangha**—the teacher, the teaching, and the organization of followers, respectively. In a core ritual of worship, the faithful Buddhist chants that he or she takes refuge in the Buddha, the dharma, and the Sangha, thus attempting to transcend this world of suffering and treat others with compassion to relieve both their suffering and his or her own.

Although not traditionally classified as a messianic religion by social scientists, Buddhism has a strong messianic element in the concept of the bodhisattva—that is, those individuals who obtain enlightenment but return to aid others rather than entering into nirvana or eternal bliss. **Avalokitesvara,** the most widely revered bodhisattva in East Asia, appears in several forms as a Buddha, an **arhat** (enlightened human), and an animal, and in many spheres (heavens and hells) to free all creatures from suffering and lead them to the Pure Land of Amita. Thus, this bodhisattva is known for compassion and mercy and is celebrated in the Lotus Sutra, in which she appears in 33 transformations to meet the needs of each audience to whom she addresses her teachings.

Avalokitesvara, the Goddess of Mercy, is one of the most famous images represented in Chinese popular culture. As such, she serves a function similar to the widely popular Virgin Mary in Christianity, especially appealing to the poor and women because of her approachability and merciful demeanor. According to legend, Avalokitesvara vowed to serve the Buddha and fled to a nunnery over the objections of her father. His agents pursued her and set fire to a temple where she had sought sanctuary, resulting in her death but also in the reward of immortality. According to the story, her father was blinded as punishment, but Avalokitesvara plucked out her own eyes to restore her father's sight, thus earning her title as Goddess of Mercy.

Avalokitesvara, known as Kuan-yin (Guanyin) or Kuan-shih-in in Chinese (see Overmyer, 2002), are not just ancient or mythical figures. They play a very real role in contemporary world politics, believed by some to be incarnated as the Tibetan Dalai Lama, now living in exile in India. The current Dalai Lama, 14th in the line of succession, has become a popular figure worldwide, especially after winning the Nobel Peace Prize in 1989 for championing the nonviolent struggle for Tibetan independence from China. Avalokitesvara's mantra, "*Om Mani Padme Hum*," is one of the most popular prayers in Buddhism.

Buddhist Rituals

Most Buddhist rituals involve efforts to obtain merit of some sort, either for oneself or for someone else, thereby transcending one's current state through the performance of the ritual. The foundation of Buddhist ritual consists of *dana* (giving), *sila* (precepts), and *kamma* (karma). Keeping the precepts includes conscious acts of service and paying respect to others. All of these ritual acts are rooted in the concept of karma, which is conceived in much the same way as it is in Hinduism: Because every act has a consequence, if one engages in meditations, chanting, and service and keeps the precepts, merit will accrue to oneself and to others.

A central Buddhist ritual is the chanting of refuge that a believer takes in the Three Jewels: the Buddha, the dharma, and the Sangha. This chant is a prelude to the performance of Buddhist rituals throughout Asia, which incorporate many ancient rites and a variety of indigenous practices. The metaphor of crossing the river helps to explain the meaning of these sacred symbols: An explorer—the Buddha—makes the trip first, proving that it is possible. He comes back to show us the way. The dharma, or teachings of Buddhism, is the vehicle of transport, the boat or raft we will use in our journey; and the Sangha, the organization of followers, is the boat's crew, in whom we can have confidence because of their training and discipline. Chanting that one is taking refuge in the Three Jewels helps a believer to recenter and focus attention on the path to liberation from this world of suffering:

> Veneration to the Blessed One, the Enlightened One, the Perfectly Enlightened One is as follows:
>
> To the Buddha, the [chosen] resort, I go.
>
> To [the] Dhamma, the [chosen] resort, I go.
>
> To [the] Sangha, the [chosen] resort, I go.
>
> —Ti-ratana in the Pali text[5]

Chanting is very common throughout Buddhism—especially chanting the *paritta*, the protection discourses. Chants are most significant at times of crisis or change: death, illness, possession, danger, embarking upon a new activity, entering a new house, and so on (see Cousins, 1984, p. 310).

Formal worship varies widely in the Buddhist tradition, in part because the religion has been grafted onto local traditions so is often practiced according to ancient indigenous customs. Followers often worship relics with offerings (food, water, clothing, incense, candles, etc.), bowing or prostrating, cleaning or adorning the relic, and chanting verses. Worship can also be carried out at home, although a temple or pagoda facilitates the worshippers' gain of merit, sometimes through ingenious methods. The Ten Thousand Buddhas Temple, located in Taichung, Taiwan, contains 10,000 images of the Buddha, all of which can be worshipped in a single visit, and hundreds of which can be venerated with one bow or prostration.

Sikhism

Deep within the self is the Light of God. It radiates throughout the expanse of His creation.

Through the Guru's teachings, the darkness of spiritual ignorance is dispelled.

The heart lotus flower blossoms forth and eternal peace is obtained,

as one's light merges into the Supreme Light.

—Guru Amar Das

Another tradition that challenged the religious orthodoxies in South Asia was the Sikh tradition, founded in the Punjab, in South Asia, during the 15th century CE, based on the teachings of Guru Nanak Dev Ji and 10 gurus who followed him. Although a relatively small population compared to the world's major faiths, Sikhism continues to grow, has diffused to every continent, and has an estimated 22.5 million adherents in Asia, as well as another 600,000 in North America. Its presence is often visible, especially in urban areas of South Asia and elsewhere, because of the distinctive turban and uncut hair worn by male adherents.

The Sikhs emphasized a monotheistic faith and the principles of justice and equality, cultivating a disciplined lifestyle and meditation. Guru Nanak's writings in the Punjabi language constituted the first sacred text, supplemented by the writings of his successor gurus and even those of

Hindu and Muslim saints to form the *Adi Granth* by the 1680s (Mann, 2006, p. 46). Indeed, the authority of the personal was transferred to the sacred book by Guru Gobind Singh in 1798, making it the Guru Granth Sahib. Community prayer at the center of Sikh worship began at the house of the Guru Nanak, the *gurdwara*, the name later used to designate the Sikh houses of worship.

The culture of personal discipline and the martial tradition of the Sikhs led to their becoming important partners of the British Raj, where they often served in the military and other positions of authority. The relationship with the broader society has not always been peaceful, however, especially with the emergence of a secessionist movement in the 1980s seeking an independent Sikh state. The Indian government attacked the center of the movement in Amritsar resulting in a decade-long period of violence and mayhem.

One interesting outgrowth of the emphasis on justice and the community's tension with Hindu orthodoxy was the institution of the community kitchen, which was open to anyone regardless of class or caste. Sikhs would cook and eat together, thus breaking with the Hindu caste norms against eating together to pervaded much of Indian society. The shared meals have also become a kind of community service, where the poor are welcome to come and eat free meals alongside members of the congregation.

Religious Life in China and East Asia

As Buddhism diffused east into China, East Asia, and Southeast Asia over the centuries, it found more fertile ground there than in its homeland. It became intertwined with indigenous religions of China and elsewhere, especially with Confucianism and Taoism, where the three traditions mingle with other local indigenous ritual practices to create a rich tapestry of spiritual beliefs, practices, and institutions that have evolved.

It is helpful to look at these three living religions in the context of their long and impressive history dating back to ancient China, before the time of **Confucius, Lao-tzu,** and the Buddha, to the ancient I Ching, also called The Book of Changes and The Classic of Changes. The major portion of the work is traditionally attributed to Wen Wang, founder of the Chou dynasty in the 12th century BCE, making it one of, if not *the*, oldest known sacred scriptures, although it has several additions from later centuries.

Religious belief and practice in China now consists of an amalgamation of several religious traditions, especially Confucianism, Taoism, and Buddhism, making it difficult to categorize the traditions to which individuals belong. The *Encyclopædia Britannica Online* (Religion: Year in Review 2010, 2011) identifies this complex of beliefs as "Chinese Folk Religionists" and describes its adherents as follows:

> Followers of a unique complex of beliefs and practices that may include universism (yin/yang cosmology with dualities earth/heaven, evil/good, darkness/light), ancestor cult, Confucian ethics, divination, festivals, folk religion, goddess worship, household gods, local deities, mediums, metaphysics, monasteries, neo-Confucianism, popular religion, sacrifices, shamans, spirit-writing, and Daoist (Taoist) and Buddhist elements.

Taoism is a set of indigenous Chinese religious practices, sometimes primal, that are closely related to the natural world and loosely linked to the teachings of its legendary founder, Lao-tzu. Confucianism is an official ethos and a system of ethics for public life, as first formulated by the Chinese philosopher K'ung-Fu-tzu, known in the West as Confucius, who lived in the 6th century BCE. Confucianism emphasizes filial piety (*hsiao*), respect for one's parents and ancestors. K'ung-Fu-tzu was himself more oriented toward this world than the spiritual realm, but his teachings had a clear affinity with the notion of ancestor worship. Living in a tumultuous time in which warfare was common and corruption thrived, K'ung-Fu-tzu emphasized the importance of order in human relationships, based on respect, benevolence, and reciprocity. He valued learning, ritual, and discipline as vehicles through which to overcome the ignorance and chaos of his time. Taoists offered an alternative, less hierarchical perspective on the world and often ridiculed the formal hierarchism of K'ung-Fu-tzu's philosophies.

Confucian officials considered both Buddhism and Taoism heterodox, but over time, these religions were incorporated into popular Chinese culture, perhaps *because* they were somewhat at odds with the official culture. The Buddhist emphasis on respect for all life, its general tolerance for all religious perspectives, and its ability to incorporate other traditions in its teaching helped this religion to establish deep roots in China by the 1st century CE. Buddhism coexisted alongside Taoism and Confucianism with Confucianism remaining the official state cult until the 20th century and the two other religions pervading folk culture.

When Buddhist texts were translated into Chinese, translators used the idioms of Taoism to express the ideas and rituals of the new religion. Many elements of Taoism and Buddhism became so closely intertwined that it is often difficult to separate the two traditions. Not only did Buddhism rely upon Taoist language but Taoism in turn also borrowed from Buddhist stories and concepts, especially as the Taoists began to have access to the rich collection of Buddhist texts coming from India from the 5th century on. As Ch'en (1964/1972, p. 474) noted, sometimes Lao-tzu's name was simply substituted for the Buddha's and texts were incorporated wholesale. Other times the copyist simply forgot to change the names, as in the Taoist passage that proclaims this: "Of all the teachings in the world, the Buddha's teaching is the foremost."[6]

Confucianism and the I Ching

It is said that if good people work for a country for a hundred years, it is possible to overcome violence and eliminate killing. This saying is indeed true.

—K'ung-Fu-tzu, The Analects 13:11

K'ung-Fu-tzu is known for his rationality, but his own intellectual development is rooted in the ancient and mysterious book of Chinese wisdom, the I Ching, or "The Book of Changes," thought traditionally to have been written by Emperor Fu Hsi (2953–2838 BCE), although it may have come from even earlier divination practices. We simply do not know its origins because they are so ancient. Ironically, K'ung-Fu-tzu did not attempt to initiate a new movement; rather, as Wei-ming (n.d.) put it, "Confucius considered himself a transmitter who consciously tried to reanimate the old in order to attain the new." In particular, K'ung-Fu-tzu was illuminating scholarly traditions of the sage–kings from centuries before him, perhaps culminating in the Duke of Chou, who died in 1094 BCE after developing a feudal ritual system which was "based on blood ties, marriage alliances, and old covenants as well as on newly negotiated contracts and was an elaborate system of mutual dependence" within social relationships (Wei-ming, n.d.). The Duke of Chou's contribution was supposedly based on the I Ching, which K'ung-Fu-tzu himself studied along with the duke's commentary.

The ancient practice of divination that the I Ching reflects is difficult for us to understand—rooted in efforts to discern guidelines for behavior in signs of nature—traditionally obtained by casting yarrow sticks or coins, a practice similar to those carried out today in Buddhist temples. The idea is to tap unseen aspects of reality, its substances, and processes—much as modern scientists do with experimental methods in subatomic physics and medicine. Although it may not make sense to those of us raised with the methods of modern science, it was considered useful in the hands of a creative and wise sage who could read the signs and offer advice by consulting reality at all levels, attending to the forces of heaven and patterns on earth and within the psyche. I do not pretend to understand it, but it is helpful to try to see it from the point of view of the believers. Ritsema and Karcher (1995) put it this way:

> Consulting an oracle and seeing yourself in terms of the symbols or magic spells it presents is a way of contacting what has been repressed in the creation of the modern world. It puts you back into what the ancients called the sea of soul . . . by giving advice on attitudes and actions that lead to the experience of imaginative meaning. Oracular consultation insists on the importance of imagination. It is the heart of magic through which the living world speaks to you. (p. 8)

The I Ching organizes these forces of nature into images so that an individual reading becomes possible, beginning with a problem and a question regarding how one can act in creative relation with the forces shaping a particular moment in time. Whereas the traditional method was the casting of yarrow sticks or coins, today it can be done by computer: See the website http://flytrapinteractive.com/~complimentary/iching/ and ask your own questions. The I Ching "responds" with patterns in a hexagram that have a long tradition of interpretations and commentaries. The usefulness of such ancient texts is always in the wisdom of the interpretation, of course, and the imaginative problem solving that comes from meditating on the question and receiving assistance from the right person for interpreting. K'ung-Fu-tzu himself said, "If I could study the I Ching fifty more years, I would have all the knowledge I need" (Confucius, 1998, 7:17).

If the method of inquiry he used was mysterious, the outcome of his deliberations was quite simple and clear-cut, laid out in the Lun-yü, The Analects, the most sacred scripture in the Confucian tradition and probably compiled by a second generation of his disciples and based on the master's sayings as preserved in oral and written form. At the core of his thought was the importance of educating oneself spiritually for self-realization—a lifelong effort to develop proper character. The ultimate goal of education is *ren* (or *jen*) or human-heartedness, which is the highest virtue. The path to the attainment of *ren* is the practice of *li*, that is, social norms. *Li* is not something fixed, however, but is subject to change according to individual situations and to discern it, one must rely upon the principle governing the adoption of *li*, that is, *yi*, proper character or the principle of rationality. Thus, *yi* (rationality) leads to *li* (social norms), which in turn leads to *ren* (human-heartedness).

The education of individuals prepares the way for a peaceful and well-ordered society. A ruler has the responsibility to govern by example, disciplining his or her own life to be a model for all the people. The core of this rational living with human-heartedness is a set of five relationships: (1) father and son, (2) elder brother and the younger, (3) husband and wife, (4) elder and younger, and (5) ruler and subject. In each case, the latter should show respect and give deference to the former. Although the key to a peaceful society is thus submission to hierarchy, the relationships are reciprocal and much of the content of The Analects addresses the responsibility of the superordinate to the subordinate. The elder and the ruler should show kindness, nobility, care, humaneness, and benevolence to their social inferiors.

The bottom line of the Confucian ethic in these relationships should sound familiar, as it appears in some form in most religious and ethical systems: What you do not want done to yourself, do not do to others.

Confucian thought became the bedrock of Chinese culture over the centuries, in part because of an elective affinity between his teachings and the interests of the emperors but also because of the highly respected scholarship of his disciples. According to Karl-Heinz Pohl (2003, p. 470), Jesuit missionaries brought valued Confucian teachings to Europe, where they had an impact on the enlightenment.

Not everyone was comfortable with K'ung-Fu-tzu's hierarchical advice, however. Chinese religious thought has important countertraditions in the ideas of Mo-tzu, who taught universal love rather than filial piety, and in the teachings of Lao-tzu, who insisted it was better to have people develop their own sense of the Way rather than to impose it by force. The founder of Taoism believed that ethical systems like K'ung-Fu-tzu's sometimes contribute to problems rather than solving them and that such moral exhortations prove self-defeating.

Indeed, as Stephen Karcher (1999) has noted the following:

Confucianism and Taoism were the two poles to traditional Chinese culture. . . . Confucianism, an intensely conservative, moralistic, and hierarchical teaching, emerged first. It went on to become the official philosophy of Imperial China. It defined a political and cultural elite who identified the way or *tao* with the internationalization of a particular set of social relations. Taoism, the second child, mocked social values and established power alike. An intensely individualistic teaching, it developed methods of dis-identifying with social institutions and commonly held motivations. (p. 299)

It is to the tradition founded on Lao-tzu's wisdom that I now turn my attention, to the critic of the Confucian establishment who also become a vehicle for a critique of the Western establishment in the countercultures of Western civilization 15 centuries later.

Taoism

Thirty spokes meet at a nave;

Because of the hole we may use the wheel.

Clay is moulded into a vessel;

Because of the hollow we may use the cup.

Walls are built around a hearth

Because of the doors we may use the house.

Thus wealth comes from what is,

But worth from what is not.

—Lao-tzu (1995)

Legend has it that Lao-tzu, fed up with people not listening to him, decided to leave his country. A border guard recognized him and insisted that he share his wisdom before he left, so the sage sat down and wrote the famous *Tao te Ching* (or *Daodejing*) with only about 5,000 Chinese characters. As the border guard watched over him, he inscribed these words of wisdom, contending that the cosmos is organized in a certain Way and that power or virtue (both from the same word *te*) comes naturally by acting in accordance with the universe and its nature. He handed his text to the guard, mounted his bull, and continued on his journey. According to another legend, he became Siddhartha's teacher. The *Tao te Ching*, according to Pohl (2003), "is not only the most widely translated Chinese work, but, after the Holy Bible it is the most widely translated book worldwide" (p. 481).

Just like a combination lock with a series of numbers that have to be arranged in a proper sequence and lined up in just the right way, so people will find that when they are aligned with the universe, it will be unlocked for them. If they are not properly aligned, no matter how hard they pull, it will not open. Thus, the most appropriate and even effective way to act and think is with wu-wei, sometimes translated as "effortless activity." Seidel and Strickmann (n.d.) suggested that wu-wei "is an action so well in accordance with things that its author leaves no trace of himself in his work" or quoting the *Tao te Ching*, "Perfect activity leaves no track behind it; perfect speech is like a jade worker whose tool leaves no mark" (Seidel & Strickmann, n.d.).

In his introduction to *The Way of Chuang-tzu*, the foremost spokesperson for Taoism in the 4th and 3rd centuries BCE, the Christian monk Thomas Merton (1965) explained the principle this way:

The true character of wu wei is not mere inactivity but *perfect action*—because it is act without activity. In other words, it is action not carried out independently of Heaven and earth, and in conflict with the dynamism of the whole, but in perfect harmony with the whole. It is not mere passivity, but it is action that seems both effortless and spontaneous because performed "rightly," in perfect accordance with our nature and with our place in the scheme of things. (p. 28)

In attempting to discern the nature of the Way, Lao-tzu found that opposites are complementary, symbolized in the famous yin-yang symbol now part of global popular culture. Light and dark, male and female, hard and soft, even life and death, are all part of the Way. Life, however, is flexible and death inflexible and rigid, so in life one must bow like the bamboo to survive the storms. This, said Merton (1965, p. 30), is another key to Chuang-tzu's thought. "All beings are in a state of flux," so that one should be careful of clinging to a partial view, treating it as if it were "the ultimate answer to all questions." One must remember that "What is impossible today may suddenly become possible tomorrow. . . . What seems right from one point of view may, when seen from a different aspect, manifest itself as completely wrong."

Taoism seems both complex and simple at the same time, and it is easy to be taken in and to think one understands, according to the text itself, even when one does not. That is why the *Tao te Ching* begins, "The Tao that can be known is not Tao." We should, in other words, strive to understand and follow the Way (the Tao), but it is beyond our abilities to do so. That does not mean, however, that one simply sits and accepts everything the way it is but rather that one develops a different perspective on the world and its implications for one's life. One should not conquer others with brute force but find the power within oneself and the universe that brings opposites into harmony and empowers everything to live in peace. Lao-tzu (1995) even presaged Gandhi when he wrote the following in Chapter 30:

Powerful men are well advised not to use violence,

For violence has a habit of returning;

Thorns and weeds grow wherever an army goes,

And lean years follow a great war.

. . .

For even the strongest force will weaken with time,

And then its violence will return, and kill it.

The famous teacher and founder of Taoism in China was convinced that the universe itself was ultimately peaceful and that people who followed "the Way" (the Tao) of the universe would be more effective. Although rocks seem more powerful than water, over the years a stream will carve a canyon through a mountain. People who are patient and nonviolent may appear weak but in the long run will be more powerful than those who are impatient and violent. Like Mahavir in South Asia, Lao-tzu laid the foundation for a strong tradition of nonviolence throughout ancient East Asia.

If the true Tao cannot be known, then someone who pretends to explain it in writing probably knows almost nothing about it, having missed the point altogether. So, I will move on to the final Asian tradition I wish to visit briefly in this tour, one that is at the heart of Japanese national culture.

Shintoism

Although not a global religion by any means, the Japanese impact on the modern world—like the Jews'—has been disproportionate to their numbers, so we should know something about their religion. Moreover, it is an interesting case study in the persistence of a somewhat localized indigenous tradition well into the modern and postmodern eras.

The word *Shinto* means "the way of the Gods" or "the way of the **Kami**"—that is, the sacred or "what is worshipped," literally *shin* (divine being) plus *do* (way). As such, it is more a broad set of ways of honoring the spirits of the natural world than a self-conscious, distinct religious tradition. Like many ancient cultures, the Japanese lived close to nature, organizing daily life around natural events, cycles of the seasons, and so forth. The beautiful natural setting of the islands was inspiring and powerful; it appeared to be filled with forces in addition to humans and had an overall existence.

As the idea of the Kami developed, cosmogonies emerged about the Amatsu, the heavenly Kami, Izanagi, and Izanami, who organized the material world by stirring the ocean with a jeweled spear from the Floating Bridge of Heaven. As that material coagulated, it formed eight islands (Japan, or, according to some, the whole world). Eventually, as Shinto mythology intertwined with the imperial household, some believed that the first emperor of Japan, Jimmu, was a descendant of Amaterasu, the Sun Goddess, and one of the most important of the Kami.

This indigenous tradition is bound with the broader life and culture of the people from prehistoric times. It was formally developed as early as the 7th century BCE and took on its own distinct identity only after the arrival of competing traditions such as Confucianism, Buddhism, and Taoism. Confucianism probably reached Japan in the 5th century CE and became widespread over the next 200 years along with Taoism. Gradually, a pan-Japanese practice emerged as political power centralized with the imperial household whose Kami became those of the nation. The 6th-century arrival of Buddhism also stimulated syncretistic tendencies—the Shinto Kami were sometimes seen as protectors of Buddhism and shrines to them were built within the precincts of the Buddhist temples (Hirai, n.d.). Although Christianity came to Japan with the missionary work of Saint Francis Xavier in 1549, its exclusivism set it apart from the other faiths and it was banned for more than 200 years, from the 17th to the 19th centuries, and even today only attracts less than 1% of the population.

Originally, Shinto shrines were not buildings but places of natural beauty and particular interest, such as mountains, unusual rock formations, waterfalls, and forests. Gradually, as the tradition evolved, Shinto shrines were built, although they continue to value nature and are often still places of great natural beauty even in industrialized modern Japan.

Shinto rituals include individual visits to shrines and rites and festivals, such as those marking the seasons. Some people start the day with a visit to a shrine, either a public one or a private shrine in their home (which often sits beside a Buddhist shrine as well). A number of Shinto rites of passage are observed, from the initiation of a newborn baby to youth ceremonies and weddings. Funeral ceremonies are generally carried out according to Buddhist customs.

Although the official connection of Japanese religions to the state ended after World War II, Shinto rites and ceremonies reinforce family life, connections to ancestors, and an appreciation of the beauty and significance of nature.

East Asian Rituals and Institutions

Filial piety and honoring the dead—the ancestors—are important aspects of Chinese culture and East Asian life generally and sometimes bring together the multiple religious traditions held simultaneously by a believer. The departed's son in China traditionally goes to a stream to offer his parent wine and "ghost money"—that is, symbolic money for the departed to use in the afterlife. For 7 days, the son offers sacrifices before the coffin, supplemented by Taoist exorcisms and Buddhist prayers for the dead. The funeral procession includes music, firecrackers, and banners and tablets inscribed with the services the deceased performed during life. Elaborate provisions are made for the departed's needs in the next life, including paper money, clothing, bedding, and sometimes even models of houses and cars, easing the grief of those left behind because the unknown world after death is rendered more familiar and the departed are well provisioned for their next life.

Buddhist ritual practices are often connected with the agricultural year or the significant dates of local heroes or Gods. From the Chinese New Year to the end of the rainy season, Buddhist festivals are celebrated, usually with much fanfare and often with fireworks, lights, and processions. These religious festivals are also social occasions that get people together, uniting families, friends, and members of the community. Seasonal festivals are often cultural residues from agricultural times or even earlier, when people's lives were more closely linked to the cycles of nature and people were more appreciative of the "help of the Gods" in coping with lifestyle changes from one season to another.

In many parts of East and Southeast Asia, various faith traditions exist side by side, often with a kind of division of labor. In China, for example, Confucianism provides a sort of civic ethic or civil religion, Taoism specializes in the vicissitudes of daily life, and Buddhism is associated with rites of passage. Although religious temples and sacred sites are an important part of the landscape, much of the ritual life of Asia takes place in the privacy of family homes, where ancestors, Buddhas, and other Gods often peacefully coexist and frame life with a spiritual perspective.

The most well-developed religious institutions in East Asia are Buddhist, probably because the Buddha himself paid attention to the institutionalization of Buddhist ritual. At first, Buddhist ideas were spread primarily by highly disciplined monks, members of the Sangha who lived communally and developed a collective, nonhierarchical authority structure. The Sangha remains central to the organization of Buddhist life today, although it does not have an authority comparable, for example, to the clergy in Christianity. Lay Buddhist societies are also important and relatively independent from the monks. Much as with its sibling religion Hinduism, theological tolerance and institutional flexibility go hand in hand in Buddhism. Religious practices are by no means the monopoly of monks and may be conducted in the home as well as the temple. The authority of the monks and other holy figures derives more from their knowledge and demonstrated abilities than from any official authority.

The most significant aspect, sociologically, of the organization of Buddhist religion is the diversity that emerged over time as the religion's ideas and practices were diffused throughout Asia, picking up elements from indigenous cultures in different regions. The **Theravada** ("Doctrine of the Elders") **Buddhism** branch is dominant in Southern Asia, especially Sri Lanka, Burma, Cambodia, Laos, and Thailand as well as parts of Vietnam, Bangladesh, India, and various immigrant communities (especially in the United States). The larger branch, Mahayana ("The Great Vehicle") Buddhism, is practiced in China, Japan, Korea, Vietnam, and elsewhere. **Vajrayana** ("The Diamond Vehicle") **Buddhism,** which some consider part of the Mahayana tradition, is found primarily in Tibet, Mongolia, the Himalayas and in parts of China and the former Soviet Union. It is significantly smaller than the two main branches and remains much closer to classical Indian Buddhism. Another popular variant is Ch'an Buddhism, which took root in Japan, where it is known as **Zen Buddhism.**

Huston Smith (1965) observed how the image of crossing the river distinguishes the different branches of Buddhism and their respective approaches to the search for enlightenment. The Theravada tradition says we should make ourselves a little raft and shove off across the water. The Mahayana tradition says we seek out a ferry and rely on the ferryman to take us across safely. Before the crossing, the world on the other side is scarcely visible: It is a mere line on the horizon, misty and unsubstantial. After the crossing, the shore that is left behind—which was once tangible and real—is now just as unsubstantial as was the shore on which we now find ourselves. Zen Buddhism says that once we have reached the shore of enlightenment, we no longer need the raft. We no longer need the doctrines or the dharma, the Four Noble Truths, the Eightfold Path, or even the Buddha himself: Hence, the rather shocking Zen saying, "If you meet the Buddha on the road, kill him."

Hindu and Buddhist Syncretism and Co-Optation

The syncretism and co-optation modes of adaptation can best be seen in the Buddhist diffusion into China over a period of several centuries. Because Buddhism flourished in China and the rest of Asia rather than in its founder's native India, the way in which this religion was transformed and in turn affected other religious traditions is instructive of the syncretism process that all the major faith traditions have undergone. Eventually, the worldview and ethos of Buddhism became profoundly intertwined with that of Asian culture generally and consequently grew extraordinarily diverse. Different schools of Buddhism emerged as adaptations to indigenous cultures, forming separate sects.

Emperor Ming Ti, after dreaming in 62 CE of a golden image from the West, sent representatives to India, who returned with Buddhist teachers, images, and texts. Reliable records from the 1st century CE of vegetarian feasts may reflect a Buddhist influence, given the prohibition against taking life found in Buddhist but not Chinese teachings. Many Chinese at first resisted Buddhist practices, especially because the 4,000-year-old practice of honoring ancestors was undermined by the Buddhist monks, who eschewed family ties to join the Sangha, took vows of celibacy, and failed to "keep the ancestral fires lighted," thereby endangering the ancestors' status and raising the specter of their becoming "hungry ghosts" (Bush, 1988, pp. 138–139). Buddhism was not fully embraced in China until the early 4th century CE, when much of the country was conquered by the Hsiung-nu, a non-Chinese tribe from Central Asia. The Hsiung-nu adopted Buddhism because there was already some popular basis for it but also because this religion was not Chinese and advocated a universalistic ethic that encouraged intercultural tolerance. The conquerors co-opted the monks, giving them legitimacy but forcing them into service projects, such as managing official charities and building the country's infrastructure.

3

The Tour: Judaism, Christianity, and Islam

I n the 21st century, no religious tradition lives in isolation. The major
living world religious traditions spring from the Indian subcontinent,
where Hinduism and Buddhism emerged, and the ancient Middle East
(or "West Asia" as it is often called in Asia), where Judaism began and gave
birth to the other "religions of the Book," Christianity and Islam, both of
which became global players in the following centuries.

Judaism

The Judeo-Christian tradition became intertwined with Western civilization
and diffused even further as the Europeans conquered the Americas, Africa,
and parts of Asia. Islam remained strongest in the Middle East, although it
made substantial inroads into Africa and Central Asia. Unlike Islam and
Christianity, Judaism did not seek converts, nor did it syncretize much with
other religions, as did Hinduism and Buddhism. As a result, Judaism has
remained primarily an ethnic religion with social boundaries separating Jews
from other socioreligious groups, although those boundaries have softened
somewhat in places like the United States where Jews have become more
assimilated into the broader culture. Jews became widely dispersed geographically, however, and took their religion with them wherever they went.
Moreover, despite its continued identification with a specific social group,

Judaism developed a universalistic theology. The impact of the Jewish faith thus goes far beyond its limited social and ethnic boundaries.

The God of Judaism was originally a clan God, the God of Abraham, Isaac, and Jacob. Hebrew tradition emerged out of the legendary relationship of two people, Abraham and Sarah, with this clan God. These founders of Judaism were apparently part of a large 18th-century BCE migration of people around the ancient Fertile Crescent of the Middle East, from present-day Iraq to what is now Israel. According to Jewish tradition, their migration had a divine sanction:

> Now the LORD said to Abram, "Go from your country and your kindred and your father's house to the land that I will show you. And I will make of you a great nation, and I will bless you, and make your name great, so that you will be a blessing." (Genesis 12:1–2)

This ancient tribal legend contains two elements of universalism: (1) the God is still associated with a clan but is not limited to a particular place and (2) the promise of blessing to Abraham (Abram) will benefit not only his clan but also "all the families of the earth." The movement toward universality is thus part of a decision to migrate, a journey that involves both disengagement from ancestral space and a sense that the protective clan God will travel with them. This mobility of their deity, the bond between a people and a God, and a nascent sense that the God is concerned about others outside their covenant are hallmarks of Judaism that persist through the centuries.

Judaism was institutionalized as a religion only after the Exodus from Egypt, probably in the 13th century BCE. According to the Hebraic scriptures, the descendants of Abraham and Sarah migrated to Egypt during a period of famine and became slaves of the Egyptians. Moses, a Jew raised in the palace, had a spiritual vision in which the God of his ancestors instructed him to liberate the slaves from the pharaoh. In the watershed event of Hebrew history and religion, Moses and his brother Aaron organized the slaves, confronted the Pharaoh, and eventually escaped after the Egyptians suffered a series of "plagues" because the Pharaoh refused to release them. A final divine punishment befell the land:

> At midnight the Lord smote all the first-born in the land of Egypt, from the first-born of Pharaoh who sat on his throne to the first-born of the captive who was in the dungeon, and all the first-born of the cattle. (Exodus 12:29)

The only firstborn spared were the Hebrews, because they put the blood of a lamb on the doorposts of their houses so that the destroyer would pass over it.

Released at last by the pharaoh, the freed slaves wandered in the desert and established a covenant with their God. The **Decalogue** (Ten Commandments) given to Moses at Mt. Sinai established the exclusive relationship of the Hebrew people—a *berith*—with their God—no longer simply a clan God but God of the nation of Israel. In the ensuing years, the Israelites invaded their "Promised Land," conquered the people living there, and established a monarchy. The religious tradition laid down by Moses was dominant among the Israelites, but worship of the indigenous Gods of the region was frequent as well, especially Ba'al, the God of Storms, and Anath (or Ashtoreth or Astarte), the Goddess of Love, War, and Fertility. Not surprisingly in an agricultural society, the worship of these Gods involved the central issues of sex and fertility; the mating of the rain–vegetation God with the fertility Goddess was reenacted ritually by humans playing their respective roles, much to the dismay of the priests (see Deuteronomy 23:17–18).

Ancient Judaism was a primal clan religion whose social organization took the form of a **theocracy**, a fact that put its stamp on the tradition that persists to this day. As a way of life, Judaism insisted on strict ethical standards in all economic enterprise and political activity. God demanded strict adherence to standards of justice, with special attention to the poor, orphans, and widows. This concern for the downtrodden reflects the tradition's origins among an oppressed people seeking liberation.

The **Hebrew scriptures**, according to the tradition, came from God, and the first five books (the **Pentateuch**) were written by Moses (an idea modern scholars do not think literally true). The Hebrew scriptures are constituted by the following:

1. The Pentateuch, known as the **Torah**, or Law

2. The **Prophets** (Nevi'im)

3. The **Writings** (**Kethuvim**), which include such contributions as the Psalms, Proverbs, Job, and the Song of Songs

The Torah, Prophets, and Writings together constitute the **Tanakh**, or the Written Torah, which also has a vast body of commentaries containing oral and written traditions. They include the Mishnah and commentaries on the Mishnah called the Gemara, which together are called the **Talmud**, completed 1,500 years ago. A number of other important texts exist outside of the mainstream, such as the **Zohar**, the primary text of the Kabbalah in the Jewish mystical tradition. It is similar to Sufism in Islam and mysticism in other traditions in that the Zohar's emphasis is on the intense, direct experience of the divine. The text is of such depth and controversy that one group of rabbis declared that no one under 40 should read the Zohar!

Judaism emphasizes that God spoke to the Hebrew people through the Torah, still sometimes written on a scroll, kept in a cabinet in the synagogue, and treated with great respect. Traditionally written in Hebrew on parchment, the scroll should not be touched directly—someone reading it will use a pointer (*yad*) to follow along the text without touching it. This early emphasis on the written word rather than on an oral tradition placed its stamp on Western civilization as well. The rabbinic tradition, especially after the fall of Jerusalem in 70 CE, cultivated the study of the Torah and later the Talmud (the Mishnah, along with collections of rabbinical commentaries). Centuries of rabbinical training and Judaism's emphasis on learning, reading, and writing have produced a rich literature.

Jewish Beliefs

Much like Buddhism and its Three Jewels, we might summarize the ideas of the Jewish faith with three concepts: God, Torah, and Israel—that is, the deity, God's teachings, and the community; in this case, it is the holy nation, or chosen people, of Israel. Ritual is more important than doctrine to the Jewish tradition, which except for the most fundamental confession of faith allows wide latitude in belief, especially in its reformed branch. This lack of emphasis on doctrine and the concomitant stress on ritual and community is primarily a function of the fact that a Jew is not, for the most part, someone who *believes* in Judaism—that person is either born Jewish or is not. A Jew is a member of a family, and those few people who convert to the faith are not just members of a religious community but become the "adopted children" of Abraham and Sarah (see Hertzberg, 1962, p. 21).

The Shema, the heart of morning and evening prayers, expresses the fundamental belief of Judaism as a ritual confession of faith found in the Torah:

> Hear, O Israel: The Lord our God, the Lord is One. Praised be God's glorious sovereignty for ever and ever.
>
> You shall love the Lord your God with all your heart, with all your soul, and with all your might. These words which I command you this day shall be in your heart. You shall teach them diligently to your children. You shall talk of them at home and abroad, night and day. You shall bind them as a sign upon your hand; they shall be as frontlets between your eyes, and you shall inscribe them on the doorposts of your homes and upon your gates. (Deuteronomy 6:4–9)

The Jewish Cosmogony

The Jewish understanding of the nature of the cosmos begins with the Jews' own sacred history and their special relationship with the Creator. The same

God who sacralized Abraham and Sarah's migration freed the Hebrews from slavery under the Egyptians and led them to the Promised Land created the world. Moreover, at the end of the process, God created humanity in God's own image and gave this last creature dominion over the rest of creation. Herein lies the creative tension of Judaism, which can lead to remarkable dedication to the causes of justice and peace as proclaimed by the ancient Hebrews but also to the exploitation of power by those who proclaim themselves chosen by God. It is precisely this potential for abuse of power that was denounced by the ancient Hebrew prophets and was the source of the prophetic tradition and activist orientation within Western culture, a fountain of the West's strength as well as its weakness.

The creation story in the Hebrew scripture known as **Genesis** is a major piece of literature in Western culture and reveals a great deal about the civilization's worldviews. The Christian and Islamic cosmogonies are borrowed from that of their parent religion, Judaism, and expressed in the creation stories of Genesis, in which the one God created the universe. First, this story states that although its Creator declared creation good, humans disobeyed God and thus were expelled from the paradise in which there was no pain or suffering. Second, God gave humans control over the rest of the natural world, blessed them, and said to them, "Be fruitful and multiply, and fill the earth and subdue it; and have dominion over the fish of the sea and over the birds of the air, and over every living thing that moves upon the earth" (Genesis 1:28). This feature of the Judaic cosmogeny lays the groundwork for the activist approach toward the world in general, and the scientific–technological revolutions of the modern West in particular.

Third, the patriarchal strains in the creation story are significant and have been used to legitimate male dominance. Eve allegedly initiated eating the forbidden fruit, which led to the so-called "Fall," so some assign the blame for the expulsion of humans from paradise to the female sex. There are, however, two separate Hebrew creation stories, probably coming from two distinct sources. According to the first version, "So God created man in his own image, in the image of God he created him; male and female he created them" (Genesis 1:27). In the second version, God creates Adam and then decides "It is not good that the man should be alone; I will make him a helper fit for him" (Genesis 2:18). Then God makes all of the beasts and birds and finally causes the man to sleep, takes a rib from his side, and uses it to create the woman. In the first version, God creates the two genders simultaneously as equals. In the second version, God creates the female as a helper to the male in the version often remembered and cited, thus seeming to imply her inferiority.

With more women involved in scholarly examination of the scriptures and traditions of every faith, new questions arise at every turn. As Mary Jo Neitz (2003) put it, for example,

> We must ask "What is Eve's story? What is *paradise* for her?" As told in the second chapter of Genesis, God and Adam set things up and agreed about the rules before Eve arrived on the scene. In verses 19–20 Adam names every-thing, before God even creates Eve. Did Eve agree to the rules? Did she want to name things? Did she have a choice? There is another story in Hebrew folklore about Lilith, Adam's first wife, who insisted on equality and flew away when Adam tried to tell her what to do (Gottlieb, 1995, pp. 73–78). What was her story? I would like to hear Eve's version of the story we get in Chapter 2 of Genesis. (p. 5)

Finally, the Hebrew creation narrative weaves together competing themes of freedom and domination. On one hand, humans are created in the image of God and thus have a wide range of freedom; on the other, they are disobedient and therefore in need of authoritative social institutions that keep them from engaging in immoral behavior. Contradictory worldviews are an important aspect of religious myths and appear in a wide range of examples: Opposing ideas are woven into a narrative, which then often becomes a focus for cultural conflict.

Jewish Theodicies

Whereas the Eastern religions tend to explain suffering and death on the basis of long-term continuity and cycles of history, Western religions usually rely upon a theory of suffering caused by dramatic changes in history some-times a result of human agency followed by divine punishment. The Judeo-Christian–Islamic tradition posits that humans once lived in paradise, the Garden of Eden, where suffering and death did not exist. By disobeying God—eating the forbidden fruit of the knowledge of good and evil—humans became estranged from God and were expelled from paradise, ushering in both suffering and death.

The Genesis account of the Fall also suggests that God was somehow unhappy with the possibility of human immortality, and that is at least partly why God expelled Adam and Eve from the Garden (Genesis 3:22–23). Because separation from God is the cause of suffering and death, the solution to misery is a return to God. The usual paths in Judaism, Christianity, and Islam are conversion and/or repentance, turning back to God and restoring the lost relationship with the deity, a process facilitated by the rituals and actions of the faith community. In Christianity, one confesses and loves God and neighbor. In Islam, one relies upon the Five Pillars: (1) affirming one's belief in the One God, (2) performing ritual prayers, (3) sharing one's wealth (**zakat**), (4) celebrating Ramadan (the month of fasting), and (5) undertaking a pilgrimage to Mecca. In Judaism, the community must engage in repentance and a restoration of the covenant, to do justice and walk with God.

Jewish theodicies reflect the social nature of Judaism and the ancient Hebrews' covenant with their God. The Fall in the Garden of Eden identifies the emergence of evil in the world because of alienation from God. Suffering and death themselves are consequences of Adam and Eve's disobedience, which in turn resulted in their separation from the Creator, in the same sense that a child feels pain when separated from the parent. Suffering can be overcome through individual conversion and/or repentance, in order to return to God. This theory of suffering remains a constant reminder throughout Jewish history that—as in the Eastern law of karma—the person who causes suffering or breaks the covenant with God will somehow suffer, sometimes even at God's hands.

It appears, however, that the just also suffer, a problem discussed eloquently throughout Jewish literature, in part because of the great suffering that Jews have experienced as a people over the centuries—from that expressed in the ancient book of Job to the literature of the Holocaust. A popular contemporary Jewish exploration of the problem of suffering is Rabbi Harold Kushner's (1981) *When Bad Things Happen to Good People*, in which he contends that a satisfactory explanation for suffering is possible only when people forfeit the idea of God as omnipotent or all-powerful.

Jewish Ritual

Jewish **kosher laws**, a complex set of regulations about food selection, preparation, and eating according to ancient Hebrew rituals, provide an interesting case of religious rituals that link routine social life with the broader goals of the tradition; they are organized on the principle of the distinction between the sacred and the profane so emphasized by Durkheim. Such dietary restrictions serve the obvious purpose of providing norms for approved eating practices, important in primal societies in which religious and social regulation were not clearly differentiated. As well as forbidding the eating of certain kinds of animals, kosher laws prohibit mixing meat and dairy foods, which must not even be prepared or served by the same utensils. Finally, the Talmud and later works define the exact manner in which food must be prepared and eaten: Animals must be killed in a prescribed procedure; all meat must be free of surface blood before being eaten, and so on. These boundaries of food preparation and eating undoubtedly provided a practical protection in ancient times from disease and other hazards. Many of the regulations may not sound immediately logical within the context of 21st-century theories of health and nutrition but reflect ancient public health practices and perhaps even historically specific problems that the Hebrew community encountered. A related concern is a humanitarian one for the animals themselves. Some of the preparatory regulations, such as instructions for how to kill an animal properly, mitigate the extent to which it suffers and have an interesting relevance to ethical questions surrounding industrial agriculture, which usually places profit maximization above principles of ethical treatment of animals.

Anthropologist Mary Douglas suggested that the logic of the kosher laws lies in the principle of classification itself. Establishing boundaries between purity and danger is a direct way of ordering the world. Categories of forbidden food include types of animals that do not neatly fall into one category or another in the threefold Hebrew classification of the universe as the Earth, the waters, and the firmament: "Any class of creatures which is not equipped for the right kind of locomotion in its element is contrary to holiness" (Douglas, 1966, p. 55). Through this complex set of dietary laws, Orthodox Jews are reminded with every meal and in every encounter with the natural world of the link between the people and their God, the boundaries between Hebrews and other tribes. In this sense, then, the *medium* is the message. What is important is not which foods are taboo, or even which criteria are used, but that the laws have a logic to the people who follow them and that the boundaries around food consumption reinforce boundaries around religious and hence social identities. Nonetheless, Jewish kosher practices are preserved even among many secular Jews whose (sometimes unconscious) cultural taste preferences are informed by centuries of custom—some may not have moral objections to eating pork, but find it an unattractive food choice.

The major holidays of the Jewish year reflect this religion's agricultural roots and recall significant events in the history of Israel. Three autumn holidays include **Rosh Hashanah** (New Year), **Yom Kippur** (Day of Atonement), and the **Sukkoth** (the Feast of Booths, a harvest festival commemorating the wilderness wandering following the Exodus). **Pesach**, or Passover, in the spring celebrates God's liberating activity in history, portrayed symbolically in the *seder* meal. The ritual serves as a reminder of social history: the Jews' liberation from slavery and their hope for a messiah. It is sustained by the ethnic religious community but is couched in universalistic terms for the benefit of all humanity: Just as God worked through Moses to liberate the Hebrew slaves, Jews today are reminded of their duty to work for the liberation of all who are oppressed.

With its vast literature and formidable richness, the Jewish belief system became the basis for the Christian and Islamic traditions, a historical fact that resulted in its dissemination far beyond the boundaries of the ancient Hebrew tribes. Of particular importance for Christianity was the Jewish belief that the "anointed one," the messiah, would usher in a messianic age of peace and justice for all humanity. The followers of Jesus insisted that he was that messiah.

Christianity

The religion we know as Christianity was originally a Jewish sect started by a Jewish prophet, Jesus of Nazareth, who now has 1.5 to 2 billion followers (almost one third of the world's human population); more people now call themselves Christians than identify with any other religious affiliation. Sociologically, the reason for Christianity's dominance of the religious landscape has not been a superior theology but its alliance with the power structures of Western civilization, first with the Roman Empire after the conversion of Constantine in the 4th century CE and then with the Western European powers through the Middle Ages and into the expansionism of the colonial period. In the 19th century, colonial and missionary movements spread in tandem from Europe to other parts of the world, conquering land, economic and political resources, and cultures.

As many countries joined the global community through the colonial system, Christian missionaries educated their political, economic, and intellectual elites. Like other transplanted religions, Christianity transformed dramatically as it diffused across cultures and became increasingly cosmopolitan. One characteristic of Christianity, deriving from its multiculturalism, is the tremendous variety of "Christianities" around the world. As the religious movement spread, many indigenous people grafted their own religious beliefs and practices onto the basic ideas of the Christian faith. Christian worship often includes pre-Christian religious rituals and takes on the color and many of the features of the indigenous religions it has replaced.

Before it became the official religion of Europe, Christianity was a backwater religious movement favoring the poor. Despite the diversity of its participants (see Grant, 1963/1972), the early Christian movement appealed especially to isolated, marginal people in a politically insignificant location rather than to those at the center of power. Because early Christianity was much different from contemporary practice, we have to cut through many layers of culture to understand its primary form (see Loisy, 1903/1976). Christianity was initially a Jewish movement, one of several reform movements at the time, and fell within the tradition of the Jewish prophets with whom Jesus was identified. The apostles were a group of passionate itinerant preachers who ran into trouble with the authorities wherever they went—a far cry from the contemporary Christian leaders, especially in the West, who are respected members of the community (and usually middle class).

Jesus first joined the movement started by his relative, John the Baptist, and did not develop his own group of followers until after John's execution. Although Jesus never declared himself the head of a new religion and repeatedly affirmed his faithfulness to the Jewish tradition, he was a classic charismatic authority who challenged the religious establishment of his day. He struggled frequently with religious and civil authorities and from the very beginning of his ministry people sought to kill him. In his first sermon, Jesus proclaimed the following:

> The Spirit of the Lord is upon me because he has chosen me to give the good news to the poor; he has sent me to heal the afflicted in heart, to announce freedom to the prisoners and give sight to the blind; to set free the oppressed, to announce the year of grace of the Lord. (Luke 4:18–19)

Immediately following the sermon,

> All those in the synagogue were filled with wrath. And they rose up and put him out of the city, and led him to the brow of the hill on which their city was built, that they might throw him down headlong. (Luke 4:28–29)

This sort of controversy followed him throughout his ministry until the religious and political authorities finally executed him.

After Jesus' death, a dramatic occurrence changed the character of this small reform movement: His followers became convinced that he had risen from the dead. The movement itself was dramatically revitalized, becoming a thriving force that gradually spread throughout the ancient Middle East and eventually into the centers of power as well as the margins of society.

One of the most sociologically significant characteristics of Christianity is its universalistic criteria for membership. Religious belief systems are usually tied closely to a specific social grouping, but early Christianity was deliberately universal in its recruitment. Judaism is an inherited religion: To be a Jew, one essentially has to be born into the community. To be a Christian in the early church, one needed no ethnic or tribal qualifications; one simply had to declare "Jesus is Lord" and go through a rite of passage—baptism— to become a member of the community. This radically universal nature of

Christianity, consistent with Jesus' teachings, was institutionalized by the apostle Paul, who spread the new religious system throughout the ancient Middle East.

This lack of ethnic or tribal membership criteria led to a multicultural religious community very early on. Christianity emphasized belief more than ritual. A persistent central question of the community, "What must I do to be saved?" may sound familiar to many readers of this book, but it is a rare question indeed in the world's religions and in most religions makes little sense or none at all. Christianity thus became highly intellectualized as people began to fight one another over what beliefs were required to be a member of the community in good standing. In most other religious traditions, such issues were simply not central because the matter of membership was not so problematic. Thus, the Christian belief system carries an added weight of historical importance as the defining feature of membership in this religion.

It is possible, despite the enormous number of variant Christian communities throughout the world, to isolate some basic beliefs that are nearly universal among Christians although with a very wide range of interpretations regarding them:

1. God is love.

2. Jesus is Lord.

3. One should love God and one's neighbor.

4. God intervenes in history.

5. The Bible reveals God's nature and will.

The religious movement of early Christianity was centered on the person of Jesus, believed to be the messiah, the savior or liberator sent by God to deliver people from bondage. Gradually, the church came to see Jesus himself as God, along with the Creator (or Father) and the Holy Spirit. The combined divinity and humanity of Jesus—the doctrine of the **incarnation**—is not peculiar to Christianity but is one of its central characteristics. The idea of a "Godhead," similarly, is not altogether unlike the Hindu idea that a single Supreme Soul exists in three manifestations incarnated in various ways. Altogether, the three elements, or "roles" of the Christian God, known as the **Trinity**, make up one single being, although the way in which they do has been a matter of considerable controversy throughout Christendom. The Christian concept of God incorporates the complementary (or contradictory?) ways in which the divine is experienced—as an originator of life (the Father, in its patriarchal imagery), as a liberator who comes to people's aid (the Son or messiah, Jesus), and as the sustainer or spiritual energizer of life (the Spirit).

The idea that Jesus is the "Son of God" is a complicated one, especially for Muslims, who are taught that this is a blasphemous idea. The concept is quite foundational for some Christians and complex and mysterious for others; indeed, the early church had numerous debates, councils, and declarations about the issue. The Hebrew scriptures have numerous references to Sons of God, and the idea probably does not mean (at least to most people) that God took a wife and had a son, or that he is somehow physically God's offspring. Abdullah Ibrahim (n.d.)[1] pointed out that Arabs are often called "sons of the desert" but that no one takes that in a literal sense—it is meant to be understood spiritually. Moreover, he argued the following:

> Because God is so much greater than man, He chooses to express Himself in human terms so that we can understand Him. When Surah 22, Hajj, verse 61 says that Allah sees and hears, it does not mean He has ears and eyes. Rather He is expressing a spiritual truth in such a way that we can understand that He is All-knowing. Similarly, behind the title "Son of God" is a spiritual truth expressed in human terms.

The early Christians believed that the God who created the universe—a God who had become embodied in humanity and sustained people as they passed through both the suffering and the joy of life—was a God of love. In fact, God is frequently identified with love, providing an important element of the Christian theodicy, or theory of suffering and death: No matter how bad things seem to be, one can always be confident that the universe is a friendly one and love will ultimately triumph. The second major belief—that one should love one's neighbor as well as God—derives from Jesus' own teaching and forms the basis of Christian ethics, which I will discuss in more detail in Chapter 5.

Third, because Christians believe that God acts in history, the tradition has a historical perspective on the nature of the world. Most Christians believe God will ultimately bring about a dramatic transformation of the universe to fulfill the "Kingdom of God" on Earth as it is in heaven. Christians generally imagine this kingdom as a peaceable one in which love and justice reign. Christian **eschatologies** (theories about the end of time) provide two distinct theories of how the kingdom will arrive: (1) a major cataclysmic event will occur, the Apocalypse, in which good people will be rewarded and evil people punished, and (2) a universalistic vision in which all of creation will be united in the Christ when he returns. Members of the early church were convinced that Jesus would return before they died, that the end of the world was imminent, and that the Kingdom of God, as portrayed in the parables of Jesus and the teachings of the Christian scriptures, required a radical commitment to God. Believers should shed their traditional loyalties—even to their families and jobs—and live a life of love and justice.

Finally, Christianity, like Judaism and Islam, is a religion of the Book—that is, its central beliefs are written and codified in a set of sacred scriptures. Two primary collections of writings constitute the Christian **canon** (the writings declared authoritative for a given tradition): the Hebrew scriptures (known to Christians as the Old Testament) and the New Testament (written between 50 and 140 CE), together known as the Bible. In practice, however, a smaller section—a "canon within the canon"—identifies portions of the Bible as more significant than others. The church traditionally gives primary importance to the four **Gospels** because they record the life and teachings of Jesus, but in many churches (especially more conservative congregations), the Letters (or Epistles) take a more prominent place.

Much contemporary scholarship focuses on the historical development of the Christian scriptures, believed by some to be the literal Word of God and by others to have been inspired by God but written by fallible humans and changed over time. As noted in Chapter 1, controversies about inconsistencies in the Bible are as much political as religious. Nineteenth-century pro-democracy forces in Western Europe attacked the church as a basis of the monarchy's legitimacy by raising questions about the church's theodicies and other truth claims. Most of the claims about a literal interpretation of the Bible that the Roman Catholic Church made in the heat of the battle have since been withdrawn by Rome, but many Protestant groups continue to believe that such a position is a necessary doctrine for any true Christian.

Christian theodicies borrowed from Judaism rely primarily upon the theory that separation from God—the Creator and source of all good—causes evil, suffering, and death. They are **eschatological**, asserting that the problem of suffering will be resolved at the end of time, and they tend to be messianic. Christians believe that a messiah or liberator—Jesus Christ—has already come and will come again to reunite creation with God. Most Christians are also millenarian: They believe that an ideal society will come; some are convinced that the transformation will take place in this world, whereas others believe that the change will happen in the next world.

There are two major differences between Christian cosmologies and theodicies and those of the Jewish parent religion: (1) the Christian church's emphasis on "right belief" evolved as a consequence of its universalistic criteria of membership (belonging was a matter of believing) and (2) Christianity, especially **Protestantism**, places a stronger emphasis on individual responsibility. Once more, the sociological reason for this shift in emphasis lies in the nature of the tradition: Whereas Judaism has defined suffering as something the community experiences collectively, Christianity has placed more emphasis on individual choice and its consequences.

Christian Rituals

Two central rituals found across the Christian faith, the **Eucharist** (communion) and baptism, involve communication with God in the context of collective worship. Ironically, these rituals were borrowed from the Jewish tradition and adapted by the church to denote distinctive Christian identity. Even though they are explicitly designed for communication with God, both rituals address the sociological problem of membership in the community. Because the early church broke so radically with the religious context it grew out of by denying ethnic or tribal requirements for membership, the issue was an important one. Controversies about circumcision and kosher laws recorded in the New Testament reflect the difficulties of this new form of socioreligious organization no longer rooted in ethnicity. Baptism thus became a rite of passage into the community, and the Eucharist was designed to bring members together in the body of Christ.

Several layers of symbolic meaning can be found in the Eucharist or communion service, the eating of bread and drinking of wine: First, the ritual involves a reenactment of the "Last Supper" Jesus had with his disciples before his crucifixion. Second, it also recalls the ancient Hebraic Passover meal, the ritual occasion for which Jesus and his disciples had assembled. The Hebrew rite commemorated the delivery of the slaves from Egypt and the night of the "Passover," when the Egyptian sons were slaughtered but the Hebrew children were spared. (Behind both of those rituals, it should be noted, hovers the much more ancient idea of a hierophany involving human and animal sacrifices. In some ancient traditions, the God, or the animal in which the God resides, is killed and eaten, allowing participants literally to partake of the God and incorporate it into their very being.)

Because the early Christian church saw itself as the resurrected body of Christ, the Eucharist was at the center of worship life as a way of reinforcing the internalization of Jesus into the members of the community. They reconfirmed their "incorporation" in the church, the body of Jesus, when they ate the body and drank the blood (Matthew 26:26–28). As Christianity was removed from its primal roots, in which the natural and symbolic worlds are one, the relationship between the blood and body of Jesus to the wine and bread was rendered increasingly abstract. At the Council of Trent (1545–1563), the Roman Catholic Church countered the Protestant Reformation by insisting that the bread and wine actually became, through the mystery of the Eucharist ritual, the body and blood of Christ (the doctrine of *transubstantiation*). In the modern world, the literalness of the Eucharist has been stripped away; ironically, those contemporary Christian communities that insist most vociferously on a literal interpretation ("fundamentalist" Protestants) are often the most symbolic in their interpretation of the ritual. In some churches, a further abstraction is made by introducing grape juice as a symbol for the wine that represents the blood.

The most widely diffused religious tradition, Christianity has a long history of transformation from its Jewish roots. Although the central rituals of the faith have considerable continuity, they have also been diversified and modified through contact with other cultures as Christianity was carried by various empires and colonizers. Let us look briefly at this process, which hints at what may happen in the contemporary globalization of society and culture.

Because of its detachment from social criteria for membership, Christianity was syncretistic from the beginning. Its earliest advocates acted on the universalism of the founder's teachings and took the movement from the core to the periphery of the ancient Roman Empire. Along the way, it became infused first with Greek philosophy (as in the Gospel of John), and later with a wide variety of other religious and philosophical perspectives. During the Western colonial period, Christianity became transnational and global and its practices grew exceedingly diverse. Although the colonial church eradicated many indigenous local practices in an effort to carry out its perceived mission of spreading an exclusivist religious truth, Christianity itself was dramatically changed by its encounters with various traditions. In the West, especially in New World Protestant denominations, it took on features that would seem foreign to the Christians of ancient Palestine. Elsewhere, Christian belief and practice melded with local practices in a way that most Americans would find astonishing. Although the most general sets of symbols and core beliefs remain intact, they are interpreted in quite different ways around the world, expressed in rituals and ideas that reflect the indigenous religions that Christianity replaced or with which it coexists.

One of the most striking examples of this process is in India, which is predominantly Hindu but has a Christian tradition claiming roots in the 1st century CE from the apostle Thomas, who reportedly first took the Christian gospel to the Asian subcontinent. In traditional Indian culture, the Gods each have their special festival days when they are brought out of their temples and paraded through the streets on a chariot borne or pulled by believers. Over the centuries, the Indian Christian church has adopted many of the practices of the indigenous folk religions, including these processions. In Kodaikanal, in the mountains of South India, the statue of Mary is annually removed from her resting place at Saint Mary's cathedral, put on a "chariot," and reverently pulled through the streets of the town in a daylong procession. Thousands of people come on foot, and by bus, oxcart, and car, to participate. The streets blaze with lights, and as Mary works her way through the crowd babies are lifted up to the statue for her blessing. Vendors hawk their wares: everything from religious artifacts and worship aids to food, clothing, and toys for the kids. A carnival atmosphere prevails, and people gather to talk, renew acquaintances, and enjoy the holiday. Through the process of this festival honoring Mary, life in Kodaikanal is also honored and collective life is celebrated in the midst of poverty and disease.

Another example of syncretism between Christian and indigenous Indian practices can be seen in a Catholic orphanage outside of Madurai in South India, where a statue of Mary adorns the central courtyard. That the sisters who run the orphanage would make Mary the visual focal point of their establishment is not surprising. What may be surprising is that she is standing atop a lotus flower.

Although religious traditions sometimes appear fixed and immutable, they are in fact dynamic phenomena that grow out of specific social circumstances in specific natural environments and are then transformed as the conditions in which they flourish change. This tension between continuity and change is characteristic of all human institutions, including religion. The figure of Mary is a very interesting one in Christianity as she is extremely popular especially among women and within the so-called Third World. There is a sense in which she buffers women against the patriarchy of the tradition, especially in the Roman Catholic Church where she is so prominent.

Christian Hierarchy and Rebellion

Christianity's social organization, the most formalized of all the major religious traditions, is a direct response to the universalism of its theology, which encouraged efforts to convert all nonbelievers to the faith. Whereas many traditions can draw on the social organization of the culture in which they are carried, the deliberate break made by the early Christian movement with the Jewish socioreligious order precipitated the construction of cultural and ideological, rather than ethnic or tribal, boundaries to shape the new religious movement (NRM). Such a radical break with social organization is inherently destabilizing, however, so that the history of the Christian church becomes a story of continual efforts to attach the ideas of the faith and the interests of its institutions to particular social orders.

Reading between the lines of the New Testament, we can detect many religious conflicts emerging from this universalistic experiment in social organization as the early Christian community fought to establish its identity. First came the division between the ethnically Jewish church in Jerusalem, headed by the local leader Peter and the non-Jewish congregations founded or nurtured by the more cosmopolitan Roman citizen Paul, who was dramatically converted to the movement and later transformed it (hence the Christian saying, "robbing Peter to pay Paul"). Many of Paul's letters address the relationship between what it means to be a Christian and whether or not this excludes non-Jews and whether one needs to be kosher or circumcised or otherwise connected to the ethnic roots of the faith in Judaic culture. Time and again, the early community, especially under Paul's charismatic leadership, came down on the side of universalism.

These early battles were waged over and over again as Christianity diffused and became entangled in social and political alliances, finally resulting in institutional divisions of the church that were more political than theological. The church councils of the early centuries, meetings of high church officials from the far-flung corners of "Christendom," were called to settle doctrinal disputes that had political or societal bases. The Roman Catholic Church attempted to monopolize the tradition, tracing the authority of the pope back to Peter, who, according to tradition, was anointed by Jesus to succeed him. When the church split between the Roman and Eastern Orthodox branches in 1054 CE, it was because the Holy Roman Empire itself had divided into Eastern and Western branches. Similarly, the Protestant Reformation of the 16th century reflected not only legitimate religious disputes over the need for reform in a religious hierarchy that had become infatuated with power but also the struggle between Roman ecclesiastical authorities and German political elites (see Bainton, 1950). Along with the Roman Catholic and Eastern Orthodox branches, the Reformation created a third major branch of the church, Protestantism, but the Protestant branch subsequently subdivided into an extremely large number of sects that later grew into denominations in their own right, setting the stage for further reformations.

The history of the Christian church is one of continual rebellion and recrystallization in conjunction with related historical, social, and political developments. As the church attempted to compensate institutionally for its lack of a social organizational base, it became increasingly rigid and hierarchical in sharp contrast to the egalitarian nature of the early Christian community. The vast complex of the institutional order of Christianity at the beginning of the 21st century reflects the tremendous diversity of cultural styles, sociological and political alliances, and theological differences within the church. This institutional diversity runs across denominational boundaries, but it is internally marked as well and has been radically challenged by Christianity's confrontation first with a myriad of indigenous cultures and recently with the democratic ethos of the modern world. The Roman Catholic Church, with its strict chain of command developed over the centuries, has taken much of the brunt of the modern rebellion against authority, but other branches of the church have been similarly shaken by populist and democratic movements as well.

The shape of modern democracy was influenced positively by people demanding political freedoms as a vehicle for religious freedom. Important structural innovations were developed by the founders of the American republic, for example, and models of democratic participation were created by such Christian sects as the Society of Friends, also known as the Quakers. The Quakers rebelled in the mid-17th century against the formalism of ecclesiastical institutions and practices even within the reformed Protestant churches. Rejecting even the sacramental rituals of the Eucharist and baptism, the Friends (as they called themselves) emphasized the four principles of equality, peace, simplicity, and community and cultivated a radically democratic institution. One arm of the Friends, in an effort to purge the community of empty ritualism, resorted to silent worship, interrupted only when any participant felt "moved by the spirit" to speak. Gradually, however, although they continued to emphasize simplicity in their interactions and forms of worship, even the Quakers developed their own distinctive rituals and specialized language that help to solve the problems of collective life and to create the boundaries not provided by a common ethnicity.

Islam

In the name of God, the Benevolent, the Merciful!

Praise is proper to God, Lord of the Universe, the Benevolent, the Merciful!

—Qur'an 1:1–2[2]

Judaism, Christianity, and Islam are sometimes called "three strands of a tightly woven rope" within the **Abrahamic tradition** because they trace their roots to the same Semitic religious heroes, Abraham and Sarah. The Islamic faith is one of the fastest-growing and most misunderstood religions in the world today. Approximately one out of five people call themselves Muslims. Indonesia has the largest number. At the center of Islamic belief is a monotheism inherited from Judaism, the concept of *tawhid*, the unity of God, and the belief that God spoke to humans in the **Qur'an**, the Islamic scriptures, through a prophet in the tradition of Abraham, Moses, and Jesus.

Islam is perhaps the most difficult of the traditions for a non-Muslim American to write about, because of the pervasive misrepresentation of Islam and Muslims in the media—which is where most Americans get their information about Islam—and which is so patently inaccurate. Moreover, I was reminded of the difficulty of the overall task of this book while preparing the section on Islam; perhaps Daniel Martin Varisco (2005) put it best when he insisted, "It is easy to create unity out of diversity but seldom does it serve an analytic purpose" (p. 136). In his critique of academic approaches to the study of Islam, he observed the shortcomings of the classics from Clifford Geertz (1968) to Ernest Gellner (1981) to Fatima Mernissi (1975) and Akbar Ahmed (1988), as well as Edward Said's (1978) critique of "Orientalism" in Western scholarship.

Islam's founder, Muhammad, known as the **Prophet**, was born in about 570 CE on the Arabian Peninsula. Orphaned at a young age, he grew up in his merchant uncle's household and became a well-established and respected merchant himself. He married a wealthy widow, Khadija, had children (two sons died and four daughters survived), became a respected citizen of Mecca, and then became disenchanted with his comfortable life and the materialistic meaninglessness of his culture. Muhammad experimented with spiritual contemplation in a cave on Mount Hira, just outside of the city, and spent long hours in solitary meditation. One evening in the cave, a voice spoke to him, the first of a series of revelations recorded in the Qur'an. He was petrified and ran home to his wife, who was one of the few who stood by him at the beginning of his long and difficult spiritual journey.

Muhammad began relating the revelations he had received to others in Mecca, but he was met with hostility and mounting opposition there from religious and business elites who considered him a threat because of his prophetic emphasis on justice and spiritual discipline. The tribe of Quraysh, which controlled Mecca, first tried to co-opt him and then threatened him; local merchants instituted a social and economic boycott against his entire family and then attempted to kill him. Some of the Prophet's converts fled to Ethiopia in 615 CE, and in 622 CE, Muhammad and his followers left Mecca, moving to the nearby city of Medina, where they established their religious community, the **ummah**, the foundation of Islam. The Muslim calendar thus begins with that year of the migration.

The ummah grew quickly in Medina, and its influence spread to the surrounding territories. This expansion led to military conflicts between Muhammad's followers and a Quraysh army, among others. Eventually, the ummah became the dominant religious, military, and political force in the region. In 629 CE, Muhammad and his followers returned triumphantly to Mecca, took control of the city, and established the religious practices and organizational patterns of Islam.

Islamic Beliefs

> *Walk into any bookstore, and you will initially be drawn to a stack of breathless titles that are truly frightening. These journalistic exposés reveal worlds of terrorist intrigue and plots against the United States. Alongside these instant potboilers are books with a more sober tone, delivering with masterful condescension the verdict of failure upon Islamic civilization, and the promise of an apocalyptic clash between Islam and the West. . . . How can anyone make sense of all this?*
>
> —Carl Ernst (2003, p. xiii)

> *Islam's scriptural sources stress that mercy—above other divine attributions—is God's hallmark in creation and constitutes his primary relation to the world.*
>
> —Umar Faruq Abd-Allah (2004, p. 1)

It is almost a truism in discussing Islam to begin with the Five Pillars that ritually reinforce the religious and social boundaries of the ummah, draw people into the community, and provide an important redistribution of wealth from rich to poor.

The Five Pillars of Islam

Despite the diversification of the tradition as it spread throughout much of the world, Islam's core remains remarkably constant with the following elements showing up almost everywhere Muslims worship:

1. *Shahada*, or profession of faith

2. Ritual prayers (*salat*) as part of communal and individual life cycles

3. Zakat, the sharing or giving of alms from a proportion of one's wealth

4. Ramadan, a month of fasting from daybreak to sundown to cultivate spiritual, physical, and moral discipline

5. Hajj, a pilgrimage to the sacred places in and around Mecca

The first pillar, the shahada, is the profession of faith: "There is no God but God [Allah], and Muhammad is God's messenger." The shahada is affirmed at all key events in one's life cycle (especially at births and deaths) and is included in daily prayers. To embrace Islam, one must simply make this declaration before witnesses. According to the tradition, everyone is born Muslim.

The second pillar consists of making personal prayers along with the formal ritual prayer (salat) that is repeated several times daily, normally at dawn, noon, afternoon, sunset, and late evening. The salat is used to frame the entire day, reminding believers that God should be at the center of their lives. The equality of all before God is underlined as the devout of all social classes worship together. The entire Muslim world faces Mecca and prays, turning to neighbors at the end of the prayer with the traditional greeting, "Salaam" (peace).

The third pillar, zakat, requires Muslims to give a portion of their wealth to the poor and others needing assistance. This ritual giving includes the important practice of hospitality and provides a mechanism for sustaining the social order by redistributing wealth and addressing social needs. The term *zakat* literally means "purification," suggesting that wealth is defiling unless shared with others. This practice, stemming from Muhammad's vision of justice and condemnation of the materialism of his time, now requires an annual donation of 2.5% of a believer's wealth.

The fourth pillar of Islam is Ramadan, a month of fasting from daybreak to sundown to cultivate spiritual, physical, and moral discipline. Besides abstaining from food, drink, and sex during the daylight hours, Muslims are enjoined to focus on the core values and practices of their faith during that period. After sunset, the fast is broken, followed by prayers and shared meals with family and friends. After the last day of the month, Muslims celebrate 'Id al Fitr, one of the major festivals of the Muslim year.

Finally, at least once the believer must participate in the fifth pillar, hajj, the pilgrimage to the sacred places in and around Mecca. Although not required if it is a financial burden, this pilgrimage is a highlight in the lives of the faithful as they retrace the steps of generations of their predecessors. Mecca is the axis of the Muslim world, toward which all face in prayer; it is the holy center where humans encounter the sacred. By making their own pilgrimage to Mecca, the believers trace their spiritual roots back through the centuries: They reenact the founding of Islam with Muhammad and finally join Abraham and Ishmael in encountering the God who created the world and is compassionate and merciful to all.

The ritual package known as the Five Pillars reminds Muslims on a daily basis that religion is not part of the structure of Islamic society—it *is* the structure—and that Islam is not part of the daily life of a Muslim—it *is* his or her life. The rituals associated with the Five Pillars are designed to draw believers constantly back to Allah. Performed publicly, they mark Muslims as believers wherever they may be. A religious tradition that is not compartmentalized but diffused throughout a believer's life is undifferentiated and primal in character, makes more demands on its adherents, and is sometimes less tolerant of other belief systems and lifestyles than more cosmopolitan religions.

It is, however, also more effective in providing believers with religious purpose and guiding their lives, so many practitioners understandably resist efforts to "modernize" their beliefs and secularize the other spheres of their lives.

The Islamic faith centers first of all around intense commitment to and worship of the One God, the Creator of the universe. Sharing a heritage with Judaism and Christianity, the faith of most Muslims is quite different from what is portrayed in the news media. When Susan Savage (2004) examined images of Muslims on television news programs over a 2-week period, she found 417 images of Muslims as violent compared to 23 neutral images of Muslims. During that time, she found no images of pacifist or nonviolent Muslims. She also found that television was a major source of information about Islam for non-Muslims, so it is not surprising that misunderstandings are abundant.

A second important feature of Islamic belief is that the Qur'an is God's word as revealed to the Prophet Muhammad. Thus the Qur'an is the Word of God for Muslims in the same sense that Jesus is the Word of God for Christians. The Qur'an is in the Arabic language and, among the strictly orthodox, cannot be translated into any other language; this prohibition is breaking down, however, as Islam becomes more cosmopolitan.

Third, God requires a disciplined, ethical life, as outlined in the Qur'an; Muhammad believed it was essential for believers to cultivate such qualities in a natural and social environmental hostile to spiritual growth. Fourth, the tradition is institutionalized within the context of the all-encompassing religious community, the ummah, to help people nurture a spiritual lifestyle.

Doctrinal and social boundaries are drawn around Islam and the community of believers, although Islamic rulers have often shown considerable tolerance in allowing others to practice their own religions. The uncompromising monotheism of Islam led to a strong condemnation of "idolatry" and the association of other deities with God, sometimes leading to conflict with other religious communities, as in India, where a plethora of Gods is sometimes seen as competing with Allah, although this issue may be more a matter of politics than spiritual life.

Muhammad's emphasis always seems to be on cultivating a spiritual lifestyle that is God-centered and promotes justice. This is far more important than any particular doctrinal issue. As the Qur'an (5:48) puts it,

So race to virtues

God is your destination, all

And God will inform you

On all that wherein you have differed . . .

Despite Muhammad's prophetic departures from earlier religious practices, Islam remained close to its roots in early Judaism. Muslims believe Islam is merely the latest expression of the Abrahamic tradition, and the language and ethos of their religion is similar to aspects of ancient Judaism. Muhammad is a prophet who stands in the direct line of prophets stretching from Abraham and Moses through the ancient Hebrew prophets Isaiah and Jeremiah. A sophisticated and very diverse tradition today, Islam still retains many elements of a primal religion, in which religious life is ideally coterminous with everyday life. Human conduct should always be oriented to God, according to Islam, so that daily life is sacralized by the religious experience.

If the Qur'an is Islam's soul, the family is its body. The family is the basic unit of Islam and is supposed to mirror the ideal society. Religious institutions still regulate the traditional Islamic family, in which there is little blurring of gender roles. Women play an important role, and there is much debate in contemporary Islam about the status of women in the Qur'an and Hadith. This is part of the reason for the strong reaction against the intrusion of Western ideas about the changing role of women, now a major political issue throughout the Islamic world. It is difficult to know what the authentic views of the prophet Muhammad were on the issue of women, given the historical context of the oppression of women in his time and the generations of male interpretations of his life and teaching. Clearly women played a key role in his own spiritual development, from his wife Khadija, who first gave him the courage to speak, to his daughter Fatima, who used to sit beside him when he spoke in public. Recent movements among Muslim women have challenged much of what people—both within and outside the faith—think about the role of women in Islam (see, e.g., Fernea, 1998; Shaikh, 2003; Wadud, 1999; and the discussion in Chapter 6).

With their similar origins in the creation story of Genesis, Muslim theodicies, like Christian and Jewish ones, focus on the idea of separation from God as the source of suffering and death. The Creator aspect of God is highly significant to Muslims, who place a strong emphasis on a transcendent God who is nonetheless compassionate and merciful despite an exalted position. The pain of the separation from the Creator is a central theme among the **Sufis**, the mystical branch of the Muslims. Like most mystics in other traditions, the Sufis seek an intimacy with God through devotion and prayer, like that the Prophet obtained through his long hours of solitary worship. Sufis sometimes use the metaphor of the reed to explain suffering and death. The reed is fashioned into a flute that laments being separated from its source, the reed bed, and yearns to be reunited. The human separated from his or her Creator is brought back to God through the rituals of the Islamic community.

Those who are faithful to God are rewarded with paradise. Ali Mazrui (1986) noted the following:

> Muhammad's image of heaven enchants the wretched societies of thirst. The drought-ridden areas of the Middle East and Africa are fertile ground for the message of Mohammad—to people in dry lands, what is heaven if not abundance? The Islamic paradise has all kinds of rivers flowing under it—the dream of a clean natural fountain, the bliss of a wet eternity.

Those who forget God and the teachings of the Qur'an, who fail to show mercy, will be punished for their misdeeds. The faithful, however, will be rewarded for their discipline and devotion to God, for their acts of justice and mercy, and will be reunited with God for eternity.

Abd-Allah (2004) said it is accurate to describe Islam as a religion of mercy. He observed that "Islam's scriptural sources stress that mercy—above other divine attributions—is God's hallmark in creation and constitutes his primary relation to the world from its inception through eternity, in this world and the next" (p. 1). Moreover, the most important name of God is Allah, which Abd-Allah noted is similar to the word for God in the language of Jesus, Aramaic—*Alaha* (God, the true God). The next most significant name for God is "the All-Merciful, the Mercy-Giving" ('r-Rahmani 'r-Rahim), which is at the beginning of all but one chapter of the Qur'an and is central to Islamic ritual (Abd-Allah, 2004, p. 2).

I will take up some of the implications of this belief about God as most merciful when discussing the issue of Islamic ethics in Chapter 5.

Tribal Continuities in Judaism and Islam

Whereas Christianity broke with the ethnic identities of its roots, both Judaism and Islam struggled to retain elements of their tribal organization in which religious and social organization were not clearly divided. Islam retained the universalistic message of Judaism and Christianity while tightening the social boundaries of its faith community. Islam's tribe is not one of biological inheritance like Judaism but a spiritual family, the ummah. Muslims ritualize this family by referring to one another as brother and sister.

The ancient Hebrews built an ark in which to keep the tablets with the Decalogue, or Ten Commandments, that Moses received at Mt. Sinai, and they carried it to Jerusalem, where a temple was built to house it. The temple was the center of religious life until its destruction by the Babylonians in 586 BCE, when many of the Israelites were carried off to Babylon. Fifty years later, Babylon itself fell and the Jerusalem temple was rebuilt and rededicated in 515 BCE. The religious specialists who maintained the original temple prefigured the rabbis—the teachers—who became especially prominent in Jewish life after the destruction of the temple by the Romans in 70 CE following a period of Jewish revolts against Rome. In the Rabbinic period, which dates from about the beginning of the Common Era, the two institutions of modern Judaism were clearly established: the **Temple** and the Torah. The rabbis helped to shift the focus of Jewish practice from the Temple to the Torah, especially after Jews were forbidden access to Jerusalem by the Romans. The Torah itself, the written scriptures of the faith, became something of a portable institution; it was believed to contain the presence of God and served to facilitate the rituals of worship much like the Temple, but it could be transported when members of the community were forced to migrate to other parts of the world.

The shift from an emphasis on the Temple to the Torah was significant historically as Judaism became somewhat disengaged from sacred locations, less local, and more cosmopolitan but still emphasized social boundaries around its community. Jerusalem retained sacred significance because of its historical importance, but Israel's God could be encountered wherever the rituals of the faith were performed. Gradually, all members of the faith, not just the priests, could perform every ritual practice required by the tradition. No matter where they were in the world, Jews could act in their homes and their daily lives as if they were priests in the Temple (although some rituals require a minimum of 10 Jews, a *minion*, to be properly conducted). A major reason for the decentralization of Jewish institutions was the mobility and dispersion of the community. After the Babylonian Captivity in the 6th century BCE and again following the destruction of Jerusalem and the temple by the Romans in 70 CE, Jews were scattered all over the world and carried their religious traditions with them, constituting what is called the **Diaspora.**

Because ritual is more important than belief in traditional Judaism and the religion is rooted in an ethnic identity reinforced by ritual, the rabbis became ritual experts more than theologians. The emphasis on the written revelation from God in the Hebrew scriptures, however, means that the rabbis also became experts in the written tradition, especially the Torah, and were the intellectual leaders of the community as well. The authority of the Jewish rabbis is similar to the pragmatic authority of the priests, monks, and *pandits* of Hinduism and Buddhism; it comes from effectively demonstrating their knowledge of the scriptures and the rituals of the faith. The rituals must be perceived as bringing benefit to worshippers, although it is more for the collectivity than the individual.

Just as the mosque forms the center of the Muslim city from which all life emanates, so Islam and its institutions are to lie at the center of individual and collective life. Initially a relatively small, homogeneous religious movement, Islam developed institutions that were identical to the general social organization of Muslims. The **Shari'a**, or law, provided a comprehensive judicial system but also defined the Muslim state and the responsibilities of the **caliphs**, the heads of the Muslim ummah (the community of believers). The Shari'a, also called the "Way," grew out of efforts to develop systematic instructions for all aspects of individual and collective life as prescribed by the Qur'an, and is found in a collection of books that emerged in the first centuries of the faith, such as the Hadith, records of early Islam comparable to the New Testament of Christianity, and less authoritative collections within various branches of the Islamic community. Because "early Islam made no distinction between law and religion," the word Shari'a is itself a later development (Williams, 1962, p. 92). As the community grew after the death of Muhammad, first Uthman (Muhammad's son-in law) and then Ali (Muhammad's cousin) became caliphs, but internal divisions erupted into deep and long-lasting conflicts and finally civil war. The group of Muslims who supported Ali came to be known as the **Shia** ("followers"), or Shiites, and created an institution called the *imamah* (from *imam*, meaning "leader") in which each leader appointed his successor to be the imam, whom the Shias believe is divinely inspired and can provide both religious and secular leadership. When Ali was assassinated in 661 CE, his eldest son Hasan succeeded him, followed by a younger son Husayn.

The other major branch of Islam, the **Sunni Muslims**, became the majority Muslim group, deriving their name from an emphasis on the *sunnah*, the custom or practice of the Prophet. The Sunnis developed an elaborate set of means for interpreting the Shari'a that relied upon the use of consensus and analogy by the scholars who specialized in the study of Islamic law. Not surprisingly, some broad differences of opinion in how to interpret the Shari'a led to the creation of four schools of law in Sunni Islam, each named after the scholar most responsible for its creation: Shafi'i, Maliki, Hanbali, and Hanafi. Each school has its own distinctive characteristics but recognizes the others' right to their differing views (e.g., on whether the Qur'an was created by God or is eternal), thus allowing for a considerable level of tolerance and autonomy while recognizing a common membership in the universal Islamic community, the ummah.

In many parts of the world today, the Muslim community is undergoing substantial transformation as it diffuses and intermingles with other subcultures, just as the Jewish Diaspora has. Umar Faruq Abd-Allah (2004) argued the following:

> The Prophet Muhammad and his Companions were not at war with the world's cultures and ethnicities but entertained an honest, accommodating, and generally positive view of the broad social endowments of other peoples and places. The Prophet and his Companions did not look upon human culture in terms of black and white, nor did they drastically divide human societies into spheres of absolute good and absolute evil. Islam did not impose itself—neither among Arabs or non-Arabs—as an alien, culturally predatory worldview. Rather, the Prophetic message was, from the outset, based on the distinction between what was good, beneficial, and authentically human in other cultures, while seeking to alter only what was clearly detrimental. Prophetic law did not burn and obliterate what was distinctive about other peoples but sought instead to prune, nurture, and nourish, creating a positive Islamic synthesis. (p. 4)

In both Judaism and Islam, the solidarity of socioethnic ties of the community diffused authority and gave religious elites socially derived authority rather than the formal authority of a religious institution. Divisions within each tradition emerged mostly along social cleavages and have become somewhat more complicated as the traditions were diffused around the world and their bodies of adherents became more heterogeneous.

The Social Construction of Religious Traditions

Specific religious beliefs provide a coherent worldview for people in a specific cultural and historical situation; they may or may not make sense to people outside the context in which they were created. The social organization of a people, and the religious tradition(s) to which they subscribe, are intimately and dialectically related, molding and shaping each other over time. Religious tradition explains to those within a faith community why the world and its inhabitants are here; why evil, suffering, and death exist; and how one should live one's life. These ideas are expressed both through rational discourse and through myths and legends that incorporate the contradictions of life. They are reenacted and reinforced through rituals that solve social problems, identify evil, offer guidelines for action in times of crisis, and legitimate the religious institutions that emerge out of each religious movement that survives. These ideas, rituals, and institutions do not operate only on some abstract plane; they also deeply infuse the daily lives of believers.

As people with different traditions are thrown together in the same social space, the changes the new situation precipitates are profound.

Many of the same threads, such as the ethical system that orders everyday life, run through different religions. Distinct cultural styles emerge within separate religious traditions, however, that persist over long periods of time and space; they affect how people love and fight one another and establish a general tone and lifestyle for a people (their ethos). Although enormous variations exist within each broad cultural framework, some general tendencies reach across entire civilizations. We can see these large patterns more clearly when we compare Eastern and Western cultures and religion.

One general *tendency* found in contemporary Western religions (primarily the various Judaisms, Christianities, and Islams) is the idea of one God who created the universe and regularly intervenes in history. The implicit message to believers is, "Go thou and do likewise"—that is, an active deity who is constantly shaping and creating the world expects worshippers to do the same. God often sends people on missions to do the divine work in the world and sometimes even everyday economic activity; this is defined as a "calling" from the deity. Consequently, a tendency toward a this-worldly orientation, emphasizing action in the world at the behest of the deity, is evident in most of the Western religions, although each of them possesses an otherworldly theology as well.

Adherents of Eastern religions, in contrast, tend to worship either many Gods, specialized according to different functions, or humans who become deified, such as the Buddha or K'ung-Fu-tzu (Confucius). Rather than giving commands and intervening in history, these Gods seem to prefer simply to explain how the universe works and tell people how to do their best with what they have been allotted in this lifetime. They instruct people to determine their *dharma* (religious obligation/duty) for a particular stage of this lifetime and carry it out to the best of their ability. Because of karma (the law of cause and effect that governs the universe), they will either be rewarded because of their faithful commitment to their dharma or they will be punished for failing to fulfill it. One's duty is, therefore, not to change the world, as in Western religions, but to follow its rules in order to escape its control. The usual reward, according to Hindu and Buddhist traditions, is a more favorable incarnation for the next lifetime, and the expected punishment is an unfavorable one. Thus, it is not a matter either of being particularly moral or immoral or of doing or not doing what the Gods tell one to do or not to do but simply whether or not one is harmoniously aligned with the universe.

Each of the five major world religious traditions was forged as a response to the social and existential conflicts of their founders and the social strata that carried them. Hinduism probably emerged out of conflicts between the indigenous Dravidians and Aryan invaders of the Asian subcontinent. Buddhism was formed in the era of changes sweeping Asia during the 6th century BCE and the Buddha's own representative struggle with the tensions between wealth and poverty. Judaism emerged from the context of the migration and enslavement of a clan, Christianity was created in the intercultural conflict of the Roman Empire's conquest of the ancient Middle East, and Islam came into being amidst warring Bedouin tribes and the growing materialism of the merchant class. Those varied social contexts and intercultural conflicts provided the soil in which the beliefs, rituals, and institutions of each tradition were cultivated.

Diffusion and Change in Religious Traditions

Every major religion has undergone substantial change over time as its social context changed, especially as each diffused geographically and encountered other indigenous traditions. We may summarize the major modes of religious adaptation to these new situations as follows:

1. Co-optation and syncretism of indigenous traditions, as in Hinduism and Buddhism, which developed a loose federation of organizations affording considerable autonomy to local units and knitting institutional and belief systems together into new forms

2. Conversion of people in alternative traditions, as in Christianity's alliance with Western European colonialism, using more or less coercive tactics in different situations depending on a variety of factors

3. Segregation of the traditions, as in Judaism, through strict rules prohibiting interfaith marriage and excessive interactions with outsiders

4. Integration of religious communities, as in Islam and some sectors of established Christianity, which allowed people from other religious groups to participate as minority groups in the society with some measure of religious freedom

In actual practice, of course, sectors of every religious tradition have engaged in some combination of each of these adaptation strategies, although the cultural style of a tradition often shows an affinity for a single option.

Christianity's Conversion Strategy

The distinction between religious belief on one hand and social and ethnic status on the other in early Christianity resulted in a conversion-oriented faith that shaped institutional practices. From the time of the early church,

Christians felt compelled to convert nonbelievers of all national and ethnic groups to their faith. On the ideal level, it was an effort to "share the good news" with others. As a practical matter, the conversion strategy became a tool of various social forces to conquer others, from the Roman emperors to the Western European colonialists.

The Roman Emperor Constantine converted to Christianity in 312 CE, immediately before a decisive battle in which he painted the Christian cross on the shields of his soldiers, won the battle, and became convinced that the Christian God was a great war deity who could advance his imperial ambitions. Constantine's entire army was forced to convert and the Empire and Christianity spread simultaneously over broad areas of the ancient Middle East into Europe, Asia, and Africa, converting people from their indigenous religions to Christianity as the troops went from region to region. Christianity was also changed substantially through the process, shifting from a radical oppositional community, pacifist and egalitarian in nature, to a hierarchical, establishment-oriented institution that facilitated imperial military conquest. The beliefs and rituals of the faith were transformed as well by Greek philosophy and indigenous beliefs and practices throughout the ancient world. Although the central ideas remained mostly intact in the process and local religious leaders and masses of people were sometimes forced to convert formally to the Christian faith, the link between the church and the power elite of Western civilization together with the encounter with diverse beliefs and practices created a tradition so radically different that the earliest followers of Jesus might not have recognized it.

A similar process occurred in the 18th and 19th centuries as the European colonists conquered vast regions of the world in search of trade and profit, accompanied by the Christian missionaries. The missionary movement did not always legitimate the colonial conquerors, but it often did, providing the rationale for the subjugation of populations throughout the Americas, Africa, and Asia. Local Gods and spirits became identified with the saints of Christianity, indigenous festivals were grafted onto celebrations in the Christian calendar, and the majority population of the church became not white Europeans and North Americans but people of color in the so-called Third World. As the anticolonial movement took hold following World War II, people in the former colonies that gained their political independence demanded independent religious institutions as well, and much of the control over religious institutions exercised by Europeans and Americans was turned over to indigenous leaders. The newly independent nations, often run by people trained in Christian missionary schools, asserted their own authority in their local arenas but also became part of the decision-making process at the core. Italian and other white bishops and cardinals were challenged at the Vatican in Rome, and people of color sent delegations of their own leaders to church centers in New York and elsewhere. After the mid-20th century, the shape of the Christian institutions was altered along with the ethnic and racial composition of international gatherings and legislative bodies. The global village had reached Christianity.

Judaism's Universalistic Belief and Exclusivistic Practice

Where Christians and Muslims tried to convert nonbelievers to their respective religions, the Jewish community took a very different approach. For the most part, Judaism has deliberately segregated itself from its host cultures as it diffused around the world, making no effort to impose the tradition or to convert others to join but focusing instead on coexistence with its neighbors whenever Jews left the Middle Eastern homeland.

This institutional path has an elective affinity with the needs and interests of the Jewish community, which has its roots in a nomadic tribal culture and has experienced substantial geographical mobility over the centuries. As a primal religion, ancient Judaism was simply a tribal variant of existing religions in the regions where Abraham and Sarah lived before they migrated to the region now known as Palestine. It was probably only when Moses was leading the Israelites out of Egypt and back to that region, many centuries after Abraham and Sarah, that the earliest institutions of ancient Judaism began to emerge. The first notable event was when the priesthood of the Levites sided with Moses during a conflict among religions at the time that Moses received the Decalogue at Mt. Sinai.

The division of modern Judaism into three main branches—the Orthodox, Conservative, and Reform—reflects the modern transformation of the segregation strategy and especially the emigration of large numbers of Jews to the United States, where they have become more assimilated into mainstream culture than in most of Judaism's long history. The Reform movement was first initiated in the 19th century, primarily among German Jews who immigrated in large numbers to escape waves of persecution starting in Germany in the mid-1850s. They were joined by Eastern European Jews fleeing the pogroms of Eastern Europe in the late 19th and early 20th centuries. The Reform Jews wanted to retain their traditional faith and many of its rituals but felt it necessary to respond to changing social conditions with new forms of religious practice. The Conservative movement emerged as an American Jewish response to these reforms, growing out of a concern by some that the Reform movement had given up too much of the tradition; it advocated not a return to Orthodox Judaism's more rigid order but a more flexible approach that retained the spirit of the ancient tradition. Moreover, high rates of interfaith marriages further undermined the traditional segregation strategies of the Jewish tradition, a situation exacerbated by ongoing changes in Jewish social organization around the world with the establishment of the state of Israel in 1947 and the recent lifting of restrictions on Jewish emigration from the former Soviet Union.

Segregation and Integration in Islam and Established Christianity

A final mode of adaptation to intercultural conflicts is this interplay between attempts at segregation and efforts to integrate all religious groups under a broader sociopolitical order without swallowing up minority religions in the dominant faith. The structural situation created by the linking of a religious tradition with a sociopolitical order results in an unresolved tension between efforts to segregate the established and minority faith communities, while at the same time attempting to integrate them into the civil society. Sometimes this uneasy alliance results in a flexibility and tolerance for diversity; at other times, it promotes a rigidity that results in the persecution of minorities or splits within the dominant religious community.

This dynamic is most often found in the established faiths of Islam and Christianity in those contexts where they are the official religion of a sociopolitical order and forge an alliance with the state. In the established versions of Christianity and Islam, elites from the official religion set moral and religious standards as they attempt to stretch the sacred canopy of their faith across the entire society. The reign of Christendom in medieval Europe is the clearest example of this union between church and state, though similar alliances were forged elsewhere, and it was precisely this institutional arrangement—more than anything inherent in Christian beliefs or rituals—that precipitated the intense conflict surrounding the advent of modernism in the West, as we will see in Chapter 6. The Islamic community, especially in the West, has often shown some intolerance to nonbelievers, but a tradition of hospitality to strangers and sensitivity to the need for peaceful coexistence among many Muslims has resulted in the extension of tolerance to believers from other faith traditions as long as other religions do not restrict the religious freedom of Muslims. While the Christian church was tormenting Jews during the Inquisition in 17th-century Europe, for example, the Muslims allowed Jewish communities substantial religious and social freedoms within predominantly Muslim territories.

The Elementary Forms of Religious Life

This brief overview of religious beliefs, rituals, and institutions has focused on a number of characteristics relevant to an assessment of religious life at the beginning of the 21st century. First, religious traditions are dynamic, particularly when they encounter one another—as they do with increasing frequency in the emerging global village. Usually presented as immutable truth passed down intact over the generations, religions have changed dramatically over time as the social conditions around them have transformed. Always the norm, change within the globalization process has become a tradition. There are no "pure" religious traditions preserved intact over the centuries. Orthodoxy is as much the result of a political battle within a religious institution, or between the institution and forces outside it, as an inherent feature of the belief system.

It is, in part, the tension between continuity and change that makes religious traditions such a powerful force in social life: Religions often preserve ancient wisdom, allowing a culture to recover lost knowledge and insights that may disappear from view within the ruling conceptual paradigms of any given age. These preserved traditions, even as they strive to conserve, then become the source of change. "Conservative" ideas of harmony between humanity and the rest of the natural world, for example, sometimes provide the basis for a radical critique of contemporary practices that are destructive to the environment and a call for change within the society.

Because of the dynamic nature of religious life, we can expect dramatic social and religious changes in the next decades as the various regions of the world community become increasingly interdependent. The relatively autonomous sacred canopy may be an artifact of the past. Far from disappearing, however, religious beliefs, rituals, and institutions are emerging in new forms from among which people can choose in a global marketplace of religious ideas and practices. Although religion is so embedded in social practice that people do not switch lightly from one religion to another, conversions do occur from time to time. Perhaps more important is the transformation of religious traditions themselves that is growing out of the increased contact among the faith communities, especially at the top. Ancient religious institutional forms persist in the contemporary world, but religious bureaucracies have taken their place alongside other multinational corporations. Religious elites continue to play an important role in social and political life and often provide an alternative voice emphasizing values that sometimes collide with the utilitarian rationality of modern corporate and political elites.

A second salient feature of religious life is the diversity that persists even in the face of an expanding world system and, ironically, may be enhanced by the unification process. As we will see in Chapter 6, efforts to unify culture on a global level have been countered by the revival of more localized practices in the form of religious traditionalist and other protest movements. The revitalization of specific religious traditions, ethnic identities, and other countersystem movements has intensified the complex interplay between increased unity and diversity that Durkheim (1893/1933) noted at the end of the 19th century.

Many people will object to religious diversity and make exclusivist claims to the truth for their own religious perspective, but this very insistence on the value of a single religion ironically adds to the strengthened diversity of religious life on the planet. Competing religious traditions are enhanced by arguments against religion in general or the formulations of specific traditions. Some of the most eloquent statements in defense of religious beliefs have been responses to strong attacks, and some of the most important truths of religion have been revitalized by atheists when the religious institutions that formerly promoted such truths have long ignored them.

Could it be that religious life, like natural life in the ecosphere, thrives on a system based on the interdependence of a wide variety of distinct species that serve a specific function within the larger dynamic environment? When a rainforest is cleared to make way for a single crop, the fragile ecology of the region is destroyed and even that crop—valuable as it may be—cannot readily thrive. Whether the same is true of cultures and religious communities is an open question, but the fact that global unification is being met with such strong resistance in so many ways around the world suggests that the question is important.

A third important aspect of religious life is the systemic character of beliefs, rituals, and institutions. Every element of a set of religious beliefs and practices is interdependent on a number of levels. At the most abstract level, beliefs within a single tradition are interdependent and reinforce one another. These beliefs are, in turn, intricately intertwined with the rituals and institutions of the tradition, each acting on the other and thus giving shape to the daily observance of the religion.

The sacred canopy metaphor has shown us that each thread of a tradition is woven together with the others and that the unraveling of one part threatens the integrity of the entire fabric. The problem with the canopy metaphor in the global village is that it may be too static an image to represent reality. A religious tradition may be, in fact, more like an energy field than a piece of fabric. Although an energy field has definition and its effects can be seen and felt, it is still made of a dynamic interplay of ever-moving forces. Perhaps we should simply say that the fabric of the sacred canopy is not static either; at the subatomic level, it too consists of dynamic energy fields that are in constant motion despite the canopy's deceptive appearance of solidity to the human eye.

A final characteristic of religious life is its dialectical nature. Religion is something that grows out of and yet also acts back upon the social context in which it is born and persists; the elective affinities between certain religious ideas and the interests of social forces that promote them are mutually reinforcing. The canopy (or energy field) of any given religion is socially constructed out of the life experience of a group of people in a particular environment. The principal cause of the diversity across and within religious communities is the diversity of human life situations on the planet. The primary source of their similarities is the universality of human experience regardless of social and natural environments: Every individual and community appears to be living in a world that is full of joy as well as suffering and death, and the worldview that each community constructs has implications for how people can make the best of the lives they have been given.

Rather than replacing religion, as Durkheim predicted, science has become another major means for obtaining knowledge about the world in which humans live and a tool for interpreting it. Science itself is not monolithic, of course; it remains embedded in the process of dynamic changes in its beliefs, rituals, and institutions as well and takes its place in the ecology of human affairs. Rather than destroying religious life, science has in many ways revitalized and reformulated it, forcing people of faith to rethink their ideas and practices and to reapply their religious perspectives to the changing conditions of a postmodern world.

Durkheim's (1915/1965) treatise on religious life a century ago was a major contribution to our understanding of how knowledge itself is constructed and the relevance of that knowledge to our collective life. The various theories about the world and its meaning embodied in religious traditions show an elective affinity with diverse social forces attempting to shape the nature of the emerging social order. The knowledge and values of one age as preserved by religious tradition often provide the critique of the next, just as an emphasis on justice for the oppressed among the ancient Hebrews serving as slaves under the Pharaoh became the basis for scathing critiques by the prophets of King David's monarchy. This tension between sustaining and critiquing, legitimating and challenging the status quo has characterized religious traditions throughout the centuries and persists in our time, becoming a major element of ethos construction for the global village that we will explore in the next chapter.

4

Indigenous Religions

Bonnie L. Mitchell-Green

Lester R. Kurtz

W hile the majority of the world's population might identify with one
of the "cosmopolitan" world religions, many of them also have
beliefs and rituals rooted in indigenous traditions. Indeed, since those major
traditions began as indigenous faiths, it is more accurate to map humanity's

Author's Note: I, Bonnie L. Mitchell-Green, the lead author for this chapter, have gleaned most
of my information over the course of 30 years of informal interactions with numerous
American Indian friends and acquaintances from diverse tribal backgrounds. This is in keeping
with the oral traditions of most native peoples in which the primary and most-respected
learning is via repeated face-to-face interactions rather than via printed words. The following
persons must be given credit for their oral instruction: Lena Judee, John Maestas (deceased),
Robbi Ferron, Clifford and Yetta Jake (both deceased), Mark Maryboy, Wil Numkena, Nola
Lodge, Fred Cedar Face, Rodger Williams, Lacee Harris, Dan Edwards, Jimmy and Anna
Benally, Travis Parishonts, and Alex Shephard. I am knowingly violating the norms of some
tribes by mentioning the names of the deceased, because it seems a bigger crime not to give
credit where credit is due. My early readings of works authored or coauthored by Sonia
Bleeker, Frank Waters, Black Elk, Leslie Marmon Silko, Sun Bear, Richard Erdoes and John
Fire Lame Deer, Vine Deloria Jr., Winona LaDuke, Suzan Harjo, Linda Hogan, and others
were also influential.

beliefs as a tapestry of various traditions woven together over the centuries rather than the oversimplified pie chart (see Figure 2.1) at the beginning of our sociological tour. Moreover, many in the Western Hemisphere, for example, have become nominal Christians in order to gain social status or to avoid social stigma, discrimination, or even persecution. The sad history of physical and cultural genocide by Europeans from the colonial conquests on have driven many indigenous practices underground around the world.

We will honor the oral traditions we are studying in this chapter by presenting much of the information about them in their own voice. Long histories of intense discrimination by Europeans and/or Euro-Americans against all manifestations of indigenous culture included formal attempts at both physical and cultural genocide in Canada, the United States, and Australia. Thus, even in current stratified random samples of national populations, many practitioners of native-based, religious healing rituals worldwide may continue to identify themselves with one of the larger institutionalized religions, due to the social constraints and lower status still attached to minority group members whose beliefs and rituals differ markedly from those of the dominant societies by which they are surrounded.

The residues of this hostile social climate still discourage individuals from identifying an indigenous tradition as their primary religion even in North America today. In the United States, for example, the practice of many Native American rituals was illegal until the American Indian Religious Freedom Act passed in 1978, and the right to use the hallucinogenic cactus peyote in Native American Church sacred rituals is still disputed. In most of Latin America, where precolonial populations were sometimes more dense and viewed as a potential source of labor rather than as obstacle to development, the *encomienda* system enslaved native populations well into the 1900s. In Guatemala, for instance, it was not until the Agrarian Reform Law of 1952 that peasants (largely indigenous) were given unused land in order to "eliminate all feudal beliefs, practices, and institutions embedded in and emerging from local environmental settings around the world, the spiritual traditions constructed over the millennia by people living in a particular place, having migrated (except for the original Africans) from some other location over thousands of years. On the other hand, we are looking at a complex set of traditions that encountered waves of external influences from traders, conquerors, and proselytizers beginning, perhaps, with the agricultural revolution that began reorganizing life on the planet, on through the European colonial conquest of recent centuries.

The word *indigenous* originates from the Latin *indu, endo* meaning in or within and *gignere* meaning to beget. *Indigen* may be defined as "produced, growing, living, or occurring naturally in a particular region or environment"

("Indigenous," n.d.). Current usage refers to the original inhabitants of specific places, primarily rather isolated places sometimes called "regions of refuge" (Aguirre Beltrán, 1967). These regions tend to still be peripheral to the global economic system today (Wallerstein, 1984). Millions of people in some of these regions continue to have a sense of themselves as different from the surrounding, usually larger groups of relative newcomers. Ironically the same transportation and communications revolutions that threatened indigenous cultures are now also facilitating a rising consciousness and collaboration among people fighting to preserve their traditions, facilitated by networks of social movements, nongovernmental organizations, and even the United Nations. Some 4,000 to 5,000 distinct ethnic groups living in almost 200 states constitute a group that may number 300 million worldwide (Wilmer, 2008).

Indigenous peoples retain some aspects of precontact spiritual values and religious practices, often in a syncretism that includes elements from the beliefs, symbols, and rituals of the more recently invasive populations. Thus indigenous religions exist in distinction from the major cosmopolitan religions, while having often been dramatically influenced by those macro-level religions and societies. They are often "less developed" economically and maintain an oppositional philosophical stance to the values of postmodern, capitalist societies.

Another distinguishing characteristic of indigenous religions is that they generally are not based on sacred written documents but rather rely almost exclusively on oral traditions for their maintenance and survival. Sometimes this dependence on speaking, hearing, and remembering as opposed to writing and reading has been forced upon indigenous populations, as with the Spanish burning of the vast majority of Mayan codices (Tobin, 2001).

The first wave of conquest for most indigenous traditions came from dominant traditions among exploring and conquering peoples moving beyond their homelands, searching for riches, and transforming local cultures along the way. The conquering empires of antiquity diffused their own religious forms over broad reaches of the planet. Moreover, Christianity, Islam, and even Buddhism have embedded within them an element that led to imperialist conquest spreading these religions far beyond their original geographic confines, which most indigenous religions do not. An elective affinity between a proselytizing and often exclusivist theology (you must believe as we do or face eternal damnation) combined with political and military conquest has dramatically transformed the global map of spiritual practice in recent centuries.

The Jewish history of forced Diaspora is somewhat analogous to the experience of indigenous peoples who were often forcibly removed from their original homelands, except that Jews had a venerably old, sacred *written* text, while most indigenous peoples either did or currently do not. Indigenous religions have generally stayed at least on the same continent for their recorded history. And, similar to Jews, most indigenous religions today do not proselytize for members and are not interested in spreading their particular forms of worship across the globe. While there have been attempts to revitalize indigenous religions and societies within recent centuries that did involve charismatic leaders "converting" many members from disparate tribal groups to a similar practices such as the "ghost dance," we choose to focus on current beliefs and practices for this chapter.

Indigenous Beliefs

Indigenous peoples, or their recent ancestors, lived in relative isolation from modern or postmodern society so they were much closer to nature than most people reading this book. Their remaining religious beliefs and practices are therefore more closely related to nature as well. They seek explanations of changes in weather patterns, seasons, life, and death as part of a general worldview that is more comfortable with nature than that of most moderns. Like the ancient Chinese oracles and I Ching readings, indigenous believers find guidance for their daily lives and ethical and social systems in the patterns of nature.

People living in modern cultures often deride these so-called "primitive" explanations of the world, believing scientific explanations to be superior. Many of us have difficulty believing, for example, that the Native American rituals designed to bring the sun up every morning actually cause the event as claimed. Yet which of us actually knows what forces cause the earth to spin on its axis? We have many good theories—our own scientific cosmogonies—based on considerable evidence about why such things occur but no definitive proof. Even those prominent scientists who have developed the most widely accepted theories about the origins of the universe set limits on their theories in the absence of hard empirical evidence. Religious cosmogonies rely on different sources of evidence than modern scientific ones do, but the question is always the same: How did we get here? We ask it, in part, because we just want to know but more profoundly because the answer seems to have implications about why we are here and where we are going.

One of the first things to notice about indigenous religions is their diversity; no broad generalizations can be made about them. That being stated, in the Western Hemisphere where pre-Columbian populations have survived to the present, there is arguably a commonality in spiritually significant connections to a specific portion of the earth and its flora and fauna. Zedeño, Austin, and Stoffle (1997) have argued that Native American groups have sacred landscapes, as opposed to the more spatially confined sacred landmarks of Euro-Americans. Whereas in North American context, the larger religious traditions usually have special churches, temples, cathedrals, ashrams, synagogues, or shrines, American tribal groups tend to attach sacred significance to various geographic features within an entire territory bounded by both physical and cultural barriers. The landscape is often seen to be a record of the activities of a God or Holy Beings, as passed down via oral traditions in creation or other sacred stories.

Among the Hopi and Pueblo nations, for example, there are *sipapu*, or sacred emergence sites, where humans are said to have first entered this world from a world below. Humans are usually not the first species to emerge into the current world and other species provided critical assistance to help humans get here. These stories give us clues about these peoples' respect for, and arguably social relations with, the plants and animals who helped them make it into this (often fifth) world. The following is a portion of a Navajo (Diné) version of this emersion story.

> They also saw that the reeds themselves grew upward rapidly. And in another moment all thirty-two joined together to form one giant stalk with an opening in its eastern side, it is said. . . . It is also said that the young man then told the people to enter the reed through the opening. When they were all safely inside, the opening disappeared. And none too soon, for scarcely had it closed before the people inside heard the terrible sound of the surging water outside . . .
>
> . . .
>
> It is also said that the people stayed inside the reed all night long. And when the white light of day filled the sky the next morning they looked outside. But they could see no hole in the sky.
>
> So the people sent *Atseelsoii* the Hawk to look for a way through to the other side. Out he flew and immediately began to scratch at the sky with his claws.
>
> He scratched and scratched until he was lost from sight. And sometime later he returned to say that he had dug his way into the sky overhead to see some light shining through from the other side. But he was tired now and could dig no more.
>
> Next they sent *Wooneeshch'iidii* the Locust to dig. He too was gone for a long time. (Zolbrod, 1984, pp. 75–76)

Notice the use of "it is said"—which in this case can also be translated as "they say"—that continually punctuates this narrative. The teller of sacred stories is always careful not to take credit for his or her knowledge and makes it abundantly clear that the narrative was learned from others via the oral tradition (see Beck, Walters, & Francisco, 2001). This is also our first hint at the orientation of storytellers to the well-being of the collective of their people as opposed to the self-aggrandizement typical in industrialized, capitalist societies.

When Locust returns, he tells a story involving his supernatural abilities to have arrows pass through his heart from the four directions and still live, which won him the right to stay in the fifth world (Zolbrod, 1984, p. 77). The story continues in the subsequent passages I have excerpted from the original narrative.

> As for the opening that had been made in the sky, it was still too small for many of the larger people to climb through. So *Nahashch'id* the Badger was sent to make it larger. . . .

> Finally *Altse' hastiin* the First Man and his wife *Altse' asdzaa'* the First Woman were able to lead the people through the hole to the surface of this the fifth world, it is said. (Zolbrod, 1984, p. 78)

In this creation story, we see clearly illustrated that the survival of the people was predicated upon a very helpful tubular plant that closed just in time to keep the people from drowning and upon the assistance of several animals who made the hole in the sky and enlarged it sufficiently so that humans could pass through it up into the current world. We would therefore expect to find that cooperation not only among human beings but also among humans and other species would be emphasized in Navajo beliefs and practices. This is indeed the case, as the Navajo concept of walking in *hozho'* or beauty/harmony/balance is an all-encompassing ideal about the best way to live in the world. It includes an orientation toward collective as opposed to individual well-being that is still evident in older generations of Navajo living today. Respected medicine people are revered for their ability to reestablish harmony when some aspect of social life is either out of balance or perhaps threatened by major changes. Those who are the best political leaders today are still those who do the most to benefit the larger tribal community. The concept of *hozhon* also involves not only attempting to live in harmony or balance with one's extended family, clans, tribe, friends and enemies (including those from other tribes and Euro-Americans) but also with the land and the flora and fauna it contains.

In many Navajo ceremonies, the person for whom the ceremony is performed is led to affirm their relationship to the sacred via their relationship to the Holy Beings—which can include geographic and atmospheric features such as mountains and winds—and to call upon specific animals for help and protection regarding a currently challenging situation. This is evident in the respect that is shown for even small and seemingly insignificant animals, such as the horned toad and corn beetle. These interspecies relationships hint at an environmental consciousness of the significance of even the smallest creatures in maintaining ecological balance, which is the subject of entire books (Berkes, 1999; Kawagley, 1995).

Note also that Navajo ritual chants usually include references to sacred people and geographic formations in both male and female embodiments. In the emersion story, First Man and First Woman led the people into this world and they waited until everyone—including the larger ones—could enter together before entering themselves. This would lead us to expect a more egalitarian relationship between men and women than is the norm for most of the industrialized world. This is still true among traditional Navajo where inheritance and descent are matrilineal and one's mother's clan is always mentioned first in formal introductions. The nicer car is more likely to belong to the wife than the husband. The mother is more likely than the father to have the final say in important household decisions, including major purchases. Men are more likely to be political leaders, but even in this realm there have been women principle chiefs or chairs of their tribes, whereas the United States has yet to elect a woman president. Even the rain that is so essential to growing corn is also perceived to come in male and female varieties—both good but for different purposes. In sum, in all aspects of social life, male and female are believed to have complementary and generally equally important contributions to make.

We quote extensively from the following cosmogony given by Darryl B. Wilson (1992) of the Pitt River Tribe in northern California due to its excellence for illustrating several additional sociological features:

In the legends of my people there are many events of "our history" that have precise meaning and are completely understandable in our society. Our lessons are from oral "historians." As a people we have been taught, for all seasons, to listen to these stories and to apply the lessons within the stories to everyday life. This is one way to gather spiritual and mental "power" and to maintain emotional as well as physical human strength—to somehow balance our being with the awesome velocity of the churning, continuous universe.

Mis Misa is the tiny, yet powerful spirit that lives within *Akoo-Yet* (Mount Shasta) and balances the earth with the universe and the universe with the earth . . .

The lesson establishes that *Akoo-Yet* was the first mountain created long ago and that it is the seat of power, a spiritual place. It is a mountain that must be worshipped not only for its special beauty and its unique power, but also because it holds *Mis Misa* deep within it . . .

. . . The person must be born for making and maintaining the "connection" between his/her nation and all that is—and for no other purpose . . .

. . .

. . . a child is dreamed of and born. The planning is intricate since the person "departing" must have sufficient time to train the child to maturity in order that the many lessons and songs are understood as they were created—and are learned with unaltered purity . . .

To not keep the "appointment" between the society and the powers of the world is to break the delicate umbilical cord between the spirit of the society and the awesome wisdom of the universe. To not listen, intently, to the song of *Mis Misa* is to allow the song to fade. Should the song cease, then *Mis Misa* will "depart" and the earth and all of the societies upon the earth will be out of balance, and the life therein vulnerable to extinction . . .

. . .

. . . He [Grandfather] told us of his fears of how this earth could be *itamji-uw* (all used up) if all the people of all the world do not correct their manner of wasting resources and amend their arrogant disregard for all of life.

. . .

In our custom, one is not supposed to intrude into the silence created when someone who is telling a story hesitates to either search for proper words or to allow the listener time to comprehend . . .

. . .

"It is said that the power that created the earth made one mistake. It made 'vanty.'

'Vanty' makes a person love himself and nothing else. . . . We must be good to one another. . . . We must be good to earth. We must be good to life. Do not kill life. . . . This earth is the last place. We call ir *atas-p-im mukuya* (to stop, last place). We have no place else to go. We must go back to the stars."

I had a dream . . .

. . .

. . . Perhaps it wasn't a dream. Maybe Grandfather is, at this moment laboring to make another "earth" so that we might have a place to go if life is again *itamji-wu* (used up).

. . . Earth is being drilled into and her heart and her guts and her blood are being used as private property for private gain. There is an immense vacuum where the spiritual connection between human beings and nature is supposed to be . . .

History has unveiled many battles and many wars . . .

. . .

In these conflicts mother earth was treated as a woman slave . . .

. . .

. . . For how many more seasons can these mistakes find pardon within nature? (pp. 55–61)

Now we'll do the sociological interpolations. First, the narrator uses the term *oral historian* as opposed to storyteller. Historians have higher status in industrialized societies, and readers are hereby symbolically alerted that these legends and lessons are to be taken more seriously than fairy tales or other fantasy stories. Emphasis is placed on the application of the stories to daily life; otherwise, the indigenous religions carried via oral traditions would have been lost long ago. Applying the advice embedded in the stories to real-life situations leads to a sense of empowerment. This feeling of spiritual importance can serve as a powerful antidote to the physical poverty many indigenous peoples still experience, surrounded as they often are by the material wealth of larger, dominant societies. Emotional and physical health is challenging to maintain under the generally oppressive conditions in which many indigenous peoples live and their religions have helped them survive.

Second, the narrator connects this story to a particular place on Earth, as mentioned previously. Each indigenous people may hold that their emergence place or arrival spot is the center of the earth with literally cosmic significance. In this case, Mt. Shasta holds a tiny spirit within it that must be listened to carefully in order for the whole earth to stay in balance and properly aligned. Similarly, for the Kogi in Colombia, South America, the messages their *Mamas* (spiritual leaders) receive about loss of snowpack in their mountains are believed to have importance for the whole world but most importantly for those of European descent in the Americas (Ereira, 1992). And the messages are perhaps surprisingly alike among many indigenous groups—stop mining and drilling into the earth or we may all die.

Third, the most highly revered leadership positions are established through a combination of divinatory methods, including dreams of a particularly powerful sort and intense training. To learn a story or song to the point of "unaltered purity" without the aid of a written text such that it can be passed down over many generations with minimal alterations is inconceivable to most literate persons. In this context, the requirement for dedicating one's entire life to this sacred role seems reasonable in order to learn the rituals perfectly. So also the huge respect accorded to those who held the entire history of their peoples in their minds and bodies, as well as the unimaginable grief attendant to the premature death of such a leader, are comprehensible. A whole indigenous nation could figuratively and literally die with the death of such an important individual. Plus, they may believe that all societies on Earth and existence as we know it may be jeopardized if important rituals are not performed by such an individual.

Fourth, in this story we see direct reference to the importance of silences. It is often perceived as rude to break those silences prematurely. So one of the noticeable features of many indigenous groups is that spoken interactions may seem to take place more slowly than what is comfortable for outsiders and may include longer pause times between utterances than seems normal for fast-paced urbanites. Interactions, especially those with respected elders, may be filled with long silences that are knowingly included because saying too much too quickly would mean that important pieces would not be heard or remembered.

Fifth, the reference to vanity as the one mistake the Creator made is a direct critique of the foundation of postmodern, advanced capitalist societies. Capital cannot be accumulated without taking away some of the wealth that should rightfully accrue to the workers, according to Marx. In many indigenous societies, it is unacceptable for one person to accumulate wealth in obvious excess of his or her neighbors. In addition, the reference to mother earth being "treated as a woman slave" is another rather direct critique of the abuses that patriarchal societies are currently wreaking upon the earth. So indigenous societies sometimes fill the function of providing a moral alternative to the ethical foundations of the larger nation-states in which they dwell.

Sixth, through the lens of one closely associated with indigenous peoples over decades, native tribes "live in a world of symbols and images where the spiritual and commonplace are one . . . " (Erdoes, 1972, p. 109). We add that the boundaries between the real and the dreamt are also permeable. In the preceding excerpt, the narrator hints that dreams may be factual. Thus dreams may be reenacted by a whole tribe so that its meaning is physically embodied in the collective, rather than being carried only by the original dreamer. This again highlights the importance of collective activities and well-being for many indigenous peoples.

To further illustrate a difference in religious and cultural norms, we quote briefly from a Yurok woman "speaking in 1978 of a time around 1910" (Buckley, 2002):

> In the old days the [brush dance] doctor couldn't have any food, no water. She'd go up in the hills and hide all day—nobody could see her.
>
> . . .
>
> . . . You see, the feather stood up straight, the way she fixed it. But then she had to go back up and look the day after the dance, in the morning. And if the feather had fallen over then the child would die. But if the feather still stood up, through the whole dance, then the child would get well. (p. 143)

Here we see that medicine people were not just men but also women and that matters of life and death were sometimes entrusted to the females.

The Veneration of Ancestors

In many indigenous cultures, ancestor spirits are part of the ongoing life of the community. The "living dead" continue to participate, in a new way, and their descendants honor them with periodic gifts or other acts of respect. Even in the highly rational culture of modern Japan, indigenous practices of ancestor veneration persist, incorporated into the modern economy by enterprising corporations who sell, according to one estimate (Nelson 2008), household altars to the tune of $1.6 billion a year. The economic giant that is modern Japan emerged from the rubble of World War II with many of its ancient traditions intact. Nelson (2008) put it as follows:

> Most families of sufficient economic means at this time had not one but two altars in their homes: the first, called a *kamidana* 神棚, was for those beneficial but ambiguous deities known as *kami* which were associated with pre-Buddhist animism and, beginning around 1870, a reconfiguration of Shinto sponsored by the state. The other altar was the family's *butsudan* 仏壇 that served as both a stage for interacting ritually with one's ancestors and as an extension of the family's membership in a local Buddhist temple. (p. 309)

Relations to the ancestors vary tremendously in indigenous beliefs around the world, of course. Sometimes they are consulted privately or through formalized oracular rituals for advice or assistance. In some systems, ancestors are believed to reward or punish their descendants depending on their behavior as a kind of omnipresent moral police. Indigenous Zimbabwean rituals, for example, almost all have to do with seeking assistance from or appeasement of ancestors through offering of homemade beer poured on the ground. Often someone becomes possessed (via trance) by the spirit of an ancestor and gives commandments to the kin group in a different voice assumed to be that of the deceased (Cox, 1998). In others, if ancestors fail to assist the petitioner, they may be simply ignored—Melanesians reportedly wore the jawbone of an ancestor around their neck for good luck, only to discard them if the good fortune was not forthcoming (Bloch, 2010).

Some indigenous theodicies, especially in Asia, interpret suffering as a natural process and explain death as part of the cycles of nature, minimizing an individual's pain by putting it in a broad cosmic context, as in the cycles of rebirth in which the process of dying is analogous to changing one's clothes. The potential danger for the individual in such a system is in becoming disengaged from the community, which does not occur at death but rather when that person transgresses social boundaries and violates sacred taboos—a much more dangerous act than the natural process of death.

What is clear is that in many parts of the world ancient indigenous beliefs still shape contemporary events and daily lives. Many sophisticated urban dwellers in Taipei still burn symbolic "ghost money" as a sign of respect for the ancestors and major international political figures give homage to their ancestors. Nelson Mandela (1994) wrote in his autobiographical *Long Walk to Freedom* that as a young man he "learned that to neglect one's ancestors would bring ill-fortune and failure in life" (p. 10).

Indigenous Rituals

Rituals of spirituality among the world's thousands of indigenous belief systems are as varied as their settings, so difficult to generalize, so we can only give a glimpse into the myriad forms of religious practice in that sphere. They include a wide variety of what Victor Turner (1967) called "life-crisis rituals" and "rituals of affliction" (p. 6).

Life-crisis rituals "mark the transition from one phase of life or social status to another" (Turner, 1967, p. 7), such as initiation ceremonies for the passage from childhood to adulthood at puberty, as well as weddings and funerals. Traditional Navajo may still hold a *Kinaaldá* for a young woman when she reaches menarche. Paiutes still hold all-night "wakes" when someone dies. Many North American plains tribes also hold a "memorial" about a year after someone's death so that mourning can be completed. Rituals of affliction address the misfortunes—for example, illnesses, reproductive disorders, or hunting failures. In the Ndembu culture studied by Turner (1967) they come from being "caught" by spirits of the dead. Turner prefers to call these "shades," since the terms *spirit* or *ancestor spirit* suggest remote or distant ancestors, whereas the "shades" are "the spirits of those who played a prominent part in the lifetime of the persons they are troubling" (Turner, 1967, p. 10). Shades are believed to come out of the grave because they have been neglected or because of disapproved conduct. Most indigenous groups have healing rituals to deal with all types of afflictions, including distinctly modern problems such as post-traumatic stress disorder, which could be alleviated by the performance of an Enemy Way ceremony among the Navajo.

One of the most widespread and still-practiced religious rituals among indigenous groups in North America is the sweat lodge. Although there is huge variation among tribes, a few common elements can be identified. The purpose of most sweat ceremonies is to purify oneself both physically and spiritually. Water is poured at intervals over very hot rocks and produces steam, which is kept within a small, tightly enclosed round space. Persons with heart or other potentially limiting health conditions are usually warned in advance about the risks of participating in these ceremonies that may include prolonged exposure to high temperatures and high humidity. Participants generally sweat profusely during the ritual, and one is said to "sweat out" physical impurities.

Most sweats include four rounds of prayer since the number four is a pattern number for most North American indigenous groups. In some traditions, the first round of steam and prayer is focused on individual issues, the second on extended family and clan issues, the third on issues involving all other humans, and the fourth on issues concerning all life on Earth. In some cases, a sweat may be done as a preliminary purification in preparation for participation in more arduous rituals such as a Sundances. Sundances are a northern plains tradition in which dancers fast (no food or water) for 4 days in succession while dancing in the heat of the summer sun in order to secure specific blessings, usually for someone(s) other than themselves.

Another ritualized practice that may include spiritual elements is the PowWow.[1] As the term is commonly used today, a PowWow means a gathering of indigenous peoples from various tribes to celebrate native cultures.

Prize monies are usually offered to the winners of various dance contests, with contestants placed in categories by gender, age, and style of dance. In the discourses of masters of ceremonies at these intertribal events, one may hear references to the "sacred circle" of the PowWow arena, to the sound of the drum as the "heartbeat of mother earth," to intertribal dances as a way of walking in harmony with nature, and to the sacred history of some dances such as the jingle dress dance. These references may be interspersed between activities that are arguably the opposite of sacred, such as a dance in which men are playfully coerced into dancing like women or a special dance for non-Indians in which they are requested to "dance like Indians." This takes us a step beyond acknowledging that for many tribal peoples the spiritual and mundane are indistinguishable; the spiritual and the humorous may also be indistinguishable. The Eurocentric notion that sacred things need to be kept distant from mundane things does not apply to indigenous traditions of "sacred clowns." Among the Lakota, the *Heyoka* were "truly unpredictable, and could do the unexpected or tasteless even during the most solemn of occasions" (Mizrach, 2008).

Three specific practices that have occurred at PowWows also demonstrate religious content: (1) the blowing of eagle-bone whistles, (2) initiations into certain types of dances, and (3) the process of recovering a fallen feather. Common reasons for use of sacred whistles include requests for healing of persons with serious illnesses, expressions of grief regarding recently deceased persons, and supplications for strength or courage to perform a difficult task. Persons participating in an intertribal dance when a whistle is blown are encouraged to become a part of the collective prayer, to "dance with the sacred whistle" and thereby increase the power of the request of the whistle carrier. Thus, individuals may move from dancing for fun to dancing with spiritual intent immediately upon hearing the high-pitched sound of an eagle-bone whistle. Persons not previously participating in the intertribal dance may join the dance, and others may stand in respect. Each person in the arena is potentially dancing a prayer, and the more energetic the dancing, the more fervent the prayer. Also, the more persons dancing, the more powerful is the collective spiritual power. While the drum group lengthens their song, the dancer who blew the whistle as a rule performs more energetically as if to increase the spiritual efficacy of the intent.

Initiation into jingle dress dancing at a PowWow may be viewed as a sacred and pivotal event in the life of a young woman. It can signify a commitment to live by the old ways and to have strict respect for traditional norms, including abstention from alcohol and drugs. A solo dance may be performed by the initiate while a special song composed in the young woman's honor is sung. Overt encouragement to resist assimilation is readily apparent in this ritual, and this resistance is strengthened by communal witnessing of the sacred vows made.

The third PowWow practice with religious significance is the way a fallen feather is picked up. A dropped feather may be symbolic of lack of requisite attention to detail (in ensuring that all feathers are securely tied) and of something not quite right or out of balance in the person who drops it. It is sometimes equated with lack of spiritual preparedness or an inappropriate frame of mind. On the other hand, most recently (2008) I heard an emcee state that a fallen feather is just like many of the things that happen to us in life, an unfortunate event or obstacle to be overcome that is not necessarily the fault of the wearer. In any case, to restore harmony, a respected elder must perform an appropriate ritual that customarily includes at least a prayer and some special advice given to the dancer who dropped the feather.

Emphases on these religious aspects of PowWow phenomena demonstrate a powerful strategy for maintaining and strengthening an oppositional or resistant culture. The term *culture of resistance* signifies a distinct culture that has evolved from precontact times and continues to evolve under conditions of oppression (Mitchell, 1994). This term is appropriately applied to most surviving indigenous groups today.

While it is appropriate to write about indigenous religions as spiritual beliefs and practices that unite communities of adherents as per Durkheim's definition, it is also important to be aware of frequent tendencies toward very individualized interpretations and observances of religious rituals even in collectivist cultures. For example, many tribal peoples place huge emphasis on the pursuit and interpretation of individual dreams. In the imaginative realm of dreams, images and interpreted messages can be utilized to substantially modify both group belief systems and behavioral norms. This built-in flexibility has arguably contributed in positive ways to the survival of many indigenous groups into the present in spite of histories of interaction with Europeans that sometimes included attempts at both physical and cultural genocide. Warry (2007, p. 88) noted that contemporary anthropologists and Aboriginal peoples in Australia understand the term *traditional culture* as conveying "ever-changing and adaptive qualities." Indeed, many indigenous people make it very clear that if they were not very adept at adaptation they would not have survived into the present.

The book *Lame Deer, Seeker of Visions* (Lame Deer & Erdoes, 1972) describes how Lame Deer and his people—the Lakotas—were experiencing oppression in the late 20th century. Lame Deer was subjected to the boarding school experience where overt attempts were still being made to civilize indigenous populations by converting them to one form or another of Christianity. In order to reach full manhood among the Lakota, Lame Deer had to go on a *hanblechia*, or vision quest, which involved 4 days of fasting (no food or water) and crying for a vision. At the end of his quest, a large bird came and lifted Lame Deer far above the earth and gave him sacred knowledge. As a result of Lame Deer's story, other kinds of knowledge besides that gained in schools are shown to be important. Ways of knowing that are based in spiritual practices and/or come from nature may be more valid than "book learning." From his cosmic perspective, Lame Deer is able to poke fun at U.S. society and the importance *wasicus* (Euro-Americans) attach to "green frog skins" or dollar bills. This ability to retain a sense of humor even amidst severe poverty and oppression has been a critical coping mechanism for many indigenous groups.

In addition, while an individual's dream and its interpretation may exert huge influence on a tribal group, it is usually without any institutionalized coercion. Most indigenous religions have avoided the bureaucratization Weber worried about, in part due to their reliance on oral (as opposed to written) traditions but also due to their lack of financial resources that could be mobilized to create formal organizations. It is no accident that indigenous peoples tend to be some of the most poverty-stricken populations on the planet. Additionally, the intensely local and tribally specific nature of indigenous spiritual practices makes these traditions unsuitable candidates for bureaucratization or rationalization. Thus, most indigenous religions do not keep formal membership records, and it is very hard to estimate a number of adherents.

Furthermore, competing visions may exist simultaneously, and what comes through today via oral tradition may actually represent the interpretation(s) of the most charismatic person(s) at the time. The dictionary definition of a *charism* as "an extraordinary power (as of healing) given a Christian by the Holy Spirit for the good of the church" ("Charism," n.d.) is apropos here if we eliminate the European bias so that the definition reads "an extraordinary power (as of healing) given a spiritual leader by the Divine for the good of the community." We are self-consciously implying here that if no charismatic visionaries arise during times of societal crises for indigenous groups, then these groups may cease to exist both physically and culturally. Prophecies, dreams, and visions may be believed or not, and changes in spiritual practice are generally left to both individuals and groups to accept or reject.

Indigenous rituals, as in all other faiths, grow out of and then reinforce the belief structure of the religious system but also the social system, linking religious practices with social structures. Prayers recited at the family altars are conducted only by the household head on death-day anniversaries or holidays in Japan, but the altars are accessible to all family members at other times. Moreover, visitors would pay their respects before the altar before engaging in conversation or business with family members (Nelson, 2008, p. 312).

Indigenous Institutions

Spiritual leadership or authority amongst those reliant on oral tradition is generally based on the continued efficacy of services performed. For instance, a person who is at one time esteemed for powerful abilities to heal may be recognized as being "not so good" for an indefinite period and may or may not regain these spiritual gifts for the benefit of the community. This can also be seen as a direct consequence of the force of oral tradition still operating within indigenous communities.

At the same time, indigenous groups tend to retain a much stronger orientation to collective well-being than is characteristic of most Euro-American groups. Within such native traditions, charismatic individuals must continually demonstrate their commitment to social norms of sharing and providing for group well-being via such mechanisms as potlatches and giveaways. Indeed, if individual talents and subsequent accumulation of wealth are not used to serve the interests and needs of extended family, clan, and tribal communities that individual risks being informally shunned by his or her people of origin. Spiritual leaders generally operate under the same rules—too much individual accumulation of wealth is unacceptable, even if one is the best medicine person for miles around.

Indigenous peoples generally are not characterized by institutional specialization—that is, there is no sharp distinction among political, economic, religious, health care, educational, or other social institutions. A respected elder usually fulfills multiple functions that can easily include but not be limited to political leadership, linguistic and cultural/practical education of youth, spiritual guidance, physical and psychological healing, and group therapy/counseling for extended family and friends. Thus, religious leadership among indigenous peoples does not follow the Western tendency to become reified within bureaucratic structures.

Nonetheless, people with demonstrated spiritual gifts have played an important role in indigenous societies and have a certain authority, in part because of their ritual importance. Nelson Mandela (1994) recalled that as a boy he learned that "If you dishonored your ancestors in some fashion, the only way to atone for that lapse was to consult with a traditional healer or tribal elder, who communicated with the ancestors and conveyed profound apologies" (p. 10).

The nomadic Europeans tended to stigmatize indigenous peoples' reverence for sacred lakes, mountains, rivers, deserts, and so on by naming them "Devil's" whatever. The persistence of demeaning colonial names for geographic features, as with the frequent use of Squaw for places and plants in the western United States, retains huge and negative symbolic significance for still strained relationships between indigenous peoples and their conquerors today.

Similarly, Alan Anderson (1993), doing ethnographic work on the relationship between Pentecostalism and ancestor veneration in South Africa reported his converted respondents as saying they still had "visitations" but labeled them not as ancestors but as "demon spirits which need to be confronted and exorcised or idols which need to be spurned for they only lead to further misery and bondage."

Along a similar vein, the religions of indigenous populations on other continents—in Africa, Australia, and the northernmost parts of Europe and Russia—may also be characterized as containing a common element of spiritual connection to the particular place in which a people reside. The Yoruba people in southwestern Nigeria, not far from the ocean, have a creation story that begins when all is water. One version involves a mollusk that held in its shell some form of soil that then was spread around by two-winged beasts to create the earth. The Deity named this origin place Ife, and it is said to be the "cradle of existence." During the annual *Eje* ceremony, a Yoruba leader gives yams to the God of the Sea, to the ancestors, and to other local spirits responsible for crop fertility.

Indigenous Environmentalism

Because indigenous people usually live close to nature, their religious forms and rituals are usually tied to nature and often serve as models of peace and environmental harmony, although there has been considerable scholarly debate over whether indigenous peoples were ecologically sensitive or damaging (see Harvey, 2000, 2002, 2003). In the 21st century, as we face a fierce debate about how to preserve the planet's ecosystems, we have much to learn from those persisting ancient indigenous spiritual teachings that contain what we might call a species of endangered wisdom about humanity's relation to the natural world.

Winona LaDuke (1999) and all Lakota use the phrase "All our relations" to discuss the complex mesh of relationships among various peoples and between particular people and their local environments. Ali Mazrui (1986) suggested that "Africans felt close to nature—they almost had a theology of nearness—that emphasized what was near was dear." Moreover, animals were believed to have souls in traditional Africa—they "could be either good or evil, holy or profane." Although the forest was the home of the ancestors as well as the animals, the natural world has become threatened by the changes in the contemporary world.

One of the most popular figures in Native American history is Chief Seattle, after whom the city was named. He argued that happiness would come only from living in peace with nature and one's fellow human beings. Chief Seattle was born around 1790 in the part of the Oregon region now known as the city of Seattle, named after the chief because he signed the Point Elliott Treaty of 1855 that gave white settlers Indian land in exchange for the creation of a reservation. He died at Port Madison in 1866, and in 1890 the city erected a monument over his grave. As chief of the Duwamish, Suquamish, and other Puget Sound tribes, Chief Seattle befriended white settlers of the Pacific Northwest and encouraged peace between the two peoples despite his complaints about the cruel and thoughtless way in which the Europeans had treated the environment and its various inhabitants.

When the president of the United States sent an offer to Chief Seattle to buy land from him, he responded with an eloquent speech, published in 1887 and later embellished, in which he apparently proclaimed that despite their differences, "We may be brothers after all. We shall see." Although his exact words are disputed, the general sense of his message remains. He urged both his people and the European Americans to pursue peace.

> When our young men grow angry at some real or imaginary wrong, and dis-figure their faces with black paint, it denotes that their hearts are black, and that they are often cruel and relentless, and our old men and old women are unable to restrain them.
>
> But let us hope that hostilities between us may never return. We would have everything to lose and nothing to gain.
>
> Revenge by young men is considered gain, even at the cost of their own lives, but old [men who stay] at home in times of war, and mothers who have sons to lose, know better.[2]

While some may rightly question his judgment in accepting the U.S. government's offer in exchange for what turned out to be questionable promises, Chief Seattle had a larger perspective on the matter that went beyond the apparent realities of the time. He was convinced that their time as a people would be short-lived, that their demise at the hands of the whites was inevitable. He wished to minimize the damage in ensuing years. He also, however, believed that what might appear to be the end of his people would not be the final act, for even the dead would remain in the land. "And when the last red man shall have perished from the earth," he told the president the following:

> These shores shall swarm with the invisible dead of my tribe, and when your children's children think themselves alone in the field, the store, the shop, upon the highway or in the silence of the pathless woods they will not be alone. . . . The White Man will never be alone. Let him be just and deal kindly with my people, for the dead are not altogether powerless. (Chief Seattle, 1887)

Unfortunately, most of the indigenous cultures of the world have been destroyed by European and later conquerors, and many of their sacred ways have been forgotten. Not only were the indigenous populations in the Western Hemisphere reduced from about 10 million, at the time of Columbus's arrival, to about 1 million in a relatively short time, the accidental and deliberate destruction of indigenous cultures is also a global phenomenon, often carried out in the name of Christianity or some other major religious tradition.

The question remains whether religious rituals and institutions will contribute to the survival or destruction of life in the global village, issues that we will explore in more detail at the end of the book.

5

The Religious Ethos

The future of the global human community depends in large measure on the ability of humanity to forge a common ethos out of current competing traditions. The major religious traditions continue to provide guidelines for the way most people believe they should live their lives—the ethical bases for both the individual and the collective life of a society. Every social order must produce a set of ethical standards that facilitates peaceful coexistence and the emerging global village is no exception. To understand the complex issues surrounding this ethos construction, I will explore the way in which it is carried out in contemporary societies as well as the ethical inheritance provided by the major religious traditions.

Constructing a Religious Ethos

A people's ethos, or lifestyle that grows out of their worldview, serves at least three social functions. An ethos (1) facilitates the process of identity construction, (2) shapes and legitimates or challenges the stratification system of the social order, and (3) identifies taboo lines and lays out the ethical guidelines implied in a given worldview.

This chapter will examine how this process of ethos production occurs in contemporary religious practice and its implications for the question of peaceful coexistence among diverse populations and religious communities now sharing the same space. I begin by looking at the process of individual identity construction in each tradition, noting ways in which a people's understandings of the personal and the social are guided by their definitions of the sacred. The ethical implications of a worldview are built into the individual's sense of identity, the lifestyles of particular status groups and subcultures, and ultimately a culture's understanding of how people are ideally to act in the context of its various institutions.

When religious identities are linked to class, ethnicity, nationality, gender, and other social cleavages, the intensity of conflicts may increase, especially if conflicts of interest fall along the same lines. Each of the world's major religious traditions contains the potential for promoting chaos or community in the world order even though their worldviews and styles of life may differ. The fundamental ethical teachings of all the major traditions are actually very similar. They tend to begin with a basic compassion or respect for others, such as the Golden Rule from Jesus ("Do unto others as you would have them do unto you") or the Silver Rule from the Buddha ("Do not do unto others what you would not have them do unto you"). The ethical standards of each religion and the way in which its major leader deals with violations of these precepts are listed in Table 5.1.

Most traditions allow for a mitigation of the consequences of an ethical violation through a confession of guilt by the violator. Acknowledging the infraction is the key in every tradition to changing the negative consequences of one's actions. In the East, the law of karma simply explains the natural outcome of one's actions rather than declaring a God's specific judgment. Even here, however, confession seems to make a difference, as illustrated by the story of King Ajatasattu, who approached the Buddha with remorse over killing his father in order to usurp the throne. The Buddha, known for his compassion, assured the king that he could reverse the negative consequences of the horrible deed by admitting his mistake and changing his life. Confucius also had a similar teaching, a silver rule that he described in response to a question, as recorded in his *Analects* (Confucius, 1998, p. XV.24): Zi Gong asked, saying, "Is there one word which may serve as a rule of practice for all one's life?" The Master said, "Is not reciprocity such a word? What you do not want done to yourself, do not do to others."

Similarly, Moses and Muhammad both insist on repentance as a condition for escaping dire consequences; they are ready to forgive the remorseful sinner but prophetically warn of harsh consequences for those who do not repent. Moses intercedes with God on behalf of those who have sinned, however, and argues with God about sparing their lives. God agrees but kills those who refuse to admit their mistakes and reaffirm their faithfulness. In both Jewish and Islamic traditions, the punishment comes from God (although as in Christianity, sometimes individuals who punish others claim to be acting on God's behalf).

In the Christian ethos, Jesus also calls on people to repent but shows a compassion similar to the Buddha's in dealing with sinners and insists that people not pass moral judgment on each other (Matthew 7:1–6). Moreover,

Table 5.1 Ethical Teachings and Treatment of Those Who Violate Them

Founder	Key Ethical Teaching	Treatment of Violators
Buddha	Silver Rule: Don't do unto others . . . Show compassion to all creatures 10 immoral actions, 5 precepts	Karma: reap what you sow Repent, it can be okay (King Ajatsattu)
Confucius	Silver Rule Filial piety, respect	Reciprocity
Jesus	Golden Rule: Do unto others . . . Love God and neighbor Be perfect	Let the one without sin cast the first stone Judgment: give food, drink, visit the prisoner
Moses	Decalogue: 10 Commandments—do not kill, steal, envy, etc. Love God and neighbor Follow the law	Pleads with God for mercy Punishes unrepentant
Prophet Muhammad	Love God Follow the law Show mercy to all of creation Hospitality Strict personal discipline	Compassionate with repentant sinners Judgment for those who will not repent Reciprocity of mercy—show it, you get it

Jesus makes it quite clear that punishment for sin should be left to God, as demonstrated by the story of an adulterer caught in the act. By law, she is to be stoned to death, but Jesus challenges the crowd by saying that the person without sin should cast the first stone. Gradually the crowd, recognizing their own guilt in other areas, disperses without harming her.

Ironically, although Jesus may have had a remarkable compassion toward violators of ethical standards, Christian institutions have sponsored some of the most intolerant practices in human history, from the Crusades and the Inquisition to witch hunts and excommunications. Such a paradoxical intolerance of the institution coupled with the tolerance of the founder often leads to contradictory behaviors among people within the tradition, an issue I will take up again in Chapter 8. The central issues here are significant for life in the global village. First, the human community must find a way to create an ethos that permits peaceful coexistence, and the basis for that lies at the foundation of every religious tradition: Respect and compassion for others is a central obligation. Less consensus exists on how violations of the ethical standards should be treated or how to institutionalize a system of sanctions.

The religious ethos is not a simple list of acceptable and taboo behaviors, however; ethical standards do not exist in isolation but as part of a complex set of elements associated with individual and collective identities, the legitimacy of the social order, and the taboo lines and ethical standards for all spheres of life. In most life situations, conflicts emerge between abstract ethical ideas and the interests of religious elites and various social strata.

Despite both the similarities and the internal diversities of the major religious traditions, each has a distinctive ethos. The ethos of the Hindu worldview, based on the laws of karma, is rooted in the conviction that one must find one's *dharma*, or duty, for one's station in the universe and position in one's life cycle and fulfill it to the best of one's ability. Each individual is responsible for maintaining the order of the cosmos within his or her own sphere and must therefore carry out his or her duty, show respect for elders and superiors, and avoid harming living creatures. Buddhism shares with Hinduism an emphasis on explaining the nature of the world (the law of karma) and expecting individuals to determine their own duty within broad general guidelines. One is then punished or rewarded in the form of reincarnation at higher or lower statuses or in being released from the cycles of rebirth altogether. The Buddhist ethos focuses on the expression of compassion for all other creatures as a way of building "good karma."

The Jewish ethos, rooted in the tribal identity of the ancient Hebrew tradition, takes an activist stance toward the universe within God's principles such as justice; the "Chosen People" remain faithful to God and perform acts of kindness. According to rabbinic tradition (see Hertzberg, 1962, p. 186), each person should clothe the naked (as God did for Adam and Eve), visit the sick (as God did Abraham), and comfort the mourners (as God did Isaac after his father Abraham's death).

Because Islam is intended to be all-encompassing, its ethos is designed to embrace all elements of Muslim society and the daily life of its participants. The difficulty of maintaining the intensity of that commitment to God, which is shown in the internal and external struggle (jihad) of believers, pervades Muslim lifestyles and culture. An all-embracing religious ethos is frankly somewhat difficult to maintain in a multicultural society.

Despite the common symbols and cultural roots its various branches share, the diversity of forms Christianity has taken renders impossible efforts to describe a coherent ethos for this religion. Members of the various Christian communities, especially the elites, increasingly interact with one another, however, and substantial cross-fertilization is taking place.

Moreover, many elements of the Judeo-Christian tradition now play an important role in efforts to construct a contemporary global ethos because of the powerful position Christian cultures and institutions hold in an emerging world order that is dominated by Western elites. Although the ethos of contemporary Western culture differs dramatically from the ethos of the early Christian Church (it is often a religion of the rich rather than the poor), the impact of Christianity on Western norms and values, and hence on the rest of the global village, is formidable.

I will begin this exploration of religious ethos by looking at the relationship between religious traditions and the process of identity construction.

Religion and Identity Construction

The concept of identity construction—of the social processes involved in an individual's creation of a sense of identity—has been a major topic of research and discussion in recent years (see, e.g., Coy & Woehrle, 2000; Willaime, 2004; cf. Mol, 1978; Roof, 1978). At least five aspects of the identity-building process are related to religious traditions:

1. *Standards or models.* A belief system may provide the standards and models for personal and collective identity construction, especially heroic and demonic models of behavior and the criteria for evaluating good and evil. On a more subtle level, even the tone of a culture—and by implication, personalities valued or disparaged within it—may be highly influenced by a religious tradition. A religious group or subculture may encourage certain traits in the socialization process and discourage others (aggressiveness or passivity, diligence or submissiveness), because the Gods approve or disapprove of certain kinds of people. Similarly, the same tradition may promote one personality type among men and another among women—one among lighter-skinned and another among darker-skinned people.

2. *Religious social networks as reference groups.* Religious institutions often provide individuals a social network, a significant reference group of people who shape self-concept (Cochran & Beeghley, 1991). An individual turns to his or her most significant others for positive or negative reinforcement in the identity–construction process. The religious reference group may include not only those who are immediately present in a congregation or social network but also significant others who are geographically distant (e.g., important religious personages) or even the dead (e.g., the saints and ancestors). As Assimeng noted (1978) in discussing religion in Ghana, "The living believe that they are also dialectically linked in some metaphysical oneness with the

ancestors of the community 'who have lived before and gone beyond'" (p. 99). Specific advice, or the general tone of personality development, may thus be related to ongoing interactions not only with living members of one's community but also—through prayer, mediums, and now mass media—with persons physically present and absent, living and dead.

The Internet has dramatically expanded the ways in which these networks might work, although it is too early to tell just what the impact is. One immediate and obvious consequence is the possibility of searching for love on the Net, within one's own faith community. Muslims from anywhere in the world can go to http://www.muslimfriends.com/?tid=122 and connect with other Muslims. Hindus can try http://www.singlesindia.com/ among many other sites, whereas Christians can look at hundreds of sites including http://www.cherish.com/christian-dating-online.dating. A "Gays for God" website (http://www.rslevinson.com/gaylesissues/gfg/blgfg.htm) connects people with gay- and lesbian-friendly faith communities.

3. *Interaction with a deity.* The process of engaging in religious rituals, and thereby actually interacting with the Gods, may have an impact on one's sense of identity. Whether the interaction—or even the deity itself—is real or imagined is not important here, nor is it sociologically ascertainable. As W. I. Thomas (1966) observed, situations defined as real are real in their consequences; what is significant is the *perception* that one is interacting with God or the Gods. When people believe that they are acting on behalf of a deity, for example, they may consider themselves (individually or collectively) as "chosen" and therefore subject to different norms and standards than others. This phenomenon cuts more than one way, of course, since the person acting under "God's instructions" may be either a Gandhi or a mass murderer. Ellison (1993) and Pollner (1989) observed that the role-taking process in religious rituals may involve assuming the role of the "divine other" in much the same way that people engage in interaction with concrete social others. Thus, an individual may define his or her own life with reference to religious figures and heroes in their tradition and may even "begin to interpret their situations from the point of view of the 'God-role' (i.e., what God would expect and want)" (Ellison, 1993, p. 1030). Routine devotional practices may also instill certain values with reference to one's identity—for example, in a Christian context, that each individual is special and that God hates sin, not sinners or in a Hindu context, that showing compassion to others is meritorious. In its more developed forms, devotional activities like prayer, study of the scriptures, rituals of obeisance, and the consumption of religious media may play a vital role in forming a sense of self (see Ellison, 1993).

4. *Mechanisms for sustaining identity changes.* Religious traditions play an important role in the process of changes in identity, whether in terms of dramatic events such as conversion or the standard changes in one's life cycle. Rites of passage, for example, offer people a repertoire of thoughts and actions for difficult times of identity change. The frequency of religious behavior is often closely linked with the life cycle of individuals, in large part because of the rituals that sustain life changes. In the United States, for example, many young people raised in a religious family will stop participating in their churches, synagogues, or mosques when they first leave home and go out on their own to work or college. After they have their own children, they may return to regular religious observance because participation in religious rituals is part of the process of raising children, according to their own experience.

Similarly, religion often plays a role in dramatic cases of what Jones (1978) called *identity alteration*—that is, the "disentanglement from one particular pattern of identity and the process involved in the adoption of another" (p. 60), referred to by a number of terms such as conversion, brainwashing, or alternation and often perceived as a consequence of a hierophany. Cases of dramatic identity change are often associated with religion because they are difficult to achieve without the powerful legitimation of a new identity that comes from religious communities.

Most people may not have exceptional religious experiences or visions—nor are the life changes of those who have them always permanent—but human history is full of such stories. Conversion tales are the stock in trade of many forms of religious enterprise, such as Christian evangelists on street corners or on some television stations. The great divide between the former life and the new one is dramatized to show others that there is a way out of their current misery and a better life is just around the corner.

5. *A system of meaning and security.* Finally, a religion's nomos (a meaningful order or worldview) provides a "shield against terror," said Berger and Luckmann (1967); a belief system helps people feel secure, especially at society's margins, and most notably in times of crisis and death. "Every nomos is an area of meaning carved out of a vast mass of meaninglessness, a small clearing of lucidity in a formless, dark, ominous jungle" (Berger, 1969, p. 23). This nomos becomes a tradition that, as Tevye put it in *The Fiddler on the Roof*, helps us keep our balance. When life seems to be falling apart, we can rely upon socially constructed images of the world, of what is true, and of what is appropriate to help us keep our wits about us. In a homogeneous society, the tradition is usually given and taken for granted, but in a multicultural society individuals and groups of people "shop" for a nomos from a

variety of available options, making it more difficult to create a common set of values. A gap between theory and reality as it is experienced can, of course, undermine a religion's believability, and the escalation of competing world-views in the religious marketplace of the global village does pose a special threat to every belief system, as we will see in Chapters 6 and 7.

Variations in Identity Construction

Although every religious tradition plays a major role in its adherents' identity construction, the nature of the process varies substantially. Let us briefly examine some of the major similarities and differences among and within the traditions. The great divide on this issue is between those religious communities that root the individual's identity within the community itself—especially in Judaism, Islam, and some branches of Christianity—and those that emphasize more individual autonomy, at least in spiritual and ethical matters—such as Protestant Christianity, major branches of Hinduism and Buddhism, and the modern marketplace of religious affiliation. Admittedly, this divide is somewhat arbitrary, but the differences in empha-sis are real and have substantial consequences for the ways in which people live their lives.

Community-Based Identities

Jewish identity over the centuries has been linked to membership in the socioreligious community, underscored by the rituals of family and com-munity life that emphasize the special character of one's heritage and one's responsibilities to God and neighbor. Strict regulations of interaction—especially restricting marriage across the boundaries of the community—reinforce the socialization process by which Jews come to see themselves primarily as Jews. This ethnic consciousness has been enhanced by bitter persecution over the centuries, and a Jewish sense of identity has often thrived in the sense of adversity.

The adaptability of the Jewish ethos to the world of modern society has been both a blessing and a curse for Judaism. Many of the traits valued and cultivated by centuries of Jewish socialization—such as advanced education, critical rational thinking, adaptability to new environments, egalitarianism, and the pragmatic rootedness of Jewish ethics—have facilitated Jewish involvement in societies in which they have been allowed to participate, notably the multicultural North American setting. Their entry into the main-stream has placed tremendous strains on the tradition, however, as many of the boundaries between Jews and non-Jews have broken down, and Jewish identity has become undermined in the same way as have other ethnically bound religious and cultural constructs.

Muslims construct their religious identities much like Jews, in form at least if not in content, and especially with regard to the deep roots of religion in the community. The Five Pillars of Islam provide a visible and regular reminder of the believer's duty to uphold God and the *ummah* at the center of his or her life. Signs of the ever-present religious community—such as the punctuation of daily life by prayer and religiously prescribed modes of dress, especially for women—have high visibility in public life. The Islamic family and the ummah provide the context from which one's meaning in life and protection from its dangers are to come. Ultimate security comes, of course, from Allah, who watches over the community. Forged in a context of raiding Bedouin tribal life, Islam paid close attention to matters of security in both belief and practice. The traditional Islamic home was a mini fortress and life occurred—especially for the women—within the confines of an enclosed courtyard. A system called **purdah** segregated women physically from the male society outside the household. On those rare occasions when they ventured forth from the security of their homes, women brought their fortification with them in the form of their traditional dress and veil. This effort to divide the male and female worlds completely is the subject of considerable controversy in contemporary Islam.

In a traditional Islamic society, it is difficult to have an identity other than the religiously centered construction provided within the family and mosque. In this sense, Islam retains a large element of its primal character from the days of its founding. Daily life experience with other family members and society constantly reinforces the values and concepts of Islam. In a multicultural society, however, an individual's religious identity is often challenged and sometimes ridiculed, providing tremendous pressure, especially for young people and most definitely for young women, to look beyond the traditional religious community for new role models and values.

Early Christianity, like Judaism and Islam, had a strong community orientation, and its participants probably perceived of themselves primarily as members of the church. The strong **individualism** of contemporary Protestantism is a recent historical development, a product first of the rebellion against the official church during the Protestant Reformation and then in the individualization of Western culture associated with the Enlightenment and the industrial revolution. Ideas long present in Western culture that emphasized the value of individual freedom and independence were elaborated during the social transformation to modern society. Individualism had an affinity with industrialization, because it encouraged people to break free from bonds of kinship and feudalism and with Protestantism, which constituted a rebellion against feudal Christianity.

Christian traditions around the world substantially influence personal and social identities, although in different ways because of the wide range of groups drawing upon the general tradition. First, major personages and heroes of the faith provide the role models on which to base personal identity: a host of "ancestral" figures from the scriptures—especially Mary—and a large collection of saints (although their role has been diminished in Protestantism), as well as local heroes and models. Christian identities are also shaped by the process of interaction with the deity, facilitated by the representation of Jesus as one element of the Christian Trinity. Routine devotional practices and prayer traditionally provide individuals with not only a quiet time for reflection but also a specific relationship with the deity. The content and impact of that relationship are not simple, however; apparently, it can result in confidence (or, in some cases, arrogance) provided by the perception of an intimate relationship with the Creator, or it can instill a strong sense of guilt and shame if one becomes too aware of the gap between one's own life and that expected by the deity. (The comedian Emo Philips once suggested that the most universal prayer is to ask God to disregard the laws of the universe for our own personal convenience.)

Across the variety of Christian communities, the person of Jesus remains a focal point of religious belief and practice and thus a central node for the meaning system of believers. According to the tradition, God's purpose in taking on human form was to speak to humans in such a way that they could understand more fully the substance of God and how they should order their lives. Thus, it is no accident that the image of Jesus in Christian subcultures around the world may say more about the cultural context of the image than it does about the historical Jesus or the God he is intended to communicate. In many parts of the world, the image of Jesus tends to mirror the indigenous culture. To some individualistic Americans, Jesus is a friend who will aid the believer in solving personal problems; to impoverished peasants in Solentiname, Nicaragua, he is a poor revolutionary who will mobilize the people to stand up to the establishment. To a Christian paleontologist like Pierre Teilhard de Chardin, he is a cosmic force toward whom all creation will eventually be drawn. To those who find even Jesus too distant to identify with, many versions of the faith encourage a relationship with his mother Mary or one of the saints who seems more approachable.

Individualistic Identities

Even the strongly communal early Christianity probably seemed individualistic to Jewish observers of the time because individuals could break loose from their ethnic and clan groups and join the new religious movement (NRM) independently. The Protestant Reformation of the 16th century extended the individualistic potential of the faith substantially by removing the institutional guarantees of salvation and putting the responsibility for religious belief and decisions squarely in the hands of the individual. This notion of the relatively autonomous individual in Protestant Christianity, combined with the notion of the "calling" in the Puritan branch of the church, cultivated elements of modern individualism, perhaps even—as Weber argued—helping to lay the groundwork for the modern capitalist order.

The Protestant emphasis on a personal relationship with God independent of the institutional structure of the church did not eliminate the communal element of Christianity. In fact, the idea that the individual was responsible for his or her own personal relationship with God sometimes resulted in a vulnerability to social pressures, especially within small homogeneous religious communities in which little deviation from the norm was tolerated. Marx (1843/1972a) anticipated this sort of dynamic when he observed of the Reformation the following:

> Luther, without question, overcame servitude through devotion but only by substituting servitude through conviction. He shattered the faith in authority by restoring the authority of faith. He transformed the priests into laymen by turning laymen into priests. He liberated man from external religiosity by making religiosity the innermost essence of man. He liberated the body from its chains because he fettered the heart with chains. (p. 18)

Ironically, some Protestant sects, with their emphasis on personal salvation, constitute a tight social organization with a control over their members that the Vatican never dreamed of having, in large part because participants internalize the norms, values, and rituals of the community.

Moreover, Ellison (1993) noted identity creation within the religious community in the United States relates to interaction within networks from the religious community:

> Like other social institutions, religious groups are network-driven (Paris, 1982; Taylor & Chatters, 1988; Williams, 1974). Participation in church-related activities brings individuals together with others who have similar status characteristics as well as common religious beliefs. For church members, regular interaction with these like-minded others may reinforce basic role identities and role expectations. Through formal and informal involvement in their church communities, these persons may gain affirmation that their personal conduct and emotions with regard to daily events, experiences, and community affairs are reasonable and appropriate. (p. 1029)

Although most Asian societies place individuals in a tightly knit social organization, spiritual matters in both Hinduism and Buddhism have a strong individualistic element that is related to theories of dharma and karma in which each individual is responsible for his or her own actions and the consequences they produce. As S. Gopalan (1978) contended, "The *dharma* of an individual is, in its psychological sense, his innate nature—the law of his being and development" (pp. 128–129). A person who discovers and acts according to his or her actual being is also therefore ethical. Nevertheless, Gopalan (1978) continued, "It cannot be the same for all people . . . or the same for one individual all the time" (p. 129). Thus, one has to continue searching, taking one's cues from family and religious networks and above all from *pandits* and gurus, seeking the path appropriate to one's natural inclinations and abilities in this lifetime.

Religious rituals play a role in the Hindu identity-construction process by providing patterns of behavior upon which individuals and groups draw from the larger culture. The techniques of Yoga, for example, aid people in the process of "identity consolidation or self-integration," according to Sinha and Sinha (1978, p. 134). The "basic assumption of Yoga," they contended, "is that an integrated self is essentially sacred. The purpose of Yogic techniques is to unify the spirit" (Sinha & Sinha, 1978, p. 134). Once individuals discover their dharma and develop the spiritual and physical discipline to act on it to their best ability, they bring themselves into synchronicity with the universe, according to the Hindu ethos. Hindu culture, therefore, relies a great deal upon traditional practices, even in the most modern circles. Horoscopes are drawn to position the individual within the larger cycles of the universe, marriages are still arranged by families in many cases, and the Vedas can be consulted (through a pandit) for key life decisions as well as for minor ones. Traditional patterns of life and gender- and family-based interaction networks guide an individual from infancy to the grave in a well-established trajectory that continues to reinforce ancient habits of the heart as well as systems of stratification.

Buddhists find meaning in life and, as in Hinduism, construct an individual identity within the broad framework of dharma and karma. The striking difference between the ethical systems of these two closely related religions lies in the Buddha's rejection of the Hindu caste system of his time. Consequently, Buddhism (along with Christianity) has been an attractive alternative to many of the lower-caste and outcast groups in India. Given the assault on the caste concept in modern Hinduism, it is thus likely that over time the two traditions will become increasingly similar in this respect.

Discussion of caste and class lead to the next arena of identity construction, namely, social stratification. Whether a religious tradition emphasizes collective or individualistic elements of identity, an individual's sense of self is never fully autonomous and is always linked to the social order in which it is constructed. Out of the many aspects of the social order that could be examined, I will focus briefly on the types of stratification systems legitimated by various religious traditions because a tradition generally puts a strong stamp on the cultural style and thus social organization of societies in which it is dominant. Moreover, the type of social organization dominant in a specific era and region shapes religious ideas so that the ethos of a religion and the forms of social organization in the society in which it is dominant are two sides of the same coin.

Muslim Identity After September 11

In the wake of the terrorist attacks on the World Trade Center in New York and the Pentagon in Washington, D.C., the question of Muslim identity has become a major issue for both Muslims and non-Muslims alike. Because Muslims became immediately identified in the minds of many not only with savage acts of violence but also with America's new mega-enemy, hate crimes against Muslims escalated. Sometimes the victims were Sikhs, because they were mistaken for Muslims. One group of South Indians, including actress Samyuktha Varma and members of her family, were speaking in Malayalam in an airplane flying into New York City. They were immediately taken into custody and interrogated for several hours because of their suspicious behavior: They were talking excitedly in a foreign language, changing seats, passing notes, and looking out the window at the city.

Virtually every mainstream Muslim group in the United States immediately condemned the terrorist attacks, many of them insisting they were in violation of the Qur'an. It was a difficult position for many Muslims to take because, like many Christians, Jews, and members of other faith traditions, they believe that violence is acceptable under certain rare occasions. In this case, the death of innocent civilians made the condemnation of the acts relatively straightforward for mainstream leaders and organizations. Yet the problem remains of the injustices that the terrorists claimed to be addressing.

American Muslim scholar Omid Safi (2003) said that he, like other fellow Muslims in the wake of the terrorist attacks, insisted that "Islam is a religion of peace. The actions of these terrorists do not represent real Islam" (p. 24). Yet, he continued the following:

> There is something pathetically apologetic about turning the phrase "Islam is a religion of peace" into a mantra. It is bad enough to hear Muslim spokespersons repeat it so often while lacking the courage to face the forces of extremism

in our own midst. It is just as bad to hear a United States President reassure us that he respects Islam as a "religion of peace" as he prepares to bomb Muslims in Afghanistan or Iraq, or support the brutal oppression of Palestinians. (p. 24)

The ethics of war and peace, of violence and nonviolence, are complicated indeed, as we shall see again in Chapter 8 when we discuss these issues more fully. What is clear is that the events of September 11 and its aftermath have made Muslims more politically active. A poll (*Muslims in the American Public Square*, 2004) found that American Muslims "have undergone massive political shifts, and have become a relevant part of the political landscape" (p. 9). Of those surveyed, 86% said it was important for them to participate in politics—and that they were dissatisfied with the way things were going in American society.

The issue is complex for the question of identity politics as well, as Ismail (2004) noted, using the example of whether or not the public space can accommodate religious symbols such as the *hijab*:

> The problem here is that the *hijab* may be an expression of identity politics, used to deliver a message in the public sphere: a message that is not about religion per se but about difference and a right for public recognition. Identity politics in this respect asserts difference in terms of distinctions in tastes, lifestyles and modes of representations in the public sphere. These distinctions are affirmed as politically viable, and not just as culturally tolerable. (p. 614)

The political and the personal, as the women's movement puts it, are two sides of the same coin. Identity is a publicly negotiated process that affects both the everyday lives of individuals and the broad cultural and political currents of the global community.

Religion and Stratification

Religious identities are, for many if not most people on the planet, an indelible part of daily life and apparently affect everything from one's health to life chances and position within the social stratification system. Although there is still much we do not know about the link between the two, many studies provide evidence that increased religious involvement may result in better physical and mental health. It does so by "regulating health-relevant behaviors that reduce the risk of disease" (Ellison, 1999, pp. 692–693) by helping people develop social support networks and ways of coping with "certain types of stressors such as loss events (e.g., bereavement), unexpected calamities (e.g., serious accidents), and health problems (e.g., chronic pain, long-term disability)" (Ellison, 1999, p. 693). Recent research has focused on health practices, social support, psychosocial resources, and belief structures as potential mediators of the religion–health relationship (George, Ellison, & Larson, 2002).

Although our attention focuses on the world of politics, because of its important role in modern culture, much of religion's impact on people's lives takes place quietly in the background, in the sphere of daily life. Religious congregations in the United States, for example, may engage in some social services and political action but are primarily involved in providing religious "services," corporate worship, and religious education (see Chaves, 2004).

The effect of religious institutions and practice on the social system of a society is therefore often subtle and indirect. The tension between authoritarianism and egalitarianism exists in every tradition, though each religious community tends to develop an affinity for one or the other. Since the advent of the modern era, the shift from isolated societies each with a single dominant sacred canopy to an increasingly cosmopolitan global religious marketplace has produced contradictory trends within religious traditions.

First, individualizing and egalitarian values, which tend to go hand in hand, have become a dominant modern motif infusing every tradition. Each tradition has its own version of these values, although in some religions they are more dominant than in others and they come in a variety of forms. Some traditions may emphasize equality within the community but also the superiority of insiders when compared to outsiders, for example. Second, authoritarianism may emerge in response to the egalitarian motif; such a response is an almost automatic reaction to an attack on the community. When they are not currently under attack, religious elites sometimes invent enemies to justify the institution's authoritarian structure.

The authoritarian motif in religious traditions is usually defended with the argument that the structure of inequality within a society is a reflection of inequality in the cosmos as a whole. In general, this position supports a status quo attitude toward the ruling powers unless, of course, they are radically democratic or opposed to the religious tradition. Statements of support or opposition to a particular regime cannot be taken at face value by sociologists because they are sometimes explicitly designed to bring about a change. It is intriguing, for example, that St. Paul exhorted members of the early Christian Church to obey the political authorities because they were given their power by God, but he often did so from the jail cells in which those same authorities put him on charges of sedition. Two extreme examples of the support of religious traditions for hierarchical social systems are in the Hindu caste system and the medieval class system legitimated by Christianity, each of which I will now examine briefly.

The Hindu Caste System

Vociferously criticized in the modern, democratic ethos of India, the caste (or **varna**) system has been a fundamental element in Hindu theology and theodicy. In tone and outcome, very like the idea of a "calling" in Christian Puritanism, the notion of caste divisions gives religious justification to a social hierarchy and legitimates a given social order by locating an individual's dharma within the strict parameters of the caste into which he or she is born.

In one sense, this system is simply good sociological analysis: It suggests that the social context into which one is born will determine one's life chances and, to some extent, educational and occupational opportunities, social networks, marriage possibilities, and so on. The caste system is an attempt to explain the universe as it appears to be and to help people to make the best of their apparent fate. In another sense, this kind of analysis becomes a sort of self-fulfilling prophecy as people assume that the status quo is the only possibility and simply adjust their aspirations to their situations, producing a kind of fatalism that makes social change seem fruitless and legitimates the worst kind of exploitation. It is a model *of* and *for* social life.

Gandhi attacked the caste system and made his reformation of it— particularly the notion of "untouchability"—an integral part of his freedom movement. Gandhi's general condemnation of the system was institutionalized in the independent state of India, established in 1948, where law thereafter forbade caste divisions. Ancient principles of social organization die hard, however, and caste divisions are about as absent from India today as racial discrimination is in the post–civil rights era of the United States.

The traditional system identifies four castes:

1. *Brahman:* priests, teachers, and political advisers

2. *Kshatryia:* rulers and warriors

3. *Vaishya:* merchants, agriculturalists, and traders

4. *Shudra:* servants of the other three, manual laborers; also gardeners, musicians, and artisans

Besides these four castes, there are a wide range of subcastes, the *jatis* or sects, which are largely occupational- and clan-oriented and often serve as guilds, protecting the interests of their members and prohibiting unwanted outsiders from joining the occupations.

Finally, outside the four castes are the street sweepers, scavengers, and leatherworkers, the "untouchables" or "outcastes." When Gandhi attacked the system, he renamed this group the **Harijans,** or "Children of God," to object to their religiously sanctioned pariah status. So denigrated were the untouchables, at least in recent times, that their very presence was considered polluting. In some circumstances, as when sweeping the streets or going to clean latrines, they were required to cry out to warn others that they were approaching. Some contemporary Vedic scholars claim that the ancient scriptures were referring not to social castes that should be observed in this way but to the mental states that people developed because of their previous actions.[1]

What is remarkable is not that such a system exists but the extent to which it functioned consistently over the centuries, essentially providing a system of slavery, with no major rebellions undertaken by the lower castes against the upper ones until the system's moral and legal underpinnings were removed in the mid-20th century. Reinforced and legitimated by the theory of transmigration of souls, the social system remained stable because punishments for revolt were believed to be incurred not only in this life but also in the next. The theological basis for the caste system was dealt a fatal blow, however, by the Indian freedom movement. As Indians argued for their freedom from the British as a God-given right, subordinates in the caste system made the same claim for themselves and Hindu authorities withdrew their support, for the most part, from the caste system as a type of social order.

Christianity and Western Power Elites

The teachings of Jesus have a strong egalitarian flavor, but the church as an institution has been ambivalent about the distribution of authority and resources, with competing themes of egalitarianism and authoritarianism. During the long history of its alliance with Western power elites, virtually every branch of the church has legitimated status quo systems around the world. After the conversion of the Emperor Constantine in the 4th century CE, Christianity became intertwined with the Roman Empire, then with the power structures from the tsars in Russia to the aristocracy of Western Europe throughout the Middle Ages. The church then legitimated European colonialism, providing a theological justification for conquest by brute force and superior technology and an altruistic account for a process motivated by greed. At the same time, however, Christianity as a belief system and an institution has played a major role in contemporary efforts to reform or revolutionize hierarchical systems from Latin America to Asia, Africa, and Eastern Europe.

Although Jesus empowered women in a number of ways that challenged the gender hierarchy of his day and Paul declared that Christians should not have divisions within their community based on gender, Christianity contains a strong patriarchal motif that has been used to legitimate various forms of gender stratification throughout the centuries. This dual standard was present from the beginning; although women played an important role in the early church, there is no record of their presence in the inner circle or in Jesus' private moments with his close followers, such as the Last Supper. It is possible, however, that the male chroniclers of the events simply did not note their presence.

Moreover, the authoritarian structure of the church itself as it developed over the centuries seems to contradict claims of an egalitarian Christian ethos. The charismatic early movement was formalized into institutional structures that solidified after Christianity became the official religion of the Roman Empire. Although various reform movements have resulted in some models of democratic governance and lay empowerment, considerable tension between clergy and laity persists and much of the leadership in many branches of the church has no democratic pretenses or aspirations. Although at times the social control of the laity by the clergy is blatant and overt, at other times it is subtler, as in the years of silence on the part of parishioners who were sexually molested by clergy.

Egalitarianism in the World's Religions

It is not surprising that the dominant motif in the world's major religions has been a hierarchical one—the ruling powers of most societies understandably promote authoritarian religious ideologies and suppress the egalitarian beliefs. Early Chinese culture, for example, had two competing traditions: (1) that of K'ung-Fu-tzu, which emphasized the need for strict social hierarchy and respect for elders and political authorities, and (2) that of Mo Ti, who promoted an egalitarian ideology and ridiculed the followers of K'ung-Fu-tzu for their "exaggerated" emphasis on authority. The first tradition was institutionalized as Confucianism and became the official state religion of the emperors, whereas the second precipitated a relatively unstable popular movement that was almost lost over the centuries.

Surprisingly, religious traditions have also provided the basis for movements toward equality, primarily because of the ethical dimension of religious beliefs that could overcome efforts by ruling elites to completely suppress the subordinate equality motifs. One of the most significant teachings about equality is that emanating from the Christian tradition, as discussed earlier, which persisted through the centuries despite the tendency toward authoritarianism favored by the political and economic elites aligned with church authorities first in the Roman Catholic and Orthodox churches, and then with the various branches of Protestantism. The Christian emphasis on equality has its roots in the parent religion of ancient Judaism, which was distinctive in its time for its consistent emphasis on justice. The apparent equality of the early Christian community, which encouraged membership by all social classes and ethnicities, remained an element of the tradition that laid the groundwork for contemporary democratic theory despite efforts to stamp it out over the centuries.

The Emerging Egalitarianism of Ancient Judaism

Judaic systems of stratification, though rooted in their ancient understanding of God as a God of justice and community, seem very modern. The early Hebrews developed an image of their deity that shares a number of traits with other high Gods in the ancient Near Eastern pantheons. They were individuated and elevated; active in the world; conceived by natural and human analogies; powerful, just, and merciful; in bond with a people or a region; and interpreted by human representatives (see Gottwald, 1979, p. 676ff.; cf. Smith, 1952, 1973).

The Jewish God also has distinctive characteristics, however, not shared by other Gods of the ancient Near East, "just as Israel's egalitarian intertribal order is unlike the other ancient Near Eastern social systems" (Gottwald, 1979, p. 693). This uniqueness is sociological: "Yahweh forbids other Gods in Israel as Israel forbids other systems of communal organization within its intertribal order. The social-organizational exclusionary principle in Israel finds its counterpart in a symbolic-ideological exclusionary principle in the imagery of deity" (Gottwald, 1979, p. 693).

The idea of Israel as a nation of chosen people only makes sense sociologically, Gottwald (1979) argued, otherwise, it is either "irrelevant supernaturalism or exclusivist racism, or both together" (p. 693). That is to say, the ancient Hebrews perceived of Israel as different because it *was* different, primarily as an egalitarian social system surrounded by stratified societies. Although Gottwald's assessment may be somewhat idealistic given the extreme patriarchy of ancient Judaism, it is nonetheless fair to say that ancient Hebrew society was relatively egalitarian considering its sociohistorical context.

The "Separate but Equal" Policy

The social ramifications of Islam, like those of the world's other major religious traditions, are extraordinarily complicated by layers of contradictory ancient and contemporary practice. Highly egalitarian in many aspects, the Qur'an emphasizes justice and attention to the widowed and orphaned, an ethical outlook that derived from the parent Hebrew religion and was adapted to the harsh conditions of the pre-Islamic Arabic world. As noted earlier, the religious institutions of Islam are not rigidly hierarchical, and all classes worship together as equals in the mosque. There is no priesthood per se in Islam, but the *ulama* ("the learned") and the *fuqaha'* ("lawyers"), the scholars and custodians of the law, have acquired an informal authority within the community that is comparable to the status held by professionals in Western religions.

Because Islam regards all people as children of the same God, an undercurrent of egalitarianism pervades the religion's entire ethos. It is clear that the Prophet Muhammad and his earlier followers—notably his own wife, Khadijah—made tremendous advances in protecting the rights of women in a manner that was highly radical at the time (see Haddad & Esposito, 1998). Most of those protections centered on family life in a context of exploitative polygamy and female infanticide, practices that were abolished by the Muslim community. Women were further protected in the event of divorce and were allowed to inherit their husband's property.

Difficult questions arise, however, especially in the 20th century, concerning the role of women, who have a "separate but equal" status in some Islamic societies. According to traditional Islam, like many other traditional societies, women's place is in the home, and they are not afforded the same equality in other spheres, simply because it was not a question at the time of the Prophet. Consequently, a number of people—especially women—have rebelled against efforts to segregate the genders outside the courtyards of traditional Muslim houses. It is also unclear how much of the status of women in Islam is from the original religious leadership of the Prophet and his followers and how much from traditional Arab culture that is often carried with the religion as it diffuses.

Religious Taboo Lines and Ethical Systems

I will now explore the ethical systems of each of the five major religious traditions, with some attention to the way in which those religious ethics shape and are shaped by social organization and action in various spheres of daily life, economics, and politics. Perhaps the most significant aspect of religious traditions for purposes of this survey is their guidelines for the nature of collective life and the ethical standards that facilitate peaceful coexistence among peoples. Every tradition has an ethical system, often growing out of the relationship with a deity, which emphasizes social relations, with regulations about treating one another with compassion and justice, at least within the community, and often across social boundaries as well.

Hindu Ethics

The Hindu ethical system centers on the concept of dharma, identifying and carrying out one's appropriate duties. In addition to religious duties, individuals must learn to discipline themselves to transcend the profane aspects of life. Various acts are not moral or immoral in and of themselves but appropriate or inappropriate given one's dharma.

A more general code of ethics for all believers stresses the importance of engaging in religious rituals, honoring one's status superiors, giving charity to the poor, avoiding harmful acts against others, not lying, and the like. Because the consequences of engaging in unethical behavior are built into the structure of the universe, one reaps what one sows. According to one rendering of the ancient scriptures, for example, if a non-Brahman kills a Brahman, the penalty is reincarnation as a worm in one's own feces for 10,000 reincarnations. This seems like a stiff penalty, and it raises a number of difficult empirical questions for believer and observer alike: Does such a threat actually deter the killing of Brahmans by non-Brahmans? Certainly such acts have taken place, but perhaps more would have occurred if the threat were not in place. Second, is the punishment really inflicted? (This question is even more difficult to answer.)

An element of Hindu ethics that deserves special attention is sexual behavior. The relationship between religion and sex has been particularly significant in India, where the entire range of sexual activity has played a central role in the search for the sacred. These extremes appear in stories of Shiva, who was sometimes an extreme ascetic who abstained from sex but at other times intensely erotic (see Pattanaik, 2006). On one hand, intense religious practice and everyday morality have long valued *brahmacharya*, sexual abstinence, as a technique for achieving karmic merit. On the other hand, the oldest and most persistent image of the Gods themselves are of the *lingam* fixed within a **yoni**, the representations of the phallus and vagina respectively, a coupling found both in contemporary temples and in 1500 BCE Indus plains archaeological sites.

Hindu mythology and art over the centuries are replete with stories about the exploits of Shiva, a God of both sexual virility and asceticism (see O'Flaherty, 1973). One popular tale has the Gods visiting Shiva and Parvati only to find them engaged in intercourse. They continued despite the presence of their divine visitors; Vishnu laughed, but others became angry and cursed the couple, causing them to die in that position. Shiva declared that the lingam would be his new shape, which men must model and worship, and the yoni would be Parvati; this was the origin of all things (Parrinder, 1980, p. 8). In the temples dedicated to Shiva, it is no accident that the lingam, still erect in the yoni, is covered by devotees with plain white yogurt and ghee (clarified butter). At the same time, however, women and men are forbidden to touch one another or show affection to members of the opposite sex in public.

The Indian example is instructive because it shows the ambivalence and apparent contradictions of religious teaching. Religion plays an important part in regulating sexual behavior because sex is such a central, pervasive force in human life. Collective life requires sexual activity for the continued survival of the species, the society, and the family, but unregulated sexual activity usually undermines a given social order and the authority of elites to control social processes. Religious legitimations of the guidelines established in any given culture give those rules a powerful authority, especially if violations of the taboos are believed to result in long-lasting or cosmic negative consequences.

In addition to positive statements about what one should do, act, and think, religions also identify taboos, or interdictions against what should not be done. One of the most significant functions of a religious ethos in everyday life is to define taboo lines between ethical and unethical behavior. As Freud (1913/1950) pointed out, however, taboos do not prohibit things people do *not* wish to do; the list of taboos in a culture provides a catalog of its social problems. Sometimes religious taboos appear antiquated or misguided—and sometimes, in fact, they are. Nevertheless, what often appears illogical from outside a belief system or society may be entirely sensible when viewed from within. Such is the case of the sacred cow in India, which wanders the streets protected by religious taboos, even in the busiest cities. To most visitors, these cows appear bizarre or inconvenient at best, and a scandal at worst, because they may not be slaughtered for sustenance—ironic, given the fact that so many people are malnourished in the same cities and villages. Moreover, cow worship by Hindus has sparked off riots between Hindus and Muslims, who prefer to eat cows rather than revere them.

Anthropologist Marvin Harris (1974) found an explanation for this phenomenon in the small-scale, low-energy economic system of traditional India, based as it has been for centuries on animals. Cows and oxen still provide low-energy substitutes for tractors; for example, a team of bulls pulls around large mowers until they cut a considerable section of grass. The grass is then raked into a pile and consumed by the cow, which converts it into energy for mowing another section. Moreover, India's cattle annually excrete about 700 million tons of recoverable manure, about half of which is used as fertilizer and the remainder as cooking fuel or household flooring and siding. Wandering cows scour the environment for waste products and stubble unfit for human consumption, which they then convert into milk, energy, and other useful products. These ubiquitous cows are not wild animals that wander aimlessly; they are actually owned and tended by individuals who can identify them. Periodic droughts and famine in India threaten the livelihood of nearly everyone, but the Zebu cattle have energy-storing humps on their backs, are efficient, and are capable of existing for long periods of time with little food or water. They provide labor, dung, and, of course, milk and calves (a source of cash for a poor family). During crisis periods, a farmer may be tempted to kill or sell livestock, which might make immediate sense but would be disastrous in the long term. Religious taboos against killing cattle thus help to protect against irrational decisions in difficult times. Compare this rational protection of the cow in India, Harris suggested, to the irrationality of U.S. culture, where the beef industry feeds two thirds of the country's grain to cattle while people go hungry.

Despite its explicitly spiritual rationale, the religious tradition of protecting cows thus has a concrete economic function. This lesson cannot be lost in efforts to construct an ethos for the global village, and it is also the conclusion of Weber's (1904/1958) *The Protestant Ethic and the Spirit of Capitalism.* Rationality is always drawn from a single viewpoint, and what is irrational from one perspective may be rational from another. Before we condemn any religious beliefs as irrational, we need to explore their hidden logic and benefits, not only from a limited viewpoint but also from multiple perspectives.

Buddhist Ethics

The moral code of the tradition, summarized in the Buddha's Five Precepts of right behavior, provides a means for escaping suffering: Live in such a way as to transcend one's fate and avoid inflicting suffering on others, engage in acts of compassion toward other creatures, and rejoice in their good fortune. The Buddha thus advocates an ethical system that is a mirror image of Western utilitarianism: In Buddhism, one's own interests are served by serving others whereas in Western utilitarianism everyone's interests are enhanced by pursuing one's own interests. The Five Precepts are as follows:

1. Do not kill.

2. Do not steal.

3. Do not lie.

4. Do not be unchaste (with different meanings for monks and laity).

5. Do not drink intoxicants.

Vietnamese monk Thich Nhat Hanh (2008) provided an interesting update on the meaning of these teachings:

1. Aware of the suffering caused by the destruction of life, I vow to cultivate compassion and learn ways to protect the lives of people, animals, plants, and minerals. I am determined not to kill, not to let others kill, and not to condone any act of killing in the world in my thinking and in my way of life.

2. Aware of the suffering caused by exploitation, social injustice, stealing, and oppression, I vow to cultivate loving kindness and learn ways to work for the well-being of people, animals, plants, and minerals. I vow to practice generosity by sharing my time, energy, and material resources with those who are in real need. I am determined not to steal and not to possess anything that should belong to others. I will respect the property of others, but I will prevent others from profiting from human suffering or the suffering of other species on Earth.

3. Aware of the suffering caused by sexual misconduct, I vow to cultivate responsibility and learn ways to protect the safety and integrity of individuals, couples, families, and society. I am determined not to engage in sexual relations without love and a long-term commitment. To preserve the happiness of myself and others, I am determined to respect my commitments and the commitments of others. I will do everything in my power to protect children from sexual abuse and to prevent couples and families from being broken by sexual misconduct.

4. Aware of the suffering caused by unmindful speech and the inability to listen to others, I vow to cultivate loving speech and deep listening in order to bring joy and happiness to others and relieve others of their suffering. Knowing that words can create happiness or suffering, I vow to learn to speak truthfully, with words that inspire self-confidence, joy, and hope. I am determined not to spread news that I do not know to be certain and not to criticize or condemn things of which I am not sure. I will refrain from uttering words that can cause division or discord, or that can cause the family or community to break. I will make all efforts to reconcile and resolve all conflicts, however small.

5. Aware of the suffering caused by unmindful consumption, I vow to cultivate good health, both physical and mental, for myself, my family, and my society by practicing mindful eating, drinking, and consuming. I vow to ingest only items that preserve peace, well-being, and joy in my body, in my consciousness, and in the collective body and consciousness of my family and society. I am determined not to use alcohol or any other intoxicant or

to ingest foods or other items that contain toxins, such as certain TV programs, magazines, books, films, and conversations. I am aware that to damage my body or my consciousness with these poisons is to betray my
ancestors, my parents, my society, and future generations. I will work to
transform violence, fear, anger, and confusion in myself and in society by
practicing a diet for myself and for society. I understand that a proper diet
is crucial for self-transformation and for the transformation of society.

Source: Reprinted from *For a Future To Be Possible* (1993) by Thich Nhat Hanh with
permission of Parallax Press, Berkeley, California, www.parallax.org.

Buddhist monks are bound by an extensive set of rules, but the observance
of only four (taken from the Five Precepts) is necessary to avoid expulsion from
the community: the prohibitions against (1) sexual intercourse, (2) theft,
(3) murder, and (4) dishonest claims to spiritual attainment.[2] Laity are expected
to (1) follow most of the same rules as monks, (2) provide the monks with food
when they make their morning rounds with their begging bowls, and (3) social-
ize their children into the Buddha's teachings. Traditionally, all Buddhists
received some instruction at the monasteries during daily and life cycle rituals
and from storytellers, because until recently most of the population was illiter-
ate. Contemporary Buddhists, when they are allowed to, provide copies of the
Buddha's teachings in public places such as restaurants as well as in the temples.

Buddhism has a highly practical side as well, as shown in the folk religions that became associated with Buddhist religion. The Gods are considered responsible not only for physical security but also for providing
guidance for everyday life in the community. In Jordan's (1985) account of
Taiwanese religion, a village is guarded at all four corners by supernatural
protectors who ward off evil. If the guardians are not properly worshipped,
evil may enter the village, and special rites are invoked to exorcise it. In the
village of Bao-an, a child drowned in the pond, having been "pulled in by a
ghost." The dead child's maligned ghost lingered in the pond and could not
be permitted to stay, so two altar tables containing small "divination chairs"
for the Gods were placed in front of the temple. The Gods came and advised
that people should stop speaking bad words to one another, because the
death had disrupted the harmony of the village. Second, they suggested that
people should keep children away from the pond.

In Bao-an, as elsewhere, religious rituals and the ethos of a religious tradition provide concrete, as well as general, guidance for a culture's lifestyles.
Just as *kosher* laws directly or indirectly keep Jews from contracting trichi-
nosis from bad pork, the Bao-an Gods' words bring harmony to the village
by stopping harsh words and thus maintaining order in the community.
Other children would be saved from drowning by the divine injunction to
keep them away from the pond.

The rituals themselves may appear nonrational to people outside the belief system, but they seem to provide a sense of security and in some ways actually helped to protect the people of Bao-an. Is it mere illusion, mere false consciousness? In traditional Chinese folk religion, those who die are always still present in the form of Gods and ghosts; they continue to have needs in the next world and so are given gifts regularly. One's fate in the next world is thus related to how well one's descendants provide for one after death. One's status in the next world, however, is also determined by the merit accumulated in a terrestrial life. One villager put it this way:

> When we men are good, we have a good report; and when we are bad, we have a bad report. The idea is always the same. Gods are those who have done good deeds as men, those who love virtue and study the ways of the buddhas and after death join the buddhas. . . .
>   Men of a good nature become gods; men of virtue become gods, and those without it [virtue] become ghosts. (Jordan, 1985, pp. 35–37)

It is not clear, at least in Jordan's (1985) account, how much of one's fate is determined by one's own merit and how heavily one's descendants' actions count, but the rational calculus is not important. Because gaining one's own merit and honoring one's ancestors are essential components of the ethical system, both are connected logically to one's fate in the afterlife.

Although traditionally emphasizing the individual's finding his or her own path to disengage from suffering, Buddhists also have a long tradition of cultivating compassion in the public sphere, such as King Ashoka's famous public works in ancient South Asia. A recent development is the emergence of a movement of "Engaged Buddhists" who argue for the importance of social activism as an implication of the Buddhist ethic. The Engaged Buddhists are involved in everything from promoting justice and human rights (see Ledgerwood & Un, 2003) to opposing war and excessive consumerism (see, e.g., Barnhill, 2004).

Jewish Ethics

According to rabbinic tradition,

> A heathen once came to Shammai and said, "I will become a proselyte on the condition that you teach me the entire Torah while I stand on one foot." Shammai chased him away with a builder's measuring stick. When he appeared before Hillel with the same request, Hillel said, "Whatever is hateful to you, do not do to your neighbor. That is the entire Torah. The rest is commentary; go and learn it." (Shabbat 31a in Hertzberg, 1962, p. 109)

This parable summarizes the core of Jewish teaching in several ways:

1. Judaism consistently emphasizes social ethics, guidelines for interacting with others.

2. The tradition honors its teachers (rabbis) but also challenges them to think clearly.

3. All necessary knowledge is in the Torah, but the process of understanding it involves continuous revelation, debate, and interpretation.

4. Though religion is to be taken seriously, a good sense of humor can help a person understand the sacred.

The foundation for Jewish ethics is the Decalogue, or the Ten Commandments, traditionally believed given to the ancient Hebrews by God shortly after their escape from slavery in Egypt (ca. 1300 BCE). In the third month after their escape from the Egyptians, the Hebrews camped before Mount Sinai. Considerable social unrest emerged at the time, perhaps because of the difficult existence the Jews experienced as refugees, after the relative security of their slavery. At this key moment, according to Jewish tradition, Moses went up onto Mount Sinai and received a set of stone tablets from God, on which were inscribed the laws that would regulate the community's life. The commandments were as follows:

1. Thou shalt have no other gods before me (Exodus 20:3).

2. Thou shalt not make unto them any graven image, or any likeness of any thing that is in heaven above, or that is in the earth beneath, or that is in the water under the earth (v. 4).

3. Thou shalt not take the name of the Lord thy God in vain . . . (v. 7).

4. Remember the Sabbath day, to keep it holy (v. 8).

5. Honor thy father and thy mother: that thy days may be long upon the land which the Lord thy God giveth thee (v. 12).

6. Thou shalt not kill (v. 13).

7. Thou shalt not commit adultery (v. 14).

8. Thou shalt not steal (v. 15).

9. Thou shalt not bear false witness against thy neighbor (v. 16).

10. Thou shalt not covet thy neighbor's house, thou shalt not covet thy neighbor's wife, nor his manservant, nor his maidservant, nor his ox, nor his ass, nor any thing that is thy neighbor's (v. 17).[3]

Note that eight of the Ten Commandments are expressed in the form of taboos, or injunctions; only two are positive (remember the Sabbath and honor your parents). Most of them (commandments 4 through 10) concern social relations rather than direct relations with the deity. This religious tradition did not intend simply to mediate between the people and their deity; it provided a foundation for their collective life.

The events surrounding the introduction of the Decalogue are full of social drama and are equally instructive. After a preparatory period in their camp at Mount Sinai, Moses established boundaries beyond which only he, his brother Aaron, and later selected leaders from the community could go. The people were not even to touch the border of the mountain where Moses was to talk with God or, they were warned, they would be put to death. The delegation was allowed to see God (which was usually taboo), but not to climb to the summit with Moses, who stayed there for 40 days. The people left behind became restless, made a golden calf out of their jewelry,[4] and began to dance in front of it, turning from the worship of God to the indigenous religious practices of the region. Both Moses and their God were furious; Moses shattered the tablets containing the Law, and God threatened to destroy the Hebrews. When Moses asked the Israelites to choose sides, the tribe of Levi went with Moses, who reported God's instructions to his followers:

> Thus says the Lord God of Israel, "Put every man his sword on his side, and go to and fro from gate to gate throughout the camp, and slay every man his brother and every man his companion, and every man his neighbor." And the sons of Levi did according to the word of Moses; and there fell of the people that day about three thousand men. (Exodus 32:27–28)

Moses gave the Levites (as the tribe related to Levi was called) a blessing and intervened with God on behalf of the survivors. God decided not to destroy the remaining Israelites and provided Moses with new tablets. At that point, Moses and the Levites, having eliminated or intimidated all opposition forces that had rebelled by worshipping the golden calf, became priests in full control of their society.

Contained in the story of the ancient Hebraic ethical code is thus also a lesson about the nature of their God (who insists on loyalty and justice, subject to negotiation), the ruthlessness of their leaders' control, and an explanation for the Levites' special authority among the ancient Hebrews. The remarkable violence of the story, which stands in sharp contrast with the taboo against killing in the Decalogue, is almost lost in the narrative. Thus, inherent contradictions are taken for granted as the reader is swept along by the story.

In the third period of what Judaism considered its "salvation history" (after the call of Abraham and Sarah and then the Exodus from Egypt), the prophetic movement (which may have begun as early as 1050 BCE) took the establishment to task for not keeping the covenant with God. Instead of polytheism or idolatry, the prophets exposed injustice and the general faithlessness of the people as well as the emptiness of their religious rituals. As the prophet Amos (ca. 750 BCE) put it, God complained that the people "trample upon the needy, and bring the poor of the land to an end" (Amos 8:4). Consequently, their rituals of worship were no longer pleasing to the deity:

> I hate, I despise your feasts, and I take no delight in your solemn assemblies.
>
> Even though you offer me your burnt offerings and cereal offerings, I will not accept them. . . .
>
> Take away from me the noise of your songs; to the melody of your harps I will not listen.
>
> But let justice roll down like waters, and righteousness like an ever flowing stream. (Amos 5:21–24)

Because the nation had broken the covenant with God by its injustice, Amos said that God would destroy them—a sentiment echoed by other prophets such as Hosea, Micah, and Isaiah. God was no longer perceived as a God who favored only a particular clan or nation but one who insisted on justice for all people and would punish violators of these principles even if they had a special relationship with their deity. This development signaled a significant break with the particularism of both clan and national religious expressions and exhibited a universalism similar to that found in Buddhism. Such a decoupling of the belief system from a particular social structure enabled the system's survival. In 922 BCE, the kingdom split into the northern kingdom of Israel and the southern kingdom of Judah. In 721 BCE, however, the northern kingdom was conquered, and in 586 BCE, more decisively, Jerusalem was crushed by the Babylonians. The temple was destroyed, and much of the nation's elite was carried off into exile in Babylon. Ironically, not only did Judaism survive this debacle but it was even strengthened by the suffering of this "exilic" period.

Three consequences of the exilic period were as follows:

1. A clear monotheism was established that had only been hinted at in early Judaism (e.g., the Decalogue declares that no other Gods are to be worshipped before the Jewish God), because monotheism seemed to be a

plausible explanation for the defeat by the Babylonians: The Jewish God was not weaker than the Babylonian God but was punishing the Hebrews for their sins.

2. The Torah was finally written down, probably for the first time, and the community ceased to rely on oral tradition. The theological and ethical ideas of the exilic and post-exilic tradition were then written back into the earlier history of the people, who reconstructed a sacred past now recorded in the Torah.

3. The very notion of Israel was constructed after the northern and southern kingdoms were destroyed. When Jews began to question their own religious identity, it was more clearly defined. By the time the exiles returned (when the Persians defeated the Babylonians in 539 BCE), the Jews had defined precisely who they were: members of the nation of Israel who had a bloodline from their ancestors and worshipped the One God, whose word was revealed in the Torah.

Thus the conflicts precipitated by dramatic (in some cases quite violent) conflict between Judaism and other socioreligious orders resulted not in the destruction of the tradition but in the re-creation of the faith. Out of these conflicts the notion of monotheism was forged, the scriptures were written down, and Judaism became less bound to specific geographic locations. Over time, the stories of this history held the Jewish community together through centuries of adversity and dispersion throughout the world.

The example of Judaism demonstrates how it is possible to see the way in which religious traditions are constructed to meet sociological and psychological needs. In this sense, the classical theorists such as Freud, Marx, Comte, and Durkheim were right in regarding religious beliefs and practices as projections of human psychological needs; these theological explanations for the crises of life put a cosmic frame on personal troubles. Just because belief systems are constructed by humans to answer human needs and explain human perceptions does not mean, however—as the early sociological theorists claimed—that the referents of religion do not exist. It simply means that if the Gods do exist that they are perceived by humans in a human way.

Economics in Ancient Judaism

Because of its ethnic–tribal basis, ancient Judaism was closely linked with the economic life of the socioreligious community. The Jewish ethos places its distinctive stamp on economic activity in three ways: (1) Economic activity, like all elements of life, should be conducted on a highly ethical basis, with special attention to problems of injustice within the community. Everyone should be cared for and all should be treated fairly, even those outside the community. (2) Because the earth is a gift from God, humanity is given stewardship over it and should be thankful. Success in business is somehow related to God's grace, and a portion of the profits should be given as an offering. (3) Finally, the emphasis on rational thought and a disciplined life in the Jewish ethos has cultivated an entrepreneurial spirit that has served the community well over the centuries. Ironically, some of the entrepreneurial skills cultivated within the community were an inadvertent consequence of prohibitions against Jews owning land in many parts of Europe until well into the modern era. Anti-Semitic groups over the centuries have exploited these historical circumstances for their own purposes by making unfounded claims about Jewish financial acumen.

Christian Ethics

At the center of the Christian ethical system is the proposition that God is love. In its more radical forms, the Christian ethic requires one also to love one's enemy, a concept I will explore more fully in the discussion of religion and social conflict in Chapter 8. This emphasis on loving others comes from Jesus' teachings as recorded in the New Testament. When asked what constituted the greatest commandment, Jesus responded with this:

> You shall love the Lord your God with all your heart, and with all your soul, and with all your mind. This is the first and great commandment. And a second is like it, You shall love your neighbor as yourself. On these two commandments depend all the law and the prophets. (Matthew 22:37–40)

This link from the Jewish tradition between love of God and love of one's neighbor became a hallmark of the early Christian Church, which cut across class and ethnic boundaries and maintained a strict pacifism. Many in the early community sold their worldly possessions, fed the poor, and lived a communal life focused on loving one another as children of God. This radical ethical universalism, as translated into interethnic relationships, must have been shocking to many 1st-century Jews who had been taught that God required strict observance of boundaries between their community and others.

Different interpretations of these ideas appeared over time according to their affinities with different social groups. In conservative versions of Christianity, sin is a central and important aspect of human nature. The concept of "original sin," popular in some circles of Christianity especially since the Calvinist movement in 17th-century Europe, suggests that humans are born sinful because of the long lineage of disobedience that can be traced back to the original sinful act of Adam and Eve. We are all, according to the famous American Puritan preacher Jonathan Edwards (1741), "sinners in the hands of an angry God."

The idea is not popular among most progressive Christians in the 21st century because of its emphasis on inevitable guilt, which, some then argue, implies a need for tightly bound systems of social control to keep people from sinning. The concept does have some sociological basis, however: Individuals face the consequences of their parents' "sin" from the moment they are born because their own values and life situations are shaped by their social context. This phenomenon can be seen in the cycles of violence in human society—children who suffer abuse from their parents, for example, often abuse their own children, and interethnic feuds can be transmitted across the generations.

The classic consequence of sin in Christian theology is to be damned and sent to hell for punishment in the afterlife. Debates about hell have raged through the centuries in Christian theology. There is little hint in the Gospels about Jesus' own understanding of the concept, and our most famous picture of hell was not provided until 13 centuries later by the Italian poet Dante in his famous *The Divine Comedy*. In the Christian tradition, evil is personified in the figure of the devil, or Satan, an angel created by God whose pride led to his downfall. By the mid-20th century, the devil had become a mere metaphor to most people in Western countries, yet popular interest in this figure had been increasing as the end of the second millennium approached and movies like *The Omen* and *The Exorcist* drew large audiences. In the 1980s, Christian traditionalists in the United States saw the workings of Satan behind dangerous cultural trends and international events. The **charismatic movement** revived the practice of exorcism to drive out demons. As a sign of the times, Fuller Theological Seminary—the nation's largest Protestant seminary—introduced a course on how to oppose the devil (see Woodward & Gates, 1982).

Widely discredited in secular intellectual culture, then, Satan remains a ubiquitous figure in popular culture and even performs a certain sociological function: The representation of this figure reminds us that evil is a force that exists, like many social forces, above and beyond those individuals who seem to be their instruments. Because demons represent an objectified element of the social construction of reality—that is, something that exists outside of individuals—sociologists must at least remain agnostic about these figures as well as about their positive counterparts, angels. Certainly the concepts of demons and angels correspond to many people's experience whether or not they exist objectively. As Peter Berger (1969) put it, people who reject the idea that Detroit is infested with demons miss a certain element of the reality of Detroit!

Christianity and Economic Ethics

The classic studies of the impact of Christian ethics on the economic functioning of Western society are Max Weber's works in the comparative sociology of religion, beginning with *The Protestant Ethic and the Spirit of Capitalism* (1904/1958). Weber's research on his lifelong interest in Western rationality and religious phenomena was inaugurated with this study, which examined the impact of a particular religious system's approach to ordering everyday life conduct. According to Weber, the Protestant Reformation produced a religious crisis when people were no longer able to depend upon the formal ecclesiastical institution of Catholicism and its priestly representatives to assure them that they were saved. In Calvinism and Puritanism, a belief in the omniscience (all-knowing characteristic) of God led logically to a belief in predestination. If God knows everything, then God must know whether we are part of the elect, who will be rewarded in heaven for eternity, or part of the damned, who will suffer in hell. Accordingly, people began looking for new kinds of signs that they were part of the elect.

The anxiety that this dilemma created became linked with the idea of a calling in Puritanism and Calvinism: God calls people to a particular vocation; when one follows God's calling and works diligently at it, one will be rewarded. Worldly economic success consequently becomes a secure sign that a person is saved, thus laying the groundwork, Weber argued, for the systematic life control and inner discipline in the workaday world that promoted the rise of capitalism. Religious ideas, Weber concluded, do play a formative role in history, but the consequences of their impact are not always anticipated. We may also note that although the Protestant ethic has its distinctive elements, the value placed on an individual's working for the good of the collective is a major theme in religious life that runs across traditions.

Religion is often closely tied to economics in a utilitarian way. People, for example, will often appeal to a God for assistance or even intervention in economic activities. From the rain dance and mating rites to facilitate agricultural production to prayer in corporate boardrooms, candles to help people win the lottery, and offerings made at the altar of Kuan Kung, humans have invoked the Gods to assist the process of economic acquisition. Sometimes the appeal is direct and unambiguous, as in the Reverend Gene Ewing's Church by Mail. The Reverend Ewing, a master of the mass-mail market, uses it to entice people to ask God to help them as well as to send money to his organization. His mailings are filled with testimonies of people who have followed his advice and been blessed with "spiritual, physical, and financial blessings." Someone got a check in the mail for $5,000; another person got a new job; someone else was healed of cancer. Modern technologies of television and mass mailings provide a great temptation for people willing to manipulate other people's religious beliefs for personal gain. The Reverend Ewing may be sincere himself, but his methods are open to exploitation.

Membership Tests and Social Control

Jesus contended that only God, not other humans, should judge another person's actions (see Matthew 5), but a religion's ethos creates guidelines for daily life. Elites thus use the religious ethos to legitimate a system of social control that does make judgments so that those who violate normative and legal boundaries can be punished and the existing social order upheld.

The history of the Christian Church is replete with the construction of criteria for such judgments. When church and state are closely linked, religious ethical infractions are often punished by law. Often, however, informal controls are just as important to the ordering of daily life. Within a religious community, and especially in Christianity—in which the boundaries of the community are doctrinal more than social—qualification for membership becomes socially significant, and tests are developed for judging an individual's eligibility. In the early Christian Church, all a person had to do was give a profession of faith ("Jesus is Lord") and be baptized in order to join the community. The community quickly established stricter standards, however, engaging in considerable debate over what did and did not constitute appropriate signs of membership. Meeting these additional membership tests was supposed to follow naturally as a voluntary response to the ethical standards of the religious community, including such things as giving away worldly goods to the poor, living communally, and not committing any acts of violence. Soldiers who converted could remain in the military but could not kill anyone.

Such Christian tests of membership, salvation, merit, and the like sometimes led to great disparities between theory and practice. In contemporary Protestant Christianity, for example, a central tenet of the belief system is that each individual has a personal relationship to God with no mediation by church officials (as contrasted with the Catholic practice). Rhetorical and behavioral formulas with little if any biblical basis have emerged, however, that constitute very rigid tests of whether or not one is "truly Christian."

In circles where personal faith is most highly valued, conformity to rhetorical standards is strictly enforced: To prove that one has an intense personal relationship with the deity, one must express it in the approved manner that reproduces other believers' equally intense, radically personal encounter with the sacred. In some communities, a Christian is someone who abstains from smoking, drinking, or fornicating (although killing others may be a religious duty, if the act is approved by the right government). In other Christian communities, however, people who approve of killing and disapprove of fornicating are judged as not truly Christian.

Because the ethos of a culture reflects its worldview and a pluralistic culture inevitably engages in struggles over the single acceptable way to define the world and the norms of everyday life, the definition of the family has taken on particular religious importance in the United States. In the patriarchal Judeo-Christian tradition, the structures of the cosmos and the family typically reflect each other: The father who presides over the family mirrors God the Father in control of the universe. The implications of this perspective for the nature of authority in family and society are instructive. In direct (although somewhat obscured) opposition to the democratic norms of modern political culture, this model of authority requires respect for superordinates at both familial and cosmic levels and harsh sanctions against those who do not follow the norms.

The conservative Protestant movement in the United States is attempting to sustain this metaphor against the protests of the gender equality movement. At issue is what Wald, Owen, and Hill (1989) called "authority-mindedness"—that is, a model of social life in the family and the congregation that advocates an unequal division of authority. As Ammerman (1987) put it, "They come to expect groups to be divided between sheep and shepherds. The shepherds are entitled to deference and rewards, while the sheep are entitled to love and care" (p. 128). Within this belief framework, corporal punishment becomes an important part of maintaining the family's structure, just as God punishes wrongdoers.[5] Ellison and Bartkowski (1997) contended the following:

> Conservative Protestants are convinced that physical discipline communicates a positive spiritual lesson to their children. In brief, they argue that many children develop and express their understandings of God in parental images, and therefore that children will infer God's view of them based on the treatment they receive from their parents. . . . In this view, parents should teach their children by example that God is loving, merciful, and forgiving. At the same time, however, because God's punishment of sin is understood as both inevitable and consistent, it is vitally important for parental discipline to embody these characteristics as well. (p. 52)

Other Christians contend that this authority-minded model is antithetical to their faith, which posits the brotherhood and sisterhood of all humanity.

Such debates take on heightened importance in the contemporary world, in which religious communities are forced to coexist alongside other groups with differing values and standards. Many of these issues can be seen in sharp relief in the strained relationships between the sibling traditions of Christianity and Islam. The ways in which these problems are debated and resolved are central to the process of constructing an ethos for the global village.

Islamic Ethics

Umar Abd-Allah (2005) claimed that according to the Qur'an, God "designs the world and rules the universe in his aspect as the All-Merciful" (p. 2). The ethical implication of this is that "Islam enjoins its followers to be merciful to themselves, to others, and the whole of creation, teaching a karmalike law of universal reciprocity by which God shows mercy to the merciful and withholds it from those who hold it back from others" (Abd-Allah, 2005, p. 1). Ultimately, everything that happens in the universe— "even temporal deprivation, harm, and evil—will in due course, fall under the rubric of cosmic mercy" (Abd-Allah, 2005, p. 2).

The founder of Islam was known, even by his enemies, as a "prophet of mercy" (Abd-Allah, 2005, p. 1). Abd-Allah (2005) put it this way:

> Although like many prophets, Muhammad engaged in much conflict and even warfare, he was remarkably quick to seek peace and refused to be vengeful when victorious. After the Muslims conquered the city of Mecca—with very little bloodshed—the Prophet refused to punish his conquered enemies who had sought to kill him. He told the city's residents "Go to your houses. You have been set free." (pp. 2–3)

Muslims are thus expected to follow the Prophet's example in promoting a "doctrine of universal, all-embracing mercy" that is to be applied not only to other Muslims but to believers and unbelievers and even "the animate and inanimate: birds and animals, even plants and trees (Abd-Allah, 2005, pp. 4–5). In the end, Abd-Allah (2005) wrote, "The imperative to be merciful—to bring benefit to the world and avert harm—must underlie a Muslim's understanding of reality and attitude toward society" (p. 6).

The ethical and legal code of Islam is institutionalized in the *Shari'a*, Islamic law intended to encompass all of human life from the most private areas to the public organization of society. Because early Islam made no distinction between the law and religion, the Shari'a is traditionally the law for Islamic society as well as a set of religious ethical guidelines. Its roots lie in Allah's revelations to Muhammad and the Prophet's efforts to create a disciplined, ordered society in the chaotic, warring tribal culture in which he lived. The concept of the Shari'a as all-pervasive is an ancient one in Islamic practice and is given more emphasis than most doctrines of the faith. John Williams (1962) has said the following:

> Muslims, consonant with their emphasis on the Law, have been more concerned with what men do than with what they believe, and very slow to reject any group of Muslims for wrong doctrine, unless that doctrine led the group to actively exclude itself, by its deeds, from the Community. (p. 94)

As we will see, however, interpretation of the law is often a matter of dispute.

Islamic Economics

In traditional Islam, following the example of Muhammad, a theocratic system was imposed in which polity and economy were essentially under the direction of the ummah. The Islamic movement spread rapidly during its formative period, and its military expansion led to Muslim control of the Mediterranean and as far as Southern France, North Africa, the Middle East, and well into Asia, in part because of the practice of Muslim traders, known for their skills and ethical business practices. During the period of Western colonial expansion, the authority of the Shari'a was challenged by the conquering European powers, which tried to impose their secularized legal system on Islamic countries. Whereas economic law (of more concern to the colonizers) was placed under secular jurisdiction, the one area of successful resistance in most Islamic colonies was family law, which became a symbol of the struggle with the colonialists, and is still largely controlled by the religious courts.

In postcolonial Islamic societies, many religious leaders have tried to reinstate their authority over economic matters and in some ways have succeeded in the case of smaller-scale entrepreneurs, but now they find that many of the standards of business for larger companies are set by international capitalism. Some of that influence has been bounded by such practices as traditional Islamic hospitality and concern for the poor, including the annual *zakat*, which creates some redistribution of wealth within economic society. As other religious communities have discovered, however, it is difficult to resist the powerful forces of international capital, which often emphasizes profit-making to the exclusion of other ethical standards.

Religion and Sexuality

As one of the most potent aspects of human life, sexuality has a special place in the beliefs, rituals, and institutions of the world's faith traditions. It is, in fact, at the center of Durkheim's theory of religion, which emphasizes the collective conscience and social solidarity as emerging from collective exaltation and orgiastic frenzy in religious rituals. It is difficult to generalize about the relationship between religion and sex because the range of beliefs, rituals, and institutional regulations is so broad, from the celebratory to the celibate, from the tantric and orgiastic to ascetic renunciation. Moreover, the norms about sexual morality— like most ethical teachings—vary both within and between the major traditions (see, e.g., Sherkat, 2002). There are, however, some central themes we can explore in this brief overview.

Weber (1963) understood the importance of this topic when he wrote, "The relationship of religion to sexuality is extraordinarily intimate, though it is partly conscious and partly unconscious, and though it may be indirect as well as direct" (p. 236).[6] Since Weber's time, the relationship between religion and sexuality has transformed further in a direction he saw. With the industrial revolution's differentiation of the family and the economy and the privatization of religious beliefs in heterogeneous societies, issues of sexual morality and the family often became battlegrounds for the **culture wars** of modernity. In the West, with the church losing much of its control over the political sphere, much of its attention turned to personal morality. It did not give up efforts to affect politics, but many within the institution did increasingly attend to the politics of sexuality, with abortion and homosexuality high on the agenda.

Given the power of religious institutions and authorities, of course, there is frequent occasion for abuses of power, with clergy and other authorities (usually men) claiming sexual privileges, with unequal access to members of the opposite or even the same sex. Recent scandals in American Catholicism regarding priests who took advantage of their positions to assault children sexually may be just the iceberg tip's example of centuries of such power abuse.

Weber (1963) outlined a number of aspects of this phenomenon from "the intoxication of the sexual orgy" with its "notion that sexual surrender has a religious meritoriousness" (p. 237) and the opposing significance of sexual abstinence, whether a cultic chastity as a temporary abstinence prior to the administration of the sacraments to the permanent abstinence of priests and religious virtuosi. Religion becomes especially involved in sexual matters when the behavior is extraordinary, whether in the Hindu *Kamasutra* and explicit sexual poses on temple walls or the sexual chastity, brahmacharya, practiced by Mahatma Gandhi following the tradition of Hindu ascetics of old. "Chastity, as a highly extraordinary type of behavior," Weber (1963) noted, "is a symptom of charismatic qualities and a source of valuable ecstatic abilities, which qualities and abilities are necessary instruments for the magical control of the god" (p. 238). He speculated that a decisive reason for priestly celibacy in occidental Christianity "was the necessity that the ethical achievement of the priestly incumbents of ecclesiastical office not lag behind that of the ascetic virtuosity the monks" (Weber, 1963, p. 238) along with the church's efforts to prevent the priests' heirs from inheriting the church's resources.

As Jack Miles (1995) noted in his *God: A Biography*, much of the account of God's relations with humans in the Hebrew scriptures is concerned with sexual issues, starting with the sin of Adam and Eve, the nature of which is revealed by "what happens when the forbidden fruit is eaten: 'Then the eyes of both of them were opened and they perceived that they were naked; and they sewed together fig leaves and made themselves loincloths" (Genesis 3:7). It is, as Miles (1995) pointed out, not sexual "desire, in and of itself, but knowledge of one's desire that generates shame. Animals desire, but they do not know they desire" (p. 36). Consciousness of sexuality—and God's concern with it—seems to be a foundational part of human nature from that scriptural point of view.

Since relations between humans and God are often linked to social relations, as they are in the Hebrew scriptures, it is not surprising that religious teachings are frequently about sexual morality. The regulation of sexual intercourse—an important component of social organization, especially as society becomes more complex—becomes framed as a spiritual issue, although it may have always been so, since the division between religion and culture we construct in the modern world was rare in antiquity. In today's religious economy, restrictions on sexuality are something of a barometer of the regulation of life generally by religious institutions and the amount of individual freedom granted in this sphere is comparable to that in other spheres. Moreover, such issues as chastity or abstinence from premarital sex, abortion, safe sex practice, reproductive rights within marriage, and homosexuality are the lightning rods of many cultural conflicts (see, e.g., Ojo, 2005). Ellingson and Green (2002) observed the following:

> Sexuality is the primary ground on which human relationships are sanctioned as natural and good, or unnatural and wrong. Through ideology, taboo and ritual, sexuality is channeled into those behaviors recognized as licit, as opposed to those seen as illicit. (p. 2)

Consequently, they went on to say, "Sexuality occupies the attention of many religions because it is a powerful way to organize and relate human beings. . . . [It] is a central element in the construction of religious meaning" (Ellingson & Green, 2002, p. 2). It is also, however, "centrally concerned with the use and abuse of power. This is the central dialectic uniting all religious struggles" (Ellingson & Green, 2002, p. 3). For people in power, sexuality is something to be shaped, and religious institutions and rituals are often exploited to do the shaping.

Ideally, the purpose of religious teachings on sexuality is to guide people toward a channeling of sexual—and other libidinal energies and drives—in a direction that enhances spirituality and community, rather than harming or destroying it. Of course, for Sigmund Freud (1898/1961), religion itself is an illusion constructed to compensate humans for the sacrifices they have to make on behalf of civilization, repression, or sublimating libidinal drives in the interests of the social order. The spiritual purposes of sexual morality shape the extremes (celebration and celibacy) driving them toward moderation or explaining sexual virtuosity or abstinence as a sign of special spiritual powers, creating seeming contradictions in the moral teachings of many traditions. Moral injunctions also have social functions, of course, in preserving family organization and social networks by prohibiting sexuality outside of marriage and promoting taboos against incest, which might harm relationships within the family. Similarly, prohibitions against extramarital sex also aim to protect young children by limiting the number of them who are born outside the shelter of the family during their vulnerable early years. Sexuality is also used as a weapon of war and domination (see, e.g., Ellingson & Green, 2002, p. 4).

Religion and Sexuality in the East

Hinduism offers an instructive example of the wide range of sexual mores embedded in the faith traditions. In some ways, both extremes are found in that tradition, as they are elsewhere. On the one end of the spectrum are the explicit sexual images and symbols found in sacred sites and texts; at the other end is the idea of brahmacharya, an ascetic practice that included celibacy. For the most part, sexual practices were viewed as kama, physical pleasure, considered to be of divine origin. Life itself begins with sexual desire, according to the "Hymn of Creation" in the *Rig Veda* and the *Athana Veda*, which includes formations and incantations to help or hinder lovemaking, as does the famous *Kamasutra*, the "love text," that advises on courtship, lovemaking, and positions for intercourse (see Bullough, 1976, 1995). The pursuit of sexual pleasure is quite appropriate during the householder stage of life, when one has different duties (dharma) from either the student preparation stage or later in life when one begins to move toward the end of this lifetime, possibly even becoming a *sanyasi*, who symbolically left life prior to death in order to be devoted entirely to the spiritual realm.

In the Hindu rebirth narratives, sexual intercourse is an essential part of the process, and having a son is an ancient ritual that became transformed in the 1st millennium BCE. Patrick Olivelle (2005) said the following:

> The self of the deceased person is said to go up as smoke to the sky. It finally reaches the moon and comes down as rain. The individual, now transformed into water, is absorbed into plants and finally becomes food. A man eats that food and transforms it into semen, which he deposits in a woman, giving rise to a new birth of the dead man. (p. 114)

When the idea of renunciation emerges as a major spiritual practice, having children is one of those attachments that is renounced. As it says in the Bṛhadāraṇyaka Upanishad (4.4.22), "they gave up the desire for sons, the desire for wealth, and the desire for worlds, and undertook the mendicant life" (Olivelle, 2005, p. 114).

Because there is no sharp divide in Hinduism between the inner and outer worlds and the soul itself is infinite, one's own sexual power is intertwined with the cosmic sexual energies. Bullough (1995) put it like this:

> The yoni—narrowly the vagina, but in a broader sense including pubic hair, the opening or cleft of the labia, and the uterus—is considered to have a life of its own and is a sacred area worthy of reverence and a symbol of the cosmic mysteries. The penis, called the *linga*, is also an object of veneration. (p. 450)

If one transcends passion and the carnal state, copulation itself can bring about supernatural power, and in somewhat obscure tantric sects, sexual activity facilitates spiritual union with the Gods and one is able to contemplate reality face-to-face with spiritual ecstasy. The union of male and female brings primordial male and female elements of the cosmos together in a nondual state of Absolute Reality (Bullough, 1995, p. 451). This same sacredness of sexuality leads some to experiments with chastity rather than copulation, however, and that is the path taken by the modern world's most famous Hindu, Mahatma Gandhi.

Steeped in the ancient mysteries of Hindu sexuality, but encountering many other spiritual rivers in his lifetime, Gandhi endeavored through a major part of his life with efforts to channel what was an apparently strong sex drive, by his own accounts, into spiritual and activist experiments. Gandhi decided at the age of 38 to take a vow of brahmacharya, which is a commitment to spiritual discipline that involves the renunciation of material pleasures and implies a life of poverty and chastity but has its roots in ancient spiritual practices designed to connect with cosmic powers and channel them in a life of devotion. The mystical aspect of that process is far too complex for a Western-educated American sociologist like myself, but movement organizing element makes perfect sense. Gandhi's chastity made it possible for women to become intimately involved in the movements he led without the underlying sexual tension that usually accompanies charismatic leadership. Although his experiments seem to have taken a rather bizarre turn toward the end, when he was apparently sleeping naked with young women to test himself, the practice is not entirely out of step with ancient spiritual practices (see Gier, 2007).

In Buddhism, sexuality is yet another aspect of human life from which one needs to become attached on order to achieve ultimate truth (Satha-Anand, 2001). As in Hinduism, it is not that it is wrong per se but that it is part of the journey away from attachment that the spiritual seeker goes through, as did Prince Siddhartha on his path toward Buddhahood. Suwanna Satha-Anand (2001, p. 114) points to three significant occasions in Siddhartha's life as he strove to overcome suffering having observed old age, sickness, and death, his first encounter with the truth of life. Second, he felt compelled to leave his status as a householder on the night his wife gave birth to their child, and finally, he faced the temptation of the three daughters of Mara. "The Buddha-to-be remained unimpressed by their seductive manifestations and became enlightened moments later" (Satha-Anand, 2001, p. 114).

Religion and Sexuality in the Western Traditions

The first chapter of Genesis, in the ancient Hebrew scriptures that provide the foundation for the Judeo-Christian-Islamic traditions, is rich with sexual innuendo. The first words God speaks to his new creatures before humans is repeated once *homo sapiens* comes on the scene: "And God blessed them, and God said to them, 'Be fruitful and multiply, and fill the earth and subdue it'" (Genesis 1:28). In other words, they are essentially told by God to have sex. According to the story, Adam and Eve are given free rein in paradise and they are naked; they eat of the one forbidden tree, which gives them knowledge of good and evil and makes them ashamed for the first time of their nakedness. As Mark Regnerus (2007) observed, the Hebrew term for knowledge can itself imply sexual intercourse (as in Genesis 4:1, where Adam "knew" Eve. . . .) (p. 17). From then on, in fact, throughout the Genesis story, much of the explicit interaction between God and God's human creations has to do with their sexuality. In fact, the covenant God makes with Abram, the father of the Western religions, and God's promise of fertility has a price: a piece of Abram's penis. "Circumcision is not the sign of the covenant in some arbitrary and purely external way," Miles (1995, p. 53) contended.

> Abram's penis—and the penises, the sexual potency, of his descendants—is what the covenant is about. God is demanding that Abram concede, symbolically, this his fertility is not his own to exercise without divine let or hindrance. A physical reduction in the literal superabundance of Abram's penis is a sign with an intrinsic relationship to what it signifies. (Miles, 1995, p. 53)

In Leviticus, certain forms of sexual relations are prohibited (see Bullough, 1995), and this sets the stage for Western sexual morality in the subsequent millennia—those with close relationships to protect the family (incest), those "unnatural" or contrary to physical nature as it was seen at the time (with animals or in same-sex relations), in order to foster procreation, and those contrary to law such as adultery, which again protects the family. By the time religious norms were codified in Deuteronomy, adultery was a capital offense, perhaps not so much of how God would react but because of its sociological consequences: "If a man is found lying with the wife of another man, both of them shall die, the man who lay with the woman, and the woman; so you shall purge the evil from Israel" (Deuteronomy 22:22, RSV). The punishment was for the restoration of the nation. Although we do not know to what extent it was actually carried out, the severity of the sanction was apparently intended to underscore the seriousness of the taboo.

The drawing of boundaries between the sacred and the profane so characteristic of Judaism (and the rabbi's son Durkheim's sociology of religion) applied to the matter of sexuality as well. Sex within the boundaries was sacred and outside of them clearly profane. Like the Mosaic commandments, regulations about sex had to do with the dual relations among humans and between humans and their God, and this concept carried on down the line in Judaism's descendant traditions of Christianity and Islam.

Christianity expands the Judaic restrictions on sexuality by adding an ascetic strain, at first rooted perhaps in attitudes of Paul and then the abstinence of Augustine. Biblical author and early church leader Paul of Tarsus considered sexual sin "a serious matter, more grave than most trasngressions" because "A person who sins sexually has 'sinned against his own body,' a reference to defiling or degrading what Christ has purified through his atoning death (1 Cor. 6:18)" (Regnerus, 2007, p. 18). The famous North African Bishop Augustine's pre-Christian participation in the Mannichean movement that made sharp distinctions between good versus evil and their parallel spiritual and material aspects of reality (the City of God and the City of Man) affected his conceptions of morality and helped shape Christian morality. Sex was part of the material world to be left behind when becoming godly, in much the same way as the Hindu ascetic or the Buddha renounced attachment to this world in order to transcend it. In at least 11 of 27 books in the New Testment, sexuality—*porneia* in Greek—is denounced.

Islam, according to Bullough (1995) "is a sex-positive religion as contrasted to the sex-negative aspects of traditional Christianity" (p. 447) perhaps in part because the Prophet Muhammad considered sexuality as one of the joys of life. The basic rules about sex in Islam are that it should occur within the family and that one should avoid excess. Muhammad defended the rights of women and emphasized the importance of treating wives with kindness. Although there were strict prohibitions against sex outside of marriage, some laxity emerged to accommodate the desire to do so, at least for men, with the possibility of marrying additional wives or even having temporary marriages (*Mut'a*), e.g., to accommodate traders traveling away from home (Bullough, 1995).

Religion, Sexuality, and Politics

Disputes about sexual mores and doctrines are often lightning rods for deep-seated conflicts on a wide range of issues within societies, such as the debates about abortion and homosexuality that have been so politically potent in recent American political discourse. Opposition to abortion and same-sex marriage become litmus tests for loyalty to one party in a larger cultural conflict over the nature of the nation and its future social organization. "Pro-life" and "pro-family" stances become symbols of a larger stance against the diversity and tolerance of heterogeneous modern society, an issue we take up again in Chapter 8. Such disputes also show how religious and legal institutions become intertwined even in states where there's a formal separation. Marriage is a type of social organization where, in most societies, religious and legal institutions share responsibility for regulation. Legal restrictions generally reflect cultural and therefore religious norms and values or at least those of a politically powerful group within society.

The discourse around such disputes are sometimes as much religious as they are political or legal, thus revealing how these institutions are woven together in a complex fabric that usually appears whole; it is when the existing arrangements become unraveled that we see the truth of the interplay between religion and politics in people's lives. Debates about sexual morality and the role of the state and religious institutions in regulating them are embedded in a complex context of social change, from the delinking of sex and procreation with birth control technologies, to the continually shifting nature and role of the family. Steve Bruce (2010) noted that the divorce of economic from the home, a profound cultural shift occurred that affects people in their daily lives:

> At work we are supposed to be rational, instrumental, and pragmatic. We are also supposed to be universalistic: to treat customers alike, paying attention only to the matter in hand. The private sphere, by contrast, is taken to be expressive, indulgent, and emotional. (p. 128)

One of the most divisive political issues at the turn of the millennium was the question of homosexuality, from the issue of same-sex marriage to the ordination of gay and lesbian clergy.

Homosexuality

In his introduction to *Homosexuality and Religion: An Encyclopedia* that he edited, Jeffrey Siker (2007) observed the following:

> If the three traditional taboos for polite conversation include sex, politics, and religion, then the topic of homosexuality and religion is guaranteed to provoke strong reactions, polarizing rhetoric, and a series of conflicting claims that draw variously upon peoples' experience, sacred texts, established traditions and human reason. (p. ix)

Indeed, it has been a major issue for debate in public discourse but also scholarly research in recent years (see, e.g., Boswell, 1980; Bullough, 1976, 1979; Cadge, 2007; Foucault, 1988; Halperin, 2002; Kuefler, 2006; McNeill, 1976). In the United States, where there is often more heat than light on the issue in public discourse, some mainline religious institutions have cautiously begun to support civil rights of lesbian, gay, bisexual, and transgender individuals (LGBT) but have still been indecisive about the morality of the LGBT sexual orientations or the role of openly gay or lesbian clergy in the churches. Moreover, although some church leaders have led the fight for rights, individuals and institutions within the church have been the most vociferous opponents of same-sex marriage and sometimes even LGBT rights.

Cadge (2007, p. 20) noted there has been a dramatic shift in public attitudes toward homosexuality in recent decades. Prior to the mid-20th century, it was generally defined in the United States as a sin or disease, but in the 1970s, first the American Psychiatric Association, in 1973, and then the American Psychological Association, in 1975, declared it not to be a mental disease or disorder. While the debates continue to rage in many mainline churches, public opinion quietly shifted; although close to 70% of the American public believed that same-sex relations were always wrong, only 56% agreed with that opinion in the 1998 General Social Survey, and 31% did not think it was wrong at all. Moreover, regardless of their opinion about homosexual behavior, a majority of Americans—65% in a 2000 *Los Angeles Times* poll—supported protection from discrimination for gay and lesbian citizens. Moreover, in May of 2011, Gallup reported that "For the first time in Gallup's tracking of the issue, a majority of Americans (53%) believe same-sex marriage should be recognized by the law as valid, with the same rights as traditional marriages" (Newport, 2011).

A May 2011 Pew poll showed the following:

> While the public is divided over same-sex marriage, a majority of Americans (58%) say that homosexuality should be accepted, rather than discouraged, by society. . . . Among younger people in particular, there is broad support for societal acceptance of homosexuality. More than six-in-ten (63%) of those younger than age 50 . . . say that homosexuality should be accepted. (Pew Research Center for the People and Press, 2011)

One of the most important developments in the scholarly study of homosexuality and religion has been efforts to reexamine the true history of the relationship between Christianity and homosexuality. The landmark book is John E. Boswell's (1980) *Christianity, Social Tolerance and Homosexuality: Gay People in Western Europe From the Beginning of the Christian Era to the Fourteenth Century*, which has generated considerable debate but has also set a high standard of historical scholarship in the field.

In assessing the debates that followed the subsequent 25 years, Mathew Kuefler (2006) succinctly summarized what he called the Boswell Thesis:

> There were four main points that form the narrative for the book: First, that Christianity had come into existence in an atmosphere of Greek and Roman tolerance for same-sex eroticism. Second, that nothing in the Christian scriptures or early tradition required a hostile assessment of homosexuality; rather, that such assessments represented a misreading of scripture. Third, that early medieval Christians showed no real animosity toward same-sex eroticism. Fourth, that it was only in the twelfth and thirteenth centuries that Christian writers formulated a significant hostility toward homosexuality, and then read that hostility back into their scriptures and early tradition. (p. 2; cf. Hubbard, 2003)

By placing the debate in a much broader historical context, Boswell managed to deflate some of the anti-LGBT rhetoric within American

Christianity, although his efforts to address what is also a deep affective moral issue on purely rational grounds were far from convincing to many who disputed his claims.

Another important element of the debate over sexual orientation is to look at it in a comparative perspective. Although obviously this will not persuade anyone with an exclusivist theology who also believes same-sex relations to be immoral, it does provide an important angle of perspective in our religious and culturally diverse world.

At least one branch of Hinduism, for example, has a somewhat unique position on homosexuality; according to Anil Bhanot, general secretary of Hindu Council of the United Kingdom, Hindu scriptures define homosexual condition to be a "biological one, and although the scripture gives guidance to parents on how to avoid procreating a homosexual child, it does not condemn the child as unnatural" ("Hinduism Does Not Condemn Homosexuality," 2009). Whether one is homosexual or not depends on the timing of insemination and its relation to the menstruation cycle, Bhanot claimed. On the 11th and 13th nights of the menstruation cycle, the fire and water elements are equally balanced, which results in a homosexual conception, according to Bhanot ("Hinduism Does Not Condemn Homosexuality," 2009).

The World Council of Churches has often taken controversial stances, especially in favor of LGBT rights, and it has produced considerable backfire among participating churches. One orthodox priest reportedly said, "If we allow gay churches in [churches with gay priests in them], then what's next? The alcoholic church? The gambling church?" and Patriarch Kirill of Moscow and All Russia expressed his concerns to WCC General Secretary Dr. Olav Fykse Tveit when met the Patriarch in Moscow in June of 2010 (Hutt, 2010).

Clearly the issue of sexuality will continue to be a volatile issue in religious and political circles over the coming years, especially with cultural shifts such as the practice of casual sex and "hooking up" becoming a norm on many college campuses in the United States, raising questions about the role of religious institutions in influencing definitions of morality in that area. It is, of course, just one of many controversies surrounding the intersection between religion and politics.

Religion and Politics

A final dimension of the religious ethos and its impact on collective life is the crucial sphere of politics. I will now look briefly at a few examples of how religious traditions play a role in various political orders and movements.

Hinduism and Politics

The relationship between Hinduism and the political sphere is extremely complex in modern India. Religious pluralism apparently thrived on the subcontinent without major disruption until the invasion of the Muslims in the 8th century. The exclusivist claims of Islam, however, were never reconciled with the theologically tolerant Hindu beliefs. Muslim rulers found the worship of the various Gods idolatrous and made their condemnation part of the process of conquering the region. Consequently, a virulent form of anti-Muslim Hindu fanaticism also developed over time and has become a major factor in contemporary political conflict, both inside and outside the country. Ongoing territorial disputes with Pakistan and lingering memories of atrocities on all sides following the religiously based partition of British India at the time of independence have kept the flames fanned.

Despite efforts to establish a secular independent state after a century of British colonial rule, India continues to be racked by ethnic and religious strife, and the relationship between Hindu and Muslim sectors of the population is often the source of political rifts. Until fairly recently, the Indian state was virtually a one-party democracy, ruled by the diverse Congress Party that led the freedom movement and drew on the heroic status of its former leaders (notably Gandhi and Jawaharlal Nehru) to maintain its control. A breakdown of that rule occurred following the assassination of Prime Minister Indira Gandhi and again after the assassination of her son, Rajiv Gandhi.

In recent Indian politics, the Muslim–Hindu cleavage has become a central theme recalling ancient divisions and resentments. In the city of Ayodhya is a site cherished as the birthplace of the Hindu God Ram. When the Muslims invaded in the 8th century CE, they tore down the temple there and constructed a mosque. In 1990, the prominent Hindu nationalist leader L. K. Advani mounted a campaign to raze the mosque and construct a temple in honor of Ram. The controversy brought down the Indian government in 1990, and a group of Hindus attacked the mosque in 1992, touching off a series of Hindu–Muslim riots throughout the country.

Buddhism and Politics

Buddhist tolerance of other religions and its custom of private worship tended to promote a general lack of specific alliance with political elites. An exception was Tibet, where a theocracy was established around the figure of the Dalai Lama. Even when Buddhism was aligned with the state, however, its lack of exclusivism often had a broadening, rather than constricting, effect on the political culture—that is, it encouraged a tolerant and inclusive ethos that cultivated the lowering rather than the erecting of boundaries between groups.

One of the most interesting early developments in Buddhism was the conversion of the Indian Emperor Ashoka in the 3rd century BCE. After the bloody conquest of most of South Asia, he became a Buddhist. Horrified at the consequences of the wars he had conducted, Ashoka became legendary for his support of Buddhist institutions, his efforts to lead a nonviolent life, and most of all for his "Golden Age" rule, which promoted religious tolerance and high ethical standards. Although not a strict pacifist, Ashoka was opposed to warfare and animal sacrifice and became a vegetarian. Of particular importance is his famous Twelfth Rock Edict, which declared the following:

> One should not honour only one's own religion and condemn the religions of others, but one should honour others' religions for this or that reason. So doing, one helps one's own religion to grow and renders service to the religions of others too. In acting otherwise one digs the grave of one's own religion and also does harm to other religions. Whosoever honours his own religion and condemns other religions, does so indeed through devotion to his own religion, thinking "I will glorify my own religion." But on the contrary, in so doing he injures his own religion more gravely. So concord is good: Let all listen, and be willing to listen to the doctrines professed by others. (Gard, 1962, pp. 18–19)

This edict, now almost 2,300 years old, provides a remarkable testimony to the possibility of religious tolerance in a pluralistic cultural context.

Contemporary Buddhists often struggle with modern governments just as early Buddhism often met resistance from the emperors of ancient Asia. In the past three centuries, European colonial conquerors, then the Japanese, and finally indigenous political movements in China—the Nationalist and Communist parties—have confiscated monasteries or otherwise undermined Buddhist institutions (see Welch, 1972, p. 167ff.). Neither four decades of Communist rule in the People's Republic of China nor hostile governments in other parts of the region, however, have eliminated the pervasive influence of Buddhism either in the broader cultural milieu or in popular practice. In recent years, the repression has lessened, and some temples were refurbished beginning in the 1970s, partly because of efforts to cultivate contacts with foreign governments, especially those like Japan with strong Buddhist traditions. Reforms or a change of governments in East Asia (especially in China) may result in a religious revival comparable to that now going on in the former Soviet states.

Judaism and Politics

Any discussion of modern Judaism cannot fail to mention the two inter-linked watershed events of contemporary Jewish history: (1) the Holocaust and (2) the founding of the state of Israel. The Holocaust was a historical event of unspeakable horror: During World War II, the Nazis exterminated 6 million Jews, reducing the total Jewish population of the world by approx-imately one third. The impact of this brutal violence against a people who had sustained their identity for thousands of years will doubtless affect gen-eration after generation of Jews, who have already incorporated remem-brance of the Holocaust into religious rituals such as Passover.

Few, if any, Jewish families and communities were untouched by this tragedy, which precipitated the creation of a national homeland for Jews in the ancient Holy Land so closely associated with the religious tradition. When the British ended their colonial domination of the region after the war, the United Nations voted to divide Palestine into Jewish and Arab sectors, and the state of Israel was established in 1948.

The remarkable sociological significance of these events lies in the fact that the ethnic–religious identity of the Jewish community in many countries was used by the Nazis to torture and kill a major population in an attempted genocide and then was used by members of the group itself to win interna-tional sympathy for their treatment under Hitler and to establish a modern state on the basis of a religious vision. Unfortunately, one people's homeland was purchased at the price of another's. The Palestinians, who were living in the area before Israel was created, have become refugees in their own land, pushed out by well-meaning efforts by the international community to pro-vide a homeland for the beleaguered Jewish community. Competing multiple claims by various religio-political groups on the same territory—especially the famous Holy City of Jerusalem—threaten to sustain the conflict over many more decades.

Christianity and Politics

The relationship between Christianity and political life is simple in its outline and complicated in detail. Christianity's universalistic doctrines and links with empire expansion facilitated a worldwide diffusion beyond its Jewish origins but also made it a convenient ideology for various world-conquering powers, from the Holy Roman Empire to medieval crusaders, European colonialists, and some contemporary Americans.

After Emperor Constantine's conversion to Christianity in 312 CE, the church became closely linked with Western power elites well into the early modern period. Although the alliance began to unravel with the Enlightenment of the 18th and 19th centuries, the Christian missionary movement helped Europeans solidify their colonial control, first over Latin America, and then over large sections of Africa and Asia.

Two events of the past 200 years have particular relevance to current efforts to construct a multicultural ethos: (1) the disestablishment of Christianity in Western Europe (i.e., individual declarations by nations that Christianity was no longer the official state-sanctioned religion) and (2) the separation between church and state and efforts to protect religious freedom in the United States. The second case is significant because it demonstrates how the problem of multiculturalism is forcibly addressed in political discourse when previously separate culture groups begin to share the same space.

At the end of the 18th century, the nascent American republic was religiously pluralistic, but not everyone wanted it that way. The colonies contained many religious refugees seeking religious freedom who were not about to grant that same freedom to people of other faiths living in their part of the New World.[7] Two mutually exclusive traditions from the Protestant Reformation—(1) Puritanism and (2) Pietism—coexisted uneasily. Puritans (especially in Massachusetts) wanted a theocratic state—literally, a state run by God (as represented, of course, by the political elite). For the Pietists and others who believed religion to be a private matter, the goal of religious practice was not an external social organization but internal religious experience.

Boston was essentially a theocracy at the time of its founding, in which only church members could vote; Quakers were executed for religious dissent in 1650. In nearby Salem, witches were tried, found guilty, and hanged based on such criteria as not being able to recite the Lord's Prayer in the courtroom without making a mistake (see Erikson, 1966). In neighboring Rhode Island, however, Roger Williams insisted on no political involvement for religious leaders because he believed that the state corrupts the church. In most of the southern colonies, the Church of England was the official church.

In the middle colonies, the idea of combining the religious and secular—the norm for most, if not all, of human history—was questioned. In Pennsylvania, William Penn, a religious Quaker who understood the necessity of religious tolerance, argued against the idea of an established church. This was a radical idea at the time because it raised the fundamental question of basing the state's political legitimacy on religion, a practice common to most societies. Because of competing belief systems, the idea of disestablishing religion was in alignment with the interests of many of the colonial elites. The Great Awakening, a period of strong religious revivalism in the New World in the 1730s and 1740s, challenged established religion by emphasizing individual piety.

Massachusetts still collected taxes to pay the clergy, but by 1750, people could designate which clergy would receive their taxes.

During the American Revolution, the theme of liberty became an organizing principle of political culture in the colonies. It originally meant freedom from the English Crown, but minority faiths like the Baptists and Catholics also applied it to their freedom from colonial government. Since their enemy's faith—the Church of England, with the king as its titular head—was dominant in the southern colonies, independence forced a reconsideration of the question of an official state religion. In 1784, a group of evangelical ministers in Virginia gained widespread support to obtain state funds for religious instruction, but James Madison, Thomas Jefferson, and others who feared that this would lead to a reestablishment of the Anglican Church opposed them. In that struggle, Jefferson articulated the principle of religious freedom in a way that framed later debates about religion in U.S. political culture by attempting to create a separation between church and state.

At the constitutional convention in Philadelphia in 1787, representatives artfully avoided discussion of religion because they knew it could jeopardize efforts to create a union. It was not that religion was unimportant but that the Christian denominational pluralism of the colonies prevented any consensus over which single sacred canopy could be erected. Constitutional protection for religion seemed like the most pragmatic solution to a seemingly intractable conflict. Accordingly, religion was not mentioned in the U.S. Constitution, except in article 6, which read the following: "No religious test shall ever be required as a Qualification to any Office on public Trust under the United States" ("Constitution of the United States," 1787/1938, p. 192). The First Amendment, however, tackled the issue head-on, insisting, "Congress shall make no law respecting an establishment of religion, or prohibiting the free exercise thereof." At the time, three states still had an established religion. Religion did not disappear from public political life in the United States after this clear rejection of established religion, however; it persisted as a civil religion stripped of its most sectarian elements, as we will see in Chapter 7.

The close-knit alliance between the Christian Church and secular elites began to unravel in the 19th century, and in many places the tradition became a carrier of resistance movements from Third World **liberation theology** to East European pro-democracy movements and North American traditionalist ("fundamentalist") revolts against modernity. Christianity has also served from time to time as a focal point of conflict between Western and Middle Eastern powers. As the United States emerged as a world power in the 20th century, its economic, political, and military invasions of the world were often accompanied by the church, although perhaps more subtly than in the colonial period.

The process by which religious freedom is protected by law is a significant, though not always entirely successful, experiment. Clearly, citizens of the global village have to choose some form of religious pluralism in which freedom of worship is vigorously protected by the legal structure or risk subjection to authoritarian efforts to impose a single worldview and ethos on the world's 5 billion inhabitants.

Of particular importance in recent decades is the rise of the Christian Right in American politics, an issue I will postpone until a discussion of religious fundamentalism or traditionalism in Chapter 6.

Islam and Politics

Like most NRMs, Islam was born out of conflict. It began as a reform movement by the Prophet Muhammad, who, concerned about the injustice, materialism, and lax moral standards of his society, set out to do nothing less than reform the world based on new revelations from God. Not surprisingly, he encountered much resistance. The idea of the jihad, or struggle, remains a central element of Islam—struggle within each individual and between righteous and evil forces in the larger world. Boundaries between the Muslim and non-Muslim world perpetuate the struggle today, as do the stereotypes propagated by the media, especially in the West (see Said, 1978).

The way in which people within a religious tradition relate to nonbelievers in its formative years prefigures later patterns; a distinctive style emerges that persists over long periods. Today's patterns of Muslim–non-Muslim conflict—and to some extent, intensive struggles within Islam itself—must be seen within the hostile political context of early Islam, the medieval Crusades, and the colonial experience of the early modern period.

As a prophetic movement promoting justice and resisting the status quo order of its time, it is no wonder that the community that formed around the Prophet faced persecution and was even forced into exile in Medina and the victims of military attacks by the ruling elites of Mecca whose power they threatened. The Crusades, beginning in the 11th century CE, involved an effort by the religious and secular rulers of Christian Western Europe to expel the Muslims who ruled the Middle East. It may now seem part of the distant past, but this historical episode may be even more important to Muslims now than it was at the time the Crusades took place.[8] Those invasions of Muslim territory by the Roman Catholic

Church and its Western European patrons represent a formative event in the centuries-long conflict between Christendom and Islam.

Although intense conflict between Muslims and Christians threatened much of the Islamic hegemony of the Mediterranean regions during the period of the Crusades, Muslims managed to retain (or in some places regain, after losses in the initial Crusades) control over much of the region until the 15th century. This sociopolitical hegemony was significant not only to Islam but also to the European world that later reverted to Christian control. Indeed, much of the great heritage of the West was preserved by the Muslims, including works of Aristotle and classic Greek philosophy and literature that helped to stimulate the Renaissance in Western Europe in the wake of the Crusades. The cosmopolitan centers of learning in Muslim Spain attracted Jewish and Christian scholars, influencing the thinking of Moses Maimonides and Thomas Aquinas. Both the Islamic networks and the ongoing conflict in Christendom led to expanded economic trade and diffusion that prefaced the early modern period.

Although Muslim rule ended in Europe by the 17th and 18th centuries, much of Africa and Asia remained Islamic, and the Muslim influence persists there today. The greatest impact was made during the time of the Ottoman Empire based in Turkey, which ruled Central Asia, most of the Middle East, and parts of the Mediterranean and Europe during the 16th and 17th centuries.

Another Islamic empire was established in the Asian subcontinent, where the Moguls conquered the indigenous Hindus; other small Muslim states were established in Indonesia, Malaysia, Western China, and Africa. This strong political rule of broad territories showed substantial religious tolerance in many instances, although it sometimes destroyed indigenous religious institutions and came into a number of intense conflicts. The Mogul Empire finally succumbed to the colonial powers of the early modern period after the voyages of Columbus to the Americas and Vasco de Gama around Southern Africa to India at the end of the 15th century led to the colonization of large portions of the world by the Christian Europeans. When some of the colonial entrepreneurs raided West Africa for slaves, they captured many Muslims and took them to the New World, where Islam was silent under the yoke of slavery until the emergence of the Black Muslim movement in the mid-20th century.

European colonial expansion renewed resentments lingering in the collective memory from the Middle Ages and resulted in the defeat of Islamic nations by foreigners. Of 42 Muslim countries, only 4 did not experience direct military control by outsiders during the colonial period (from the

19th century), a humiliating experience for many Muslims in which Islam was derided and Islamic political and social structures were attacked by Western conquerors. Although it may seem like ancient history to some readers, some leaders of the fight against colonial rule are still alive today—this helps to explain some of the tensions between Islam and the West and the ferocity with which some leaders like Muammar Gaddafi attempt to hold on to power. Despite Gaddafi's atrocities, he was still a hero to many even when he was challenged by a popular movement in 2011, because he was the leader who stood up to the colonial powers.

When colonial domination began to unravel after World War II, super-power interventions in Muslim countries followed. The United States took on Israel as a client state in the Middle East and supported that country in its defeat of the Arab armies in 1967, in the devastation of Lebanon in 1982, and in the occupation of the West Bank. The Soviet Union, although less adventurous in the Middle East, was nonetheless a partisan in several conflicts there and actively suppressed Islam in the Central Asian republics. Since the 8th century, Islamic peoples have been subjected to what the Ayatollah Khomeini (in prerevolutionary Iran) called "West-toxification" or "Westomania," which many believed poisoned Iranian culture under the shah, Khomeini's predecessor who modernized the country's economy and society. "The goal of the Islamic Revolution in Iran," Mark Juergensmeyer (1993) claimed, "was not only to free Iranians politically from the shah but also to liberate them conceptually from Western ways of thinking" (p. 19).

Finally, even the sanctity of the Islamic home was invaded by the West in the form of youth culture and the women's movement. This may prove the ultimate insult, especially for some Islamic men, who have seen their way of life degraded and stereotyped, their political system conquered and controlled by Western colonialists, and their religion mocked by "infidels." It is one thing to have outside powers attack one's state but quite another to have them invade one's home. This new invasion may be the source of much of the anger expressed toward the West by so many Muslims today. The mainstream of most Muslim societies has reinterpreted many of the strict regulations of traditional Islam, relaxed its segregation of women, and adapted to some of the vicissitudes of the modern world, but these developments have simply fueled the flames of those dissatisfied with the direction in which the modern world has pushed them. Many Muslims believe that the authority structure of the family is under siege and that the modesty of women and children is eroding, creating an existential crisis.

Although it is not one of the Five Pillars, the Islamic notion of the jihad is clearly a core belief that emphasizes the sharp boundaries between the sacred and the profane. Faithfulness to God does not come easily because of the sinful nature of the world and one's own sinful nature, so one must always engage in struggle. Two types of jihad are outlined in traditional Islam: first, the **greater jihad**, denoting the internal struggle against one's sinful nature. The Greater Jihad rests on the Five Pillars and the rituals of the faith to remind oneself repeatedly of what is righteous and to protect oneself from the temptations of the world. The **lesser jihad** is the external struggle for which the term is now so famous.

One Islamic view—taking its cue from ancient Judaism—divides the world into two parts: (1) the "House of Islam," where Muslim law and faith prevail, and (2) the "House of Unbelief," or "House of War." Because it is blasphemous for nonbelievers to rule over the House of Islam, it is therefore a religious duty to overthrow non-Muslim powers if they try to control and persecute Muslims. The Qur'an puts it this way:

> Fight against those who fight against you along God's way, yet do not initiate hostilities; God does not love aggressors. Kill them wherever you may catch them, and expel them from anywhere they may have expelled you. Sedition is more serious than killing! Yet, do not fight them at the Hallowed Mosque unless they fight you there. If they should fight you, then fight them back; such is the reward for disbelievers. However, if they stop, God will be Forgiving, Merciful. Fight them until there is no more subversion and [all] religion belongs to God. If they stop, let there be no [more] hostility except towards wrongdoers. (2:190–193)

The intensity of Islamic thought reflects the hostile environment in which it was formed and the single-minded intensity with which Muhammad and the early Muslims struggled to establish their religious community in the midst of strong opposing forces. Still, there is room within most interpretations of Islamic doctrine for religious pluralism as long as the ummah can thrive without persecution and the mainstream Muslim leadership has explicitly denounced terrorism as contrary to the Qur'an and the teachings of the Prophet (see Chapter 8). I will look in more detail in Chapter 6 at the Islamic revival of recent years as a protest against the control and invasion of the West. This movement is sometimes called Islamic fundamentalism, although many Muslims object to the use of that term; I will refer to it instead as **Islamic traditionalism**. It is a cry of resentment against the negative stereotypes of Muslims perpetrated by Western media and the development of secular nationalism as a mode of political organization in the West that many Muslims feel is being imposed on Islamic nations in much the same way as colonial rule was in the 18th and 19th centuries.

The Islamic revival is an effort to reassert a positive individual and collective identity, to declare the beauty of the Islamic faith in the face of non-Muslim invasions.

The Ethic of Love

In a remarkable event in January 2002, Pope John Paul II met with 200 world spiritual leaders in Assisi, Italy, the birthplace of St. Francis, where they pledged not to use religion to promote violence in the world. They prayed and worshipped together in an unprecedented act of respect for each other's faiths and commitment to a peaceful world.

In spite of the remarkable frequency with which our news reports are filled with stories of religiously inspired or excused violence, religious institutions also remain the one major space in the world where one can talk about loving others without being embarrassed. Indeed, the world's scriptures are quite consistent in promoting love not only of one's neighbor but even of one's enemies (see http://origin.org/ucs/ws/theme144.cfm; Wilson, 1991). According to tradition, the Buddha (*Dhammapada*, n.d.) declared the following:

> "He abused me, he beat me, he defeated me, he robbed me!" In those who harbor such thoughts hatred is not appeased.
>
> "He abused me, he beat me, he defeated me, he robbed me!" In those who do not harbor such thoughts hatred is appeased.
>
> Hatreds never cease through hatred in this world; through love alone they cease. This is an eternal law. (pp. 3–5)

Similarly, Jesus tells his followers this:

> You have heard that it was said, "You shall love your neighbor and hate your enemy." But I say to you, Love your enemies and pray for those who persecute you just as God sends rain on both the just and the unjust. (Matthew 5:43–45)

The scriptures of Taoism, Confucianism, Sikhism, Judaism, and Islam have similar passages. Although saying or even believing in such advice may be easier than following it, the cultural space provided by such ideas is rare in contemporary political thought and the faith traditions are bound by their own canons at least to consider the possibility. It is a consequence of such teachings that the Vietnamese Buddhist monk Thich Nhat Hanh was able to respond to a question regarding what he would say to Osama bin Laden if he met him face-to-face by saying, "I would listen."

The ethic of love provides a countervoice to the pursuit of profit as the foundation principle in the economic sphere and the protection of national interests as the basic ethic of the political realm. It may not always be heeded in the process of global decision making, but it is a force nurtured by the faith traditions.

The Ethos of the Global Village

Justice is what love looks like in public.

—Cornell West

People sharing the same geographical area must develop some agreement on rules of engagement that permit coexistence and promote justice, unless one group is going to commit genocide against the others, even if the norm is to let various cultures exist in relative isolation and autonomy. The more groups are drawn together, through either conflict or cooperation, the more comprehensive must be the norms regulating cross-cultural interactions.

At the beginning of the 21st century, the pull of economic benefits from participation in a global economy (at least for elites in various cultures and subcultures) is combined with the availability of technological means for increased interaction, thus dramatically altering the extent to which everyone's fate is shared throughout the world. As we have seen, the global village is governed not by a unifying ethos or a single sacred canopy but by a marketplace of competing worldviews and corresponding ethical systems that are colliding in many spheres of life in various regions of the world.

We have much to learn about how the process of cross-cultural ethos construction unfolds; we scarcely understand the nature of the religious marketplace in the much-studied American scene, despite important work in that area by a number of scholars (see Finke & Stark, 1988, 1992; Iannaccone, 1991; Sherkat & Wilson, 1995; Warner, 1993; among others). As Sherkat and Wilson (1995) have noted, religious preferences, like other consumption choices, are developed "not only on the basis of what we want, but how others will be affected by our choices, and how others may react to them" (p. 26). Individuals making a decision to switch Christian denominations, for example, will take into account the potential response not only of members of new groups that they might join, but also of people within their current religious communities, their parents, friends, and so forth.

If there is any truth to Marx's contention that the dominant ideas of any age are those of the ruling class, then we can expect the ethos of the postmodern world order to be disproportionately influenced by the tone of international capital, the owners and managers of multinational corporations. Certainly, the value of material gain promoted by capitalist modes of production has already pervaded much of the world's cultures—not so much because it is imposed from the top (although mass advertising certainly influences popular tastes) but because the appeal of consumer goods and mass-produced entertainment has its own internal logic.

The role of religious traditions in forming an ethos for the human community is an empirical question not easily tested. At this point, we may be able to do little more than identify the questions that must be asked. All of the world's major traditions have a basic norm of compassion for others that sets moral boundaries around behavior that is harmful to others, especially within one's own society. The way in which those ethical teachings have been interpreted and enforced have varied so widely, however, that they provide no secure basis for our common security. Indeed, religious communities in many places around the world are now, just as they have been throughout human history, using their ethical standards to justify widespread slaughter and exploitation, and half of the world's population remains malnourished. The fact that religious institutions have failed to raise global consciousness about the scandal of child malnutrition reaching Holocaust proportions is not an encouraging sign. UNICEF (2010) estimated that more than 8 million children die annually, which exceeds the total 6 million Jews murdered by the Nazi regime in the Holocaust.

Because religion is so closely linked to social life, it is important to assess the implications for ethical systems of ongoing social transformations. Durkheim's (1893/1933) observation that the world system is becoming simultaneously more unified and diverse is instructive. As the world changes, so will the various ethical systems taught by the religious traditions; the ethical transformations may be as riddled with contradiction as the social: We are coming together, but resistance to the globalization of human life is as fierce as are the forces of unification.

The related human rights and democratic movements since World War II, coupled with the dismantling of the European colonial empires, offer some measure of hope in a world torn with strife. As democratic revolutions swept the globe and the Cold War ended in the last two decades of the 20th century, these developments seemed almost inevitable. Just as important, however, are the movements of resistance to globalization whose participants sometimes see the values of human rights and democratic politics as another attempt by the West to impose a secular nationalism on non-Western cultures still chafing from the colonial bonds.

Juergensmeyer claimed that a new Cold War is emerging between the "secular nationalism" of Christendom, which promotes Western models of democracy and human rights and non-Western religious communities. Significant sectors of the world, he argued, see in these latest trends an effort to thrust Islamic, Buddhist, and other global civilizations under the aegis of Christendom.

> From this vantage point, it is a serious error to suggest that Egypt or Iran should be thrust into a Western frame of reference. In this view of the world they are intrinsically part of Islamic, not Western, civilization, and it is an act of imperialism to think of them in any other way. (Juergensmeyer, 1993, p. 19)

The current ethical crisis of the global village is not a product simply of the late 20th century, however. It began in the ancient mists of prehistory, when the ethos of today's religious traditions were born in story form and continued as various localized traditions diffused, collided, and were transformed over the centuries. Although we are always in danger of overemphasizing the importance of our own historical epoch, as the result of the dramatic transformations of society and culture precipitated by the industrial and postindustrial revolutions of the past two centuries, human history seems to be at a critical juncture in our time. To those events, I now turn my attention.

6

Modernism
and Multiculturalism

A t the turn of the 20th century, Pope Pius X (1907) declared modernism to be the synthesis of all heresies. It "lays the axe to the root, not
the branch," he said as he excommunicated a number of scholars and set up
vigilance committees to report heretics to Rome. The Vatican severely suppressed biblical scholarship in the Catholic tradition until the Second
Vatican Council half a century later.

Of course, in one sense the pope was right. Modernism *is* a synthesis of all
heresies that goes to the root of faith traditions, challenging the very notion
of dogma. As reprehensible as we might find his suppression of scholarship,
he was correct about the profound shaking of the roots that modernism
brought to Catholicism and to religion in general as it entered onto the
human stage as part of the cultural and intellectual package accompanying
the birth of modern Europe. From the pope's position at the top of the church
hierarchy, his efforts to smash modernism, beginning with those priests under
his control, was a rational decision, even if we might see it differently. The
church would never be the same.

The deep and radical changes associated with the globalization of social life
are occurring with even more rapidity as we begin a new millennium. When
social organization changes, so does religious organization; new ideas, rituals,
and societies emerge through mutual, dialectical interaction. Societies have
always changed, especially when they encountered others, but never have the
scale and scope of cross-culture encounter been so widespread or intense.
When culture groups interact, the encounter changes each of them, even if they

are of unequal power. The impact of the agricultural revolution on human life unfolded over many centuries, but the industrial revolution immediately transformed humanity in profound ways. Although considerable continuity exists between the cultures created by the modern world's industrial, scientific, and democratic revolutions on the one hand and those of the early 21st century on the other, many now argue that we are living in a postmodern era that is undergoing another transformation as profound as any in human history.

In this chapter, I examine the two great cultural upheavals of the past two centuries: (1) the twin crises for religion of modernism and multiculturalism and (2) the diverse responses they have provoked among the world's cultures and religions. They constitute nothing less than what I like to call "cultural tectonics"—like the shifting of tectonic plates deep under the earth's surface that cause earthquakes, the deep cultural shifts of our time are shaking the very cultural ground on which we stand.

From Local to Cosmopolitan

Each of the world's major religions had its roots in a local primal religion, usually connected with a particular tribe or clan and a specific geographical location. Each tradition became more cosmopolitan as it diffused, encountering and incorporating other cultural forms along the way. Even today, most of the world's population never moves more than 50 miles away from their birthplace; the cultural changes at the founding of the world's religions usually involved courageous men and women stepping outside the boundaries of their comfortable lives and moving into new territory, geographically and spiritually.

These roots did not disappear as the tradition changed over time but established the form that influenced each religion's later shape. In a similar transformation process, each of the world religions has increased in its internal diversity and its **structural differentiation**. Finally, each religion has had to struggle with the two-horned dilemma of modernism: (1) the challenges of cultural pluralism and (2) scientific criticism. All these aspects of the local-cosmopolitan shift have had profound cultural and organizational consequences for each tradition. I will look now more closely at some of the specific transformations in religion that the phenomenon called "modernism" has brought about.

Internal Diversity and Structural Differentiation

The farther a tradition travels from its roots, the more diverse it becomes. Each major religious tradition incorporates a wide range of beliefs under a broad, abstract sacred canopy and thus becomes diverse in terms of beliefs, rituals, and institutions. The reason for this development is no mystery. The primal faith of each religion's origins was constructed along relatively homogeneous lines by a group of fairly like-minded people who lived close to nature in specific ecological conditions. As culture groups migrate, carrying their religious traditions with them, or as invading tribes conquer them or they defeat their neighbors, people reevaluate and reconstruct their worldview and corresponding ethos to incorporate different perspectives and adapt to new data. These groups borrow features of other traditions and then reformulate and strengthen their tradition in direct opposition to new challenges. Sometimes the culture changes dramatically as the result of a new technology or environmental condition. Among the Siksika in North America, for example, the introduction of the horse disrupted collective, egalitarian buffalo-hunting practices by allowing individuals to obtain their own buffalo independent of the group. A hierarchy developed favoring those who had horses, and buffalo-centered egalitarian religious rituals deteriorated.

The major religious traditions often adapted to new settings through syncretism or co-optation: Chinese folk Gods became Buddhas, and local African deities became Christian saints. Even Judaism, which remained an ethnic religion with strictly guarded boundaries, adapted to local conditions so that a Palestinian and a Babylonian Talmud were produced early in the Common Era, and contemporary Ashkenazi Jews differ from their Hasidic brothers and sisters. After centuries of adaptation, each religious tradition has become remarkably diverse; patterns of rituals and beliefs vary widely around the globe and sometimes even between congregations in the same neighborhood.

The process of structural differentiation is a relatively recent phenomenon that involves the creation of specific institutions to fulfill different functions; it can occur both within the religion and between religious and other spheres of life. As societies become larger and more complex, the division of labor increases (see Durkheim, 1893/1933) so that specialized institutions carry out specific functions. Functions once performed by religious organizations are now carried out elsewhere (e.g., in public school systems), and religious institutions have taken on a specialized role, concentrating more on private than on public life.

The structural differentiation of the social order itself is a striking characteristic of modern societies and has been the subject of much scholarly and political debate. Early sociologists made much of the difference between premodern and modern society, the "Great Transformation" from preindustrial to industrial society in Western Europe during the 19th century, when countless peasants were uprooted from their family and village and flocked to the cities to work in factories. Tönnies's (1887/1957) distinction between the *Gemeinschaft*, or "community," of preindustrial life and the *Gesellschaft*, or "society," of modern social organization implies a loss of communal feeling and family ties. Durkheim (1893/1933) distinguished between mechanical and organic solidarity, expounding a more optimistic view of this transformation process than did Tönnies. Elements of that vision were revived again after World War II as modernization theorists identified parallels between cultural changes precipitated by industrialization in the West and those occurring elsewhere. Western modernization included the **secularization** of society—that is, the removal of responsibility and authority in certain spheres from religious institutions and the concomitant privatization of religion, issues I will pick up again in a moment.

Critics of early modernization theory challenged the idea that industrializing non-Western societies have to follow the same path of modernization as the West (Wallerstein, 1984). Moreover, it turns out that the characterization of a vast gap between primitive and modern culture is not historically accurate (see Macfarlane, 1979), and too much has been made of "pervasive and insidious contrasts" between the old social order and the new (see Bendix, 1978; Shils, 1981; Wallerstein, 1984). Secularization has been a major issue in the sociology of religion, of course, but reflects more of a European experience than a general phenomenon; religious belief—and even participation in public religious rituals—persists in the modern world, even in the United States, despite its advanced economy and mostly secular government. As we shall see, a major reason for the secularization of Europe is the nature of the political transition and the conflict between the church on the one hand and democratic movements on the other.

The advent of the modern era profoundly affected religious life. All the world's religions have had to do battle with the collective social and intellectual giant called modernism. I will now look more closely at this phenomenon.

The Challenge of Modernism

Modernism is the emergence of a global, scientific–technological culture since the scientific, technological, and industrial revolutions that began with the Enlightenment in Western Europe. The perception among many European intellectuals in the 19th century that religion was dying was the result of two interrelated social movements: (1) the scientific and (2) democratic revolutions. Western science itself was parented by the Judeo-Christian tradition, which put a high premium on cognitive development and scholarship. The scientific movement took on a life of its own, however, and scientists soon came into direct conflict with church authorities because of their questions about particular claims made by the Vatican and other religious authorities.

This conflict began in the 17th century, when church authorities charged the famous scientist Galileo with **heresy** because he contended that the earth revolved around the sun and not vice versa. The Jesuit inquisitors forced Galileo to recant his "heretical" ideas (quoted in White, 1896–1897, Vol. 1:139). It was not just a specific doctrine that was at stake, of course, but the reputation of the Roman Catholic Church and the entire social order legitimated by Christianity. If the church could be wrong about the immovability of the earth, perhaps it made other, more significant mistakes as well. In Bertolt Brecht's (1966) dramatized version of the Inquisition, Father Inchofer predicted that if the peasants stopped believing in the sun's rotation around the earth, they would revolt. As the conflict between scientists and church authorities escalated, their positions polarized (as is usually the case in significant social conflict). It became increasingly difficult to be both a champion of scientific inquiry and a Christian; neither side would allow it.

The second challenge to the church occurred through a similar social process: As the democratic revolution emerged in the late 18th century (especially in France), its major opponents included not just the monarchy but also the church because of its alliance with Europe's political elites that dated from the 4th century. When the monarchy of France was challenged by the French Revolution, the crown and the church stood side by side to defend the *ancien régime*. Positions tended to solidify in 19th-century Europe: Either one was pro-democratic, in favor of the development of science, and anti-Christian or one was in favor of retaining the monarch and traditional Christianity and restricting the development of science.

Because a sacred canopy is woven out of a vast complex of interdependent parts, when one aspect of it is challenged, the validity of the entire system is called into question. Thus, when scientists began to doubt some dogmas of the Christian Church, such as the process of creation and the authorship of the Bible, they cast doubt on the validity of the entire Christian tradition. A bitter conflict ensued between scientists and church authorities that persisted for centuries in Western culture. Christianity, and especially the Roman Catholic Church, bore the brunt of the modernist crisis because of its monopolistic truth claims and political alliances with the ancien régime and also because modernism came first to the West as it pioneered the industrial revolution. Science was used as a weapon in the battle to wrest control of the social order from the church, so the battle lines were drawn sharply, distorting the nature of the conflict.

The Revolt Against Religion: The Crisis Begins

The philosophical turmoil beginning in 17th-century Europe and culminating in the 18th-century Enlightenment set the stage for the "warfare between science and religion" (White, 1896–1897). In addition to the new scientific empiricism that Galileo and others initiated, many Europeans exploring the world in the 16th and 17th centuries had observed common threads in all of the world's religions. Lord Herbert of Cherbury claimed, in his 1624 *De Veritate* (Herbert, 1624/1937), that all humans had a "natural religion," a position that came to be known as **deism**. Although it may not be inherently anti-Christian, deism was an early challenger of some of the church's doctrines and encouraged tolerance of non-Christian religious perspectives.

Advocates of science also came into direct conflict with church officials over the growing development of textual criticism—that is, the scientific study of texts, including scriptures. Scholars examined such questions as the authorship, historical development, and composition of biblical writings and concluded that many of the church's claims about these texts were untrue.[1] Thomas Hobbes (1651) questioned the traditional belief that Moses authored the Pentateuch (the first five books of the Bible). Baruch Spinoza (1670/1883) contended that theologians simply used the Bible for their own purposes, pointing out inconsistencies and historical problems in the biblical texts. Some scholars using critical methods were devout Christians, but the methods were appropriated by others who opposed the church and its hold over believers. One of these dissenters was Pierre Bayle (1697–1706), whose *Dictionnaire Historique et Critique* was widely read and was supported by such influential intellectuals of the 18th century as Rousseau, Voltaire, Montesquieu, David Hume, Benjamin Franklin, and Thomas Jefferson.

In the political realm, advocates of the 1789 French Revolution attempted to destroy the Catholic church not so much as a religion but as a power in society (Tocqueville, 1862/1945). As the battle lines were drawn, people were forced to choose between the church and the monarchy on the one hand or science and the republic on the other. It was almost impossible to hold to any middle ground; people often demonstrated their loyalty to either the king or the new democratic government by choosing to send their children to the traditional church schools or to the newly established secular schools run by the state.

The Problem of Historical Contingency

By the 19th century, new rifts emerged between theorists of the burgeoning social sciences and orthodox Christian belief. Auguste Comte (1853) contended that human thought developed through three historical stages: (1) the

theological or superstitious, (2) the metaphysical, and (3) the positive or scientific. His developmental positivism was provided as a scientific basis for
colonial expansionism and a general euphoria about progress and the future
of humanity. The controversy that Comte's framework generated exploded
with the theories of evolution that spread throughout Europe, especially in
Charles Darwin's *On the Origin of Species by Means of Natural Selection*
(1859/1952) and *The Descent of Man* (1871/1952). Darwin himself claimed
that there was "no good reason why the views given . . . should shock the
religious feelings of everyone" (1859/1952, p. 239; cf. 1871/1952, p. 593),
but many church leaders claimed that his theories contradicted the Genesis
creation account. In 1865, *The American Church Review* contended that if
Darwin's hypothesis was true, "then is the Bible an unbearable fiction . . . then
have Christians for nearly two thousand years been duped by a monstrous
lie. . . . Darwin requires us to disbelieve the authoritative word of the
Creator" (quoted in White, 1896–1897).

The principle of evolution raised another issue even more profound than
specific disputes over the authenticity of the Genesis cosmogony: Did religious doctrine also evolve historically? The major problem raised by the
"scientific study of the Bible," or historical criticism, is the issue of **historical
contingency.** When religious truth is considered an infallible revelation from
the deity, the idea of the historical nature of the tradition can precipitate a
crisis because the conclusion of most historiographic science is that each religious tradition has its own history, and the immutable, taken-for-granted
truths of the faith have in fact mutated considerably over time. One example
is the disputed authorship of the Pentateuch, traditionally attributed to
Moses himself. Study of the texts created problems in Jewish, Islamic, and
especially Christian circles when scholars discovered that the sacred scriptures changed over time and were written by different scribes; later versions
of the Hebrew scriptures were written with vowel marks whereas early versions did not have them.

Evolution and biblical criticism were used by the "anticlerical" (anti-
church) movement to undermine the church's authority, and the church
(notably the Vatican) responded with a scathing attack on both science and
democracy. When Pope Pius X condemned modernism in 1907 as a synthesis
of heresies, however, in a way he was right. Conflicts about orthodoxy and
heresy, in the Christian tradition, have historically concerned the truth or
falsity of particular dogmas. In the 19th and 20th centuries, the notion of
dogma itself came under attack and the word took on a negative connotation
(Kurtz, 1986). Not specific doctrines or religious ideas but the idea of absolute religious truth itself was attacked by modern scholarship and by social
reformers who saw the church as an enemy because it supported the monarchy and opposed democracy and supported theology and opposed science.

The Problem of Relativism

Although some religions are more exclusivist in their formulations than others, virtually all of them either assert or imply that their own version of the world is true, thereby rendering competing worldviews inferior. Although this position may obtain some legitimacy in isolated cultures, it obviously becomes problematic in a multicultural context. This brings us to the second major problem of modernism and a central theological issue for contemporary religions—the issue of **relativism**.

Religious conservatives criticize such modern perspectives as humanism and liberalism on the grounds that they erode absolutes, working to destroy the moral basis of a society. Certain ethical standards, the critics of relativism argue, are universally and unequivocally true and therefore absolute because they come from God, who is absolute. This debate is particularly difficult to analyze with any intellectual reliability because first and foremost it involves a framing issue—that is, if one believes that certain truths, values, or beliefs are absolutes, then any suggestion of relativism will be simply dismissed out of hand. If one refuses to take an absolutist stand, however, a paradox emerges, for the relativist by definition cannot know with certainty that absolutes do not exist.

Georg Simmel (1907/1978) provided a useful perspective on the problem that helps to bridge the gulf between these two seemingly irreconcilable positions. A relativist position, he noted, does not inherently deny the possibility of the existence of an absolute or absolutes; it merely insists that one cannot know the absolute absolutely but only from a specific point of view, unless one is actually God. Simmel's restatement of the problem suggests that any exclusivist truth claims (such as papal infallibility) might in fact be blasphemous, according to traditional definition, because only God knows the full truth about anything. Without addressing the issue of divine versus human perspective, a sociological analysis of religion can establish that absolutist or exclusivist truth claims are not entirely rare in human history. Moreover, because a wide range of human perspectives tends to represent the world in widely different ways, it is difficult to accept the premise that any single one has a complete corner on the truth.

The Revolt Against Modernism (The Counterrevolution)

When absolutism is threatened, people often respond with fanaticism. If one facet of a tradition seems at risk, people often believe that the entire system is. The conflict between establishment religion in the West and broad cultural movements of modernism in the 19th century provides an exemplary case of the scandal modernism created in many religious traditions. During the late 19th and early 20th centuries, the Vatican came down harshly on modernism, unintentionally fanning the flames of discontent but eventually putting them out by driving some of the dissidents out of the church and silencing many more (Lyng & Kurtz, 1985).

Roman Catholicism Versus Modernism

As its own authority in European society was attacked by waves of democratic and scientific revolutions, Rome launched a systematic campaign against modernism over a period of several decades. At the First Vatican Council, convened in 1870, the pope was declared infallible in matters of faith and morals for the first time in the history of the church. In 1899, Pope Leo XIII denounced "Americanism" as a heresy in an effort to criticize both modern science and certain forms of democracy as examples of the pernicious practice of adapting religious doctrine and practice to new social circumstances (Klein, 1951; Leo XIII, 1899).

The church's external enemies, attacking its hold on European society and politics, used scientific criticism as a weapon to undermine the legitimacy of the sacred canopy woven from Christian doctrine. Little could be done by the waning papacy against such formidable enemies, but inner enemies—heretics—were identified and soundly condemned. When Pope Pius X (1908) issued decrees condemning modernism, he set up a system of secret vigilance committees in every diocese to identify and report suspected heretics to Rome.

The modernist scholars denounced by the Vatican faced remarkable ambivalence in their dual roles as Christians (in some cases, clergy) and as scholars, because the culture wars tried to force them to choose sides. The findings of their research contradicted the pronouncements of the Vatican, and Rome exploited their vulnerable positions to mask its own vulnerability. The process of identifying and denouncing heretics, of institutionalizing a campaign to root them out of the church, served as a purification ritual for the beleaguered church hierarchy and set the tone for other "counterrevolutions" to follow.

The Counterrevolution Continues

Conservative Protestant and Catholic Christians in the United States today continue to struggle with many of the same issues the Vatican confronted in the 19th century. The evangelical and traditionalist movements of late 20th-century and early 21st-century American culture reflect efforts by a large sector of the Christian community to resist reliance upon scientific and secular thought to establish the moral boundaries of contemporary life.

The modernist crisis has also spread to other religious traditions, especially in what some claim is a "postmodern period," in which all tradition allegedly ceases to function (see Robertson, 1991, 1992a, 1992b). It was not long after Christianity's encounter with scientific empiricism that questions were raised about other religious doctrines. Did God dictate the entire Qur'an word by word to Muhammad? Did Krishna really serve as Arjuna's charioteer on some historical battlefield? Have the Vedas really always existed and were not authored by humans? Does Matsu really live in the statues at the temple, or are the figures just symbolic representations of her? Or, more fundamentally, does a Goddess of Mercy Matsu really exist, or is she just a projection of some social or psychological reality?

As with secularization, the extent to which modernism becomes a problem depends on the nature of the belief system and religious institutions as well as the conflict strategies chosen, especially by religious elites. Each religious tradition faces modernism at its most vulnerable spot. For Judaism and Islam, modernism raises questions about social boundaries and creates a cultural climate in which it is difficult to sustain the rituals that reinforce them. In the Christian churches, modernism attacks the integrity of the belief structure by challenging the tradition's claims about its scriptures and by presenting alternative perspectives that answer the same questions in different ways. Modernism may not offer as much of a challenge to the Eastern religions, which have traditionally co-opted, rather than opposed, new ideas. Nonetheless, the more mystical elements of Eastern thought are sometimes problematic in the modern scientific ethos. Both the internal spiritual claims of such disciplines as Yoga and the cosmic worldviews that posit cycles of the universe lasting hundreds of thousands of years appear too fantastic and unverifiable to many modern Hindus and Buddhists, despite the effort by some to articulate their claims with scientific rhetoric.

Moreover, the **instrumental rationality**—that is, the utilitarian calculation of means to an end—of the technocratic ethos that now pervades the global village frequently undermines the value rational (rational from a values perspective) and nonrational (e.g., affective) bases of action promoted by Hinduism and Buddhism as well as other religious traditions. The technocratic ethos also values efficiency and practicality above all other values, including justice and personal loyalty. Religious and technocratic values do not always collide, but they often do.

It is clear that a countertrend has also emerged, often not merely a retrogressive or temporary movement but a solid reaction against the dominant worldview that modernism represents. I will return to this issue in later chapters, but a few points must be made here. First, the rising traditionalist theologies in Islamic societies, North American Christianity, and Hindu nationalism are powerful movements that will not go away and represent a basic revolt against modernism that has persisted long after the first major battles between religion and modernity arose 200 years ago. The most hopeful prognosis is that these movements will help to mitigate some of the more damaging effects of modernism without destroying the possibility of rapprochement among the religious communities or between religious and secular forces.

Second, mainstream cultures have begun to learn the limitations of both science and tolerance. Science and technology have been humbled in recent decades, particularly by the ecological crises brought on by unbridled technological development in some sectors of the world and the destructive consequences of scientific and technological research in the development of weapons of mass destruction. Some of the old battles have died down, and new areas of agreement have developed between scientific and religious spheres.

Finally, antimodernist movements are now facilitated by modern technologies of mass communication and transportation. Now, millions of like-minded people who might have felt isolated in their opposition to the intrusion of modernism into their culture can join together across geographical boundaries, support and encourage one another in their critiques of modernism, and build networks and strategies for collective action.

Historical Outcomes of the Modernist Crisis

Modern culture contains two simultaneous, contradictory trends: (1) increasing unity and (2) increasing diversity (Durkheim, 1893/1933). The emerging global culture has some common denominators and centripetal forces that draw people together as well as profound countertrends that challenge the mainstream—notably, the revival of traditional cultures in many parts of the world, in part as a protest against the flood of foreign influences.

What have been the direct outcomes of the modernist challenge to traditional religion? Some of the consequences were as follows:

1. The substitution of religious traditions with rationalism, **scientism**, and individualism

2. The secularization of public life and the privatization of the religious, so that people from different faiths could share a common social life

3. The revitalization of traditional forms

4. The construction of quasi-religious forms that fulfill many of the social and psychological functions of traditional religions, such as civil religion and nationalism

5. The creation of new forms of religious belief and practice created through processes of syncretism

The first two of these consequences were a logical outgrowth of the Enlightenment worldview that precipitated the modernist crisis in the first place. The latter three represent attempts to reshape, rather than replace, the world's religious traditions. I will explore the first two briefly here and the last three when I explore the crisis of multiculturalism next.

Rationalism, Scientism, and Individualism

The first Western solution to the problems raised by modern science was advanced during the Enlightenment: Replace the illusions and arbitrary authority of traditional Christianity with a rational system of norms and values based on science. Most of the Enlightenment philosophers and their heirs thus emphasized a rationalism that insisted on reason as a source of knowledge superior to religious tradition. As we have seen, the social sciences took the lead in this effort; many were optimistic that a new scientific morality could be founded on either a scientific psychology and economics— Adam Smith and the British utilitarians—or a scientific sociology—Auguste Comte and the French positivists. The German idealists exposed the "illusions" of religious thought in the West, which would be replaced with the rationality of Kant, the socialist revolutions of Marx, and the psychotherapeutic treatment of Freud. In its most extreme forms, this approach resulted in the development of scientism, a quasi-religious belief in the scientific method.

The warfare between science and theology that characterized the 19th and early 20th centuries has subsided somewhat, because both sides have pulled back their troops and because bridges have been built between the two sides. As Ian Barbour (1960) has observed, the relationship between science and religion is now one not of content differences but of methods. Barbour noted a number of similarities and differences between the two methods that are most instructive. Both science and religion involve (1) the interaction of experience and interpretation, (2) the role of community and analogy (neither perspective makes sense to the outsider who is not familiar with the symbolic language of the community), (3) the primacy of relationships rather than objects (relationships among people and with their deities in religion and among elements of matter—such as probability waves in atomic structures—in science), and (4) the use of reason to test interpretations of the way in which reality is experienced.

According to Barbour (1960), however, some differences between the two methods remain significant, including (1) science focuses on means rather than ends; (2) science aims at knowledge of reproducible relations expressible in general laws, whereas religion emphasizes configurational understanding and the significance of a unique part in relation to the whole; and (3) science promotes objective detachment as opposed to the personal involvement valued by religion. But finally, Barbour (1960) concluded, science and religion are complementary forms of investigating reality: "Either science or religion alone affords a partial view," he claims. "We need to use various categories and frames of reference. The man who says, 'Love is not real because I cannot weigh it' is confusing two frames of reference" (p. 214). The same could be said of the multiple perspectives within and between scientific and religious traditions. Every scientific field, every religious tradition has its own fields and schools, its own perspectives, which point to reality as it is experienced by people from a given point in the social and natural world.

Various religious traditions have responded differently, of course, to the challenge of modern science, as have schools of thought within each tradition. The historical context and nature of ecclesiastical institutions, as well as flexibility of doctrinal orientations, also influence the extent to which traditions resist or accommodate modern science. Most religions, like Buddhism, have both those who wish to accommodate and those who resist, although the inclusiveness of the tradition favors an incorporation of scientific paradigms into the Buddhist worldview. A few Buddhists have found science inadequate; others have emphasized the similarities between the Buddha's search for truth and the methods of modern science. Buddhist advocates of science point to a passage in which the Buddha encourages testing of his teachings through one's own powers of reason (Kitagawa & Reynolds, 1976, p. 46). Dr. Luang Suriyabongse asserted that "the Buddha was the greatest discoverer and scientist of all time" (quoted in Kitagawa & Reynolds, 1976, p. 47). Other Buddhists were less sanguine about science. G. P. Malalasekera, the first president of the World Fellowship of Buddhists, asserted that the Buddha utilized science when appropriate but that the ultimate mysteries of his teachings went far beyond the knowledge that any purely scientific approach could offer (Kitagawa & Reynolds, 1976, p. 47).

A major feature of the Enlightenment was an individualism that maintains the independence of the human individual as the source of all values, rights, and duties. This claim was a response to the legitimation crisis of the modern West—the democratic legacy to which individualism gave birth is a hallmark of the Enlightenment. Granting the individual the freedom to make doctrinal and ethical claims independent of religious communities or institutions solves the crisis of faith by making religion a matter of personal choice, but it also creates new social problems. People are not required to rely on suspect institutions, but the dilemma of how to maintain collective life is unsolved if each individual is responsible only to himself or herself.

Secularization and Privatization

The second major consequence of the crisis of modernism is constituted by the dual processes of secularization and privatization in which religious institutions are differentiated from other spheres of life. The process of secularization was the hallmark of political and religious reform following the French Revolution, when the church, in alliance with the monarchy, had such a monopoly on the organization of public life that those who wished to discard the ancien régime felt that they had to eliminate the church's control as well. It is, at first glance, a logical solution to the problem because, like Jefferson's (1802) proposed "wall of separation between church and state," it allows a diverse people to build a collective life despite their religious differences. When people enter the marketplace, the statehouse, or the classroom, they can leave their religion at home or in their hearts.

In Europe, the passing of the old regime resulted in the secularization of many spheres of European society, notably in politics and education. As the officially recognized religion in most of Europe, the Christian Church had constructed the authorized ethos, socialized and educated the youth, and pronounced judgment on moral boundaries. During the French Revolution, advocates of democracy recognized the power of the Catholic schools in promoting monarchical ideas among young people and so made the creation of state schools a major priority (see Durkheim, 1961). The schools had also been used to promote Christianity; in the secular sphere, they would promote the principles of "liberty, equality, and fraternity" and the civic virtues required by a democratic society (Dansette, 1961; Kurtz, 1986, p. 30ff.). In country after country in Europe and the New World, the official relationship between the state and the church was terminated.

As a response to modernism, secularization is closely linked to the scientific ethos, which insists that it is possible to adopt an objective viewpoint that stands above cultural biases; it lies at the heart of the Enlightenment project that would substitute religion with scientific rationality. The problem with this effort, as I have noted, is that the all-encompassing nature of religious perspectives makes it difficult, if not impossible, to "check them at the door" when entering the public arena. A religious perspective by definition involves itself in the ethical standards that apply to all other spheres of life. For most people, historically, and in most cultures, religion is diffused throughout everyday life.

Most sociologists, and many other people living in "modern" society, have trouble understanding this because they compartmentalize religion, assigning it to a specific sphere and attending to it only on an occasional basis. Although attention to the sacred has always had a cyclical nature, commonly oriented around festivals and seasonal rites or daily and weekly times of prayer and worship, such differentiation is a fairly recent development in the world's history. The compartmentalization of religion, however, represents one solution to the problem of a multicultural social reality. Here, each of many religious perspectives takes its place in a cultural marketplace alongside secular worldviews and ethical standards. If everyone in our interaction space shared the same basic worldview and ethos, we could go about our daily lives taking religious elements for granted. When we encounter a plurality of worldviews on a regular basis in the public sphere, however, we often accommodate that reality by assigning the explicitly religious to certain times and sectors of our lives in the process known as privatization.

This process is enforced by social norms of modern Western urban life: When people constantly talk of God or their religious convictions, they are either religious professionals, for whom it is acceptable (although they are alienated somewhat from mainstream society), or religious "fanatics," some of whom are shunned; others are revered but are still set apart (like Mother Teresa). If we think that the copy machine will not work because it is infested with demons or the operator has bad karma, we will probably keep our opinions to ourselves. In everyday life, people in heterogeneous urban societies usually approach the sacred with some ambivalence. Those who are religious acknowledge that their faith should inform their entire life, but they are often hesitant to engage in religious talk in a multicultural setting because they get negative feedback. Religious convictions are thus considered *private* concerns that should be kept to oneself, like one's sexual fantasies.

Sociologists of religion have made much of the secularization phenomenon, in large part because it was a major event in Europe and many of the prominent figures in the field were European and tended to generalize from their own experience. Many of the perspectives of the founders of modern social science persisted outside of Europe. In less secularized societies, however, the religious traditions people bring with them continue to shape the ways in which they define the world, the meaning of life, and the nature of ethical behavior in everyday life. A majority of people on the planet live in a much more sacralized environment than do the people who write and read books on the sociology of religion. For billions of people, religious traditions are a natural part of everyday life (cf. Davie, 2010).

Because of the highly visible disestablishment of religion in the United States, along with the establishment of state schools as well as other spheres of secularization in society, many people assumed that U.S. culture had become secularized in an analogous fashion. As Warner (1993, p. 1049ff.) noted, the secularization paradigm for American religion proposed by Berger (1969) and Parsons (1960, 1967, 1969) was increasingly challenged by an **antisecularization thesis** growing primarily out of empirical research on American religion, which found it flourishing (see, e.g., Ammerman, 1987; Christiano, 1987; Ebaugh, 1991; Neitz, 1987, 1990). Indeed, the counterevidence was so strong that it appears that "the antisecularization thesis . . . has become the accepted wisdom" (Sharot, 1991, p. 271).

As Sherkat and Ellison (1999) observed, many of the controversies in the sociology of religion "are rooted in debates between scholars wedded to secularization theories and those who explain religious behaviors and trajectories through a different lens" (p. 364).

This antisecularization thesis contends that U.S. culture, far from secularizing, has become more religious than ever; religious participation not only persisted throughout U.S. history, Warner (1993) pointed out, but also actually increased dramatically well into the 20th century. At the time of the American Revolution in 1776, about 10% of the population were church members, compared with about 60% in the 1990s (Warner, 1993, p. 1049; cf. Caplow, Bahr, & Chadwick, 1983, pp. 28–29; Finke & Stark, 1986; Hertzberg, 1962, pp. 47–50). A major reason for the lack of secularization in American society may well be that the battles there have not been as vociferous as in Europe, where they were politicized in an ethos that preferred democracy over monarchy and where church officials promoted the retention of the monarchy and fought democratization in every sphere.

Sherkat and Ellison (1999), in examining the available data on religion in the United States, noted that—contrary to the secularization theories—religion continues to play an important role in American life in terms of "(a) the distribution of beliefs and commitments, (b) trends in beliefs and attachments, and (c) predictors of religiosity" (p. 365). A mere 2.2% of American respondents to the General Social Survey (GSS), according to Melton (2003), said that they do not believe in God, whereas more than 80% said they believe that the Bible is divinely inspired. Membership and participation in religious organizations is high compared with other countries, with 61% of Americans claiming to belong to more than 2,000 religious organizations, about half of them being Christian denominations (see Melton, 2003).

At the bottom of the secularization debate is what Murray (2004) called the "rationalizing impulse," which was also a central theme of Max Weber's work on the sociology of religion. The intellectual work of the modern world was the application of reason to every sphere of life, but it had serious unintended consequences. Murray (2004) put it this way:

> The paradox of Western civilization [is] the dialectic of the Enlightenment. The rationalizing impulse that led to the liberation of the modern subject from tyrannical faith in myth, superstition, and sovereign power, and their embodiment in the objective world, is according to Adorno and Horkheimer, also responsible for its negative by reducing it to the status of that objective, or natural world from which it was trying to liberate itself. (p. 15)

The very process of liberation from the tyranny of the ancien régime may lead to a new entrapment not only the "iron cage" of the modern socioeconomic order about which Weber wrote but also a myopic intellectual inquiry that provides only certain kinds of answers and neglects other forms of inquiry that science, narrowly defined, might not understand or have the capacity to address.

Although the modernist crisis certainly affected U.S. culture profoundly and the battles have not completely subsided, the working compromise establishing some boundaries between religious and political spheres instituted at the founding of the republic, as well as the forced multiculturalism of American culture over time, have mitigated the polarization between religious and scientific perspectives and among various religious traditions.

The Modernist Crisis and the Twenty-First Century

Not only is religion alive and well in the world's most advanced industrial society but it is thriving in many other areas of the world. Along with the creation of new religious forms, we are now witnessing some dramatic revitalizations of traditional forms of religious life. The growing interdependence of the various human cultures, along with the economic and social webs woven across thousands of former boundaries, is creating an unprecedented series of changes in the nature of human theology. On the one hand, the very notion of religious belief has been called into question by the secular nature of thought in industrial society. On the other hand, the idea of a tightly woven, nearly seamless sacred canopy has clearly become obsolete (if it ever truly existed) as people from various strands of religious thought encounter ideas from other traditions. It is virtually impossible for any believer in the world today to live in isolation.

Each religious tradition faces a similar dilemma, although it is more acute in those that are more exclusivist in their theology: How can they encounter the ideas of another faith expression, and indeed interact with people from that community on a regular basis, without losing the integrity of their own faith? The idea of religious traditions encountering one another is, of course, nothing new. The process has occurred again and again over the centuries, as we saw in our tour of the world's religions. The scale and significance of those encounters are new. Members of every major religious tradition now, in a sense, meet one another on the street and must decide whether to kill each other, pass by indifferently, or somehow engage one another.

One important consequence of these encounters, however, is the rediscovery of the rich diversity each tradition embraces. From within one's own small corner of a faith community, the canopy might appear relatively uniform and seamless. When we look closely at any sacred canopy, however, we discover that it is not uniform at all but a patchwork of contradictory ideas stitched together over the centuries. The great prophets and seers of the planet emerge from the profane order of human existence at times of crisis, when the canopy is ripped apart by wars and invasions, social or economic ferment, and natural disaster. The power of the prophets and the Gods comes from being able to restore the canopy so that it can once again be taken for granted. Recognizing the affinities between their interests and the ideas of a particular religious perspective—or even an antireligious belief system—various social strata and classes struggling for a position in a new social order also attempt to seize that power.

The Western Enlightenment worldview, which in essence called for the substitution of religion with modern science, became in effect another competing sacred canopy with its own arbitrariness, contradictions, and truth claims that legitimated a new system of social exploitation. Against those secularizing worldviews and the global military–bureaucratic structures they legitimated arose a series of protest theologies, sometimes from the Left and occasionally from the Right, but almost always from the bottom up, reaffirming sacred frames for explaining and ordering the world.

Not all responses to the modernist crisis involved either efforts to substitute religion with science or simply to have religion retreat into private spheres as public life became secularized. Other efforts to cope with modern and postmodern life involved the revitalization of traditional forms, the creation of new religious movements (NRMs), and the formation of quasi-religious systems such as civil religion and nationalism. Let us look now at these permutations of religion in the wake of the challenge that followed on the heels of the crisis of modernism: the crisis of multiculturalism and competing religious beliefs, rituals, and institutions.

The Crisis of Multiculturalism

As modern cultures emerged in the West around the time of the 18th-century Enlightenment, profound changes occurred in the religious sphere of life as in all other spheres. Religious impulses pushed toward an all-embracing sacred canopy that pervaded everyday life and linked it with the broadest theories of the cosmos. The pluralism of the modern world, however, has created a contradictory impulse that limits religious spheres of influence as multiple traditions conflict with one another. New forms of religious life have taken shape out of competing tendencies of the world's religions to provide a sacred canopy for all elements of life on the one hand and resistance from alternative traditions to the monopoly of any such system on the other hand. The tension between opposing forces precipitated the privatization of religion and a differentiation of spheres of life in the modern world. Even as people's religious traditions continue to guide their private ethics, a wall of separation has emerged between religion and politics, religion and economics, faith and education.

In a structurally differentiated society, every institution has a specialized task; the task of religious institutions is to tend to spiritual and ethical issues. Religion intrudes on all other spheres, however, because its ethics generally apply to all areas of life. A modern society compartmentalizes institutions, but we cannot compartmentalize people.[2] The cultural life of the global village is a product of an interaction of religious traditions among themselves as well as between each tradition and the multitude of others within which they came into contact as the global society emerged. In itself, this is not a new phenomenon; each major cultural tradition is, after all, a product of multicultural encounters. In both form and content, moreover, our late 20th-century multicultural crisis is similar to the modernist crisis a century ago. What *is* new is the scale and scope of the process.

When cultures collide, the sacred canopy of each tradition competes in a cultural marketplace and is thus open to the scrutiny of potential consumers. When new Gods arrive on the scene, we often react in much the same way as King Kadmeus did when Dionysius was said to have appeared in Thebes in ancient Greece—we pull out our swords for battle. Multiculturalism creates a crisis but also an opportunity. We can learn from the successes and failures of the past and, with a sociological imagination, construct new religious traditions in the next centuries. It is not a question of *whether* we will do it, but *how*, and *what kinds* of new traditions will emerge. The construction of new social and cultural forms is never a simple process, however, and the crisis of multiculturalism, following in the wake of modernism, has precipitated culture wars around the globe.

Besides the responses to modernism already discussed—the substitution of religion with alternative ideologies and the secularization of public life— efforts to deal with these crises helped to forge a transformation of religion in the following outcomes:

1. Antimodernist movements such as the protest theologies of Christian and Islamic traditionalism (fundamentalism)

2. Liberation theologies from Latin America and the women's movement

3. New religious and quasi-religious forms, such as individualism and consumerism, civil religion, and nationalism

4. Religious syncretism, the development of religious movements that bring together elements of various religious traditions in a new (and often controversial) manner

I will examine each of these developments briefly in order to understand contemporary trends and future possibilities for religious life in the global village.

Culture Wars and Protest Theologies

Multiculturalism (the product of sustained encounters among the various religious traditions of the world) along with modernism (the critical force of modern secular thought) fuel ongoing culture wars in virtually every corner of the globe.[3] These wars are, in part, a result of what Jürgen Habermas (1975) called a "legitimation crisis."[4] Many religious movements struggle against the hegemony of modern cultural centers and the invasion of materialistic, relativistic, and hedonistic culture from these centers. These movements, based upon traditional indigenous cultures, have a variety of political agendas, often with diametrically opposed implications for reorganizing society in the global village. From the liberation theologians of Latin America and Africa to the Islamic and Christian traditionalists ("fundamentalists") of the Middle East and North America, however, all share a common characteristic. Religious frames empower the participants in their struggles against the oppressive structures of what Weber (1904/1958) called the "iron cage" of the modern socioeconomic order and what Simmel (1908/1971) called the growth of "objective culture." In the contemporary world, the family and religion are viewed as enclaves of "communicatively structured interaction"—that is, interaction is based on who people "are" rather than on what they are "worth," in some market sense. When even that terrain is invaded, people resist (see Habermas, 1987, p. 393). Fields (1991) contended that this social context helps explain the current rise of traditionalism in the United States; it can no doubt apply to other traditionalisms as well. "After years of withdrawal," Fields (1991) argued, traditionalist groups are now resisting:

> Thus, the major thrusts of the ideology of activist fundamentalism . . . involve a reduction of state intrusion into the economy and the family coupled with state promotion of religious doctrine as the basis of law. While seemingly contradictory, institutionalization of this ideology would produce a shift in the relationship between subsystems and the lifeworld, while political discourse would become more "substantive" than "formal" or "technical." (p. 185)

This two-pronged effort to eject the modern world from private life and at the same time to transform public life so that the gap between private values and the cultural ethos is less striking lies at the core of traditionalism. This dual development appears in various conservative subforms around the world, as well as in the emergence of liberation theology among the poor, especially in Latin America.

Cultural cleavages in the global village sometimes fall along the lines of the religious traditions. In recent decades, for example, movements in the Islamic world used centuries-old themes of Muslim–non-Muslim divisions as a vehicle for expressing their discontent about the invasions of their lives by Christians from the West, Marxists from the Soviet Union, Hindus from Delhi, and the like. Protestant–Catholic cleavages in Northern Ireland reflect economic and political divisions that parallel deep religious differences. In many instances, however, new lines of conflict cut across old ones, reflecting competing impulses toward orthodoxy and progressivism or modernism (see Hunter, 1991). These conflicts involve the question of how to set standards in all social spheres—the family, law, art, education, and politics. They concern, as Hunter (1991) put it, "allegiances to different formulations and sources of moral authority" and "how we are to order our lives together" (pp. 118, 50).

These culture wars are sparked by the interaction of diverse worldviews in the global village. Widespread dissatisfaction with the way the world is ordered and the ways in which it is changing fuels the conflicts. Moreover, interpersonal, intertribal, interethnic, interclass, interregional, and international conflicts are often framed in religious terms, intensifying them and giving them a significance to participants that transcends the mundane struggle for survival, wealth, and power. These battles, framed as religious conflicts, take on larger-than-life proportions as the struggle of good against evil.

First, the cultural dimension of the crisis involves the assault on religious traditions by the interaction of various religious orientations and scientific critiques, undermining each system's legitimation. Second, advanced capitalism is in a state of crisis, precipitated, Habermas and others argued, by the inability of political and economic systems to meet expectations about delivering material comforts and economic stability. Finally, these broader crises result in what Habermas (1984) called the "colonization of the lifeworld"— that is, the welfare state and the public realm in general have invaded the private sphere of lived experience, including the family, which is now subjected to the imperatives of advanced capitalism—"consumerism and possessive individualism, motives of performance and competition" that shape behavior. Everyday life is thus squeezed into a "purposive-rational action orientation [that] calls for the reaction of hedonism freed from the pressures of rationality" (Habermas, 1987, p. 325). Habermas was writing primarily of the world of advanced capitalism, but elements of his analysis are applicable to other parts of the world as well. The lifeworld of the poor in the Third World, for example, has been colonized by the advanced capitalism emanating from the West in such a radical manner that many have organized with the only cultural tools they had available—that is, those from their Christian tradition.

People often perceive the culture wars of the modern world as strictly moral conflicts, but they also have a basis in social organization. Orthodoxy thrives in sociologically simple (small, homogeneous) communities and modernism in complex ones (large, heterogeneous societies). In a small, rural village, where everyone comes from the same ethnic group and class differences are minimal, it is feasible to have rigid moral boundaries and a sacred canopy that unifies the ethos of the entire society. In the global village, however, competing claims of diverse groups with radically different ethnic, historical, and class backgrounds render consensus on moral codes virtually impossible.

Sociological insights into religion center on the proposition that religious and social systems are intimately connected. Consequently, they change in a dialectical fashion, influencing and shaping one another. The dramatic transformation of the social world in the 20th century is both shaped by and in turn produces remarkable changes in the nature of religious life on the planet.

As the world system emerges, many indigenous traditions are either incorporated into broader religious traditions or eliminated altogether, radically changing the global map of religious ecology. Colonialism and modernization destroyed so many local expressions of religious life that it appeared to some that few religious traditions would remain. Religious life has proven remarkably resilient, however, and we would be mistaken to assume that the shape of religion on the planet will be determined only by the central traditions. Moreover, many local variations of the global traditions incorporated many of the indigenous beliefs and practices.

The formation of NRMs has become widespread, some in the form of movements that syncretize a range of older forms, just as intense intercultural conflict created the major religions themselves in earlier times. Other movements, such as the communist and socialist systems of the 20th century, claim to be nonreligious but take on many of the characteristics of religions. They attempt to provide a worldview and an ethos, as well as a general sacred canopy to guide both the ideology and the structure of life, from the personal to the public. In the Soviet Union, the state created a relatively rich, though often cynically practiced, ritual life apparently intended to replace traditional religious practices. When I visited the Soviet Union in 1988, I was surprised to see highly decorated evergreen trees in public places. When I asked about them, one of our hosts explained that they were New Year's trees and that Father Frost, a large man with a red suit and white beard, would put presents under the tree for the children. Similarly, a humanist seder in Austin, Texas, had all the elements of a traditional Jewish Passover except that there was no deity.

Throughout the modern world, with its nation-state system, various forms of nationalism offer a surrogate religion in which identity is forged not on a religious anvil but is linked to citizenship in a nation. This development has given us the democratic political institutions so widely cherished in modern life but also nationalistic wars and new forms of ethnocentrism. A similar ambivalence pervades the individualism of contemporary Western culture. It has taken on a quasi-religious form as well, leading to both a highly ethical humanism that stresses justice, community service, and civic responsibility and, at the same time, a narcissistic hedonism in which the lofty values of individual freedom and self-actualization are translated into patterns of consumption of mass-produced goods.

Revitalizing the Traditions

The root of the contemporary culture wars is a deep-seated discontent with both traditional and modern culture, which invades virtually every corner of the global village and undermines traditional worldviews and the ethos of indigenous cultures. Instead of showing respect for elders and attending religious ceremonies, many young people from East Asia to Latin America now play brash American rock music, watch Western movies, and drive motorcycles. Instead of following the strict moral codes of their traditions, some young women of the Arabic countries are discarding the veil and demanding to work with men outside the home. These cultural movements lead to conflicts at every level of society, from the intergenerational struggles about conflicting norms and values within families to regional and international conflicts about religious, cultural, and economic issues.

Most individuals experience an ambivalence toward modern culture that pervades the global village, resulting in contradictory behaviors as people express first one and then another aspect of their love–hate relationship with the new world. Young people often love their parents and their village but are also drawn toward the lifestyles seen in Western movies. This ambivalence helps explain the penchant for many influences from rock music to postmodernist theory through which one can simultaneously emulate and attack the cultural center of the global village.

In their landmark five-volume exploration, Martin E. Marty and R. Scott Appleby (1995) concluded the following:

> Radical Shi'ite Muslims in Lebanon and Iran, militant Sikhs in Punjab, Jewish extremists on the West Bank, Hindu nationalists at Ayodhya, and Christian cultural warriors in the United States—despite being worlds apart from one another geographically, historically, and in the specific content of their beliefs and practices—were establishing "progressive," world-creating and world-conquering movements that looked to the past for inspiration rather than for a blueprint. (p. 2)

Although much has changed since that important collaborative assessment, especially since September 11 and the subsequent expansion of terrorism and the so-called war against it since that time, it is still instructive to look at the project's effort to summarize a very complex phenomenon. Gabriel A. Almond, Emmanuel Sivan, and R. Scott Appleby (1995) extracted three general findings from the project: first, that "religious fundamentalist movements are distinct from other religious movements in that they are inherently interactive, reactive, and oppositional—that is, they are inexorably drawn to some form of antagonistic engagement with the world outside the enclave" (p. 503). They are, therefore, inevitably political and look to the future rather than to the past and therefore "cannot resist being caught up in modern bureaucratic and institutional dynamics—the dynamics of change" (Almond et al., p. 504) despite their efforts to revitalize tradition.

Second, although fundamentalists movements are political, they are also "genuinely religious," which sets them apart from other protest movements. As a consequence, "Militance, coalition building, 'diplomacy'—all of the 'ordinary' pursuits of minority political movements—take on unique rhythms and patterns in fundamentalisms, due to their religious character" (Almond et al., p. 504). Finally, "while all fundamentalisms tend to be hegemonic, their world-conquering impulse is modified in practice in a variety of ways" (Almond et al., p. 504). They are constrained by the long-term trends of their host societies, their host religions, and even their antitraditional character in a way that leads to a fascinating sociological paradox. Their "willingness to manipulate the religious tradition and introduce innovation for political rather than strictly spiritual purposes" actually "delegitimates the so-called true believers in the eyes of many other believers" (Almond et al., p. 504).

David Zeidan's (2003) comparative study of themes in Christian and Islamic fundamentalist discourses finds "family resemblances" in the two kinds of movements, despite significant differences as well. He concluded that rather than being the "notorious 'clash of civilizations'" or a political confrontation "between West and East, or North and South," these ongoing conflicts are a "*kulturkampf*," a competition between two cultural tendencies *within* each civilization, one based on religious values, the other on secularism and materialism" (Zeidan, 2003, p. 280).

> For fundamentalisms of both persuasions the real question is whether the universe and human society are ultimately based on absolute values, revealed by a transcendent power, that serve as criteria and testing stones for new ideas and developments and that demand application in all areas of human life. . . . (Zeidan, 2003, p. 281)

It is, in short, framed by many as a conflict between "the Divine Principle and a secular materialist culture," as Khurshid Ahmad (1983, p. 228) put it.

As a representative battlefield, I will briefly examine the culture wars in the United States, in which traditionalist Christians see themselves as struggling with modernists and progressivists to preserve moral values, and then turn to traditionalism in Islam and religious revivals elsewhere, some of which—such as Gandhian nonviolence and Christian liberation theology—take quite a different turn and have divergent political implications.

Christian Traditionalist Protests

The culture wars occurring in the United States provide a convenient microcosm of the global conflict. American culture is organized in much the same way as culture in the global village: one broadly hewn worldview and its corresponding ethos are hegemonic, but the dominant paradigm is never fully accepted. The vast diversity of religious and ethnic perspectives brought to America by immigrants from around the world, constituting strong religious and ethnic subcultures in the country, present a strong and vocal challenge to efforts to create a hegemonic culture.

Polarized camps in the wars advocate two different styles of social authority, identified by Richard Merleman (1984) and elaborated by James Davison Hunter (1991) as "tight bounded" and "loose bounded." In tight-bounded communities, moral obligations are viewed as rigid and given, whereas loose-bounded groups view moral commitment as voluntary, contingent, and fluid. I will refer to them as the *orthodox* and *modernist* camps. Reality in American public culture, Hunter (1991) argued, is increasingly shaped by the "knowledge workers" of the modernist camp, such as "public policy specialists . . . special interest lobbyists, public interest lawyers, independent writers and ideologues, journalists and editors, community organizers and movement activists" (p. 60). These shapers of rhetoric and definers of moral standards challenge the fundamental religious tenets of many subcultures in the nation, partly because of efforts to create the proverbial "melting pot" and partly as a function of the common denominator effect of capitalism.

At stake in each area of dispute in the culture wars is the question of authority: Who is responsible for the care of the family and how much shall the state and other institutions intervene? Who can define the role of women in society, and how much authority should men retain in the family? Who defines the ethical boundaries of business, public culture, education, and other issues of public policy? Finally—and most fundamentally—who can determine when life begins and when it should end?

Specific battles of the culture wars often reveal their fundamental structure, as in the debates over gender-specific language and the Equal Rights Amendment (ERA). The orthodox party, represented by the Christian Right, claimed that the ERA would destroy the traditional family and motherhood. Modernists saw it as an essential step in the creation of a just society that ensures equality for all. The same issues underlie such seemingly unrelated public battles as homosexual marriage, prayer in the schools, and abortion. Some religious leaders claim that gay rights movements represent a vicious attack on traditional family values, whereas others defend it as an essential struggle for dignity. Modernists see collective prayer in public schools as an intrusion on the religious freedom of those not of the dominant religious perspective. The founder of the Campus Crusade for Christ, however, insisted that the 1963 Supreme Court ruling against prayer in the public schools constitutes the primary cause of the social problems faced in the United States today. Since the decision, he argued, "premarital sexual activity has increased over 200%, pregnancies to unwed mothers are up almost 400%; gonorrhea is up over 200% . . . [and] adultery has increased from 100% to 250% [*sic*]" (Hunter, 1991, p. 204).

The battle between the orthodox and the modernists is not just an American phenomenon. It fans conflicts within and between religious traditions around the world that have some common characteristics whether they are between the African National Congress and pro-apartheid forces in South African churches, secular-oriented and nationalist Hindus in India, or proponents and opponents of women's rights among the world's Muslims. Although I am referring to all of them here as "traditionalists," they are known in popular parlance as "fundamentalists," which is such a loaded term that it is problematic for academic discussion. When one group and not another is referred to as fundamentalist, it touches a nerve in the culture wars, as shown in a letter to the editor in *The Economist* by Mohammed Azim (2005), who observed that in writing about the Christian Right, a story in their journal "carefully avoided the 'f' word—the more familiar term 'fundamentalist'—in your descriptions. At what point, if ever, do these rightist *Christian* factions become fundamentalists, or is that a pejorative term reserved exclusively for Muslims and, occasionally, Hindus?" (p. 1). Moreover, it is helpful to remember, as Said Amir Arjomand (1995) pointed out, scriptural fundamentalism "is the common undercurrent of all fundamentalisms" (p. 192) regardless of the religious tradition.

Despite shared national, linguistic, and geographical bonds, the combatants of the culture wars live in different worlds and talk past one another. The orthodox and modernist camps tend to "operate from different philosophical assumptions and by very different rules of logic and moral judgment" (Hunter, 1991, p. 250). A negotiated settlement is not likely in the near future, Hunter (1991) argued, because vocal advocates "at either end of the cultural axis are not inclined toward working for a genuinely pluralistic resolution" (p. 298).

The New Christian Right

One of the most visible developments in American culture in the late 1970s and 1980s was the rise of the so-called New Christian Right, which soared into prominence with the presidential administration of Ronald Reagan. It revived again with the election of George W. Bush, who relied upon the Christian Right as a major political base, addressed some of their core issues, and appointed Pentecostal John Ashcroft as his attorney general and conservative judges John Roberts and Samuel Alito as justices on the Supreme Court.

Although it had an impact on American politics at a number of points in its history (see, e.g., Sutton, 2005), Christian traditionalism in the United States during the 1980s contained a number of interesting hybrids that used modern technology to promote its decidedly antimodern beliefs. Television brought the Christian Right into America's homes and made it possible for the movement to become a major force in American culture. In 1970, according to Arbitron, 38 religious programs had a combined audience size of 9,803,000; by 1980, 66 programs had 20,538,000 viewers, significantly transforming the religious landscape of the country (Hadden & Swann, 1981, p. 55).

Television was the most important but not the only medium used by traditionalists in mobilizing their movement: Other means used included computerized direct mail, Lear jets, and even a computer billboard created by John Marler's "Computers for Christ." Marler encouraged the development of a computer network to exchange information about everything from how to attack disturbing social problems to theological debates. Marler himself claimed to have "proof" that God dictated "each and every character and word in the Bible" (quoted in "Evangelist Uses Computer Exchange," 1984).

Many Christian conservatives overcame a natural antipathy to partisan politics, beginning in the late 1970s, because they were fed up with the drift of American culture. On the West Coast, where the two sides of the culture wars meet on the battlefield daily; in the South, where many still view the national culture with considerable ambivalence; and across the airwaves into every conservative corner of resistance, they began to talk with a new confidence about changing the country. And this time, they were not just talking about individual conversion; they were talking politics.[5]

In January 1979, a number of antigay, antipornography, and pro-family groups on the West Coast were brought together by California ministers Robert Grant and Richard Zone to found Christian Voice. Pat Robertson featured Voice on his program *The 700 Club*, and it quickly amassed a mailing list of 150,000 laity and 37,000 clergy, including Catholics and Mormons as well as Protestants. A few months later, in July 1979, Jerry Falwell founded the Moral Majority, with a strong Southern contingent and a serious computer-based fund-raising effort that gathered $1,000,000 in the first month, largely with support from the *Old Time Gospel Hour* audience. Falwell traveled to all 50 states holding "I Love America" rallies; by 1981, he claimed to have 4 million members in his organization—hardly a majority but certainly a substantial bloc.

Another significant Christian Right organization, Roundtable, sponsored workshops (including a 1980 Dallas extravaganza with Ronald Reagan) to teach clergy how to mobilize their congregations to support conservative political candidates. These organizations and others joined with the televangelists to oppose the Strategic Arms Limitation Treaty between the Soviet Union and the United States in late 1979, to protest state interference with Christian schools, to oppose abortion and endorse school prayer, and to elect a new set of political leaders. Although not very successful at recruiting people to run for Congress, the New Christian Right did play an important role in defeating Senators Robert Packwood, Frank Church, Evan Bayh, George McGovern, and John Culver, as well as several House members from conservative districts.

The Christian Right's greatest coup, though, was the election of Ronald Reagan as president. Not only did the conservatives mobilize people in the church to vote but they also raised large sums of money for political action committees (PACs). Whereas liberal PACs only raised about $1.2 million in 1977 and 1978, conservative PACs raised about $6.4 million. In 1979 and 1980, the disparity was even greater: Liberal PACs raised about $2.1 million and conservative PACs, $11.3 million (Latus, 1983).

Some students of the Christian Right predict that the movement mobilized resources to influence U.S. politics at the turn of the 21st century for the following reasons:

1. A loss of confidence in the liberal philosophy because of persistent military defeats, failed leadership, poverty, crime, drug use, and the like

2. The legitimization of a conservative cultural revolution during the Reagan era and the linkage between Christianity and free enterprise capitalism, prayer in schools, and the protest against secular humanism

3. The New Christian Right's monopoly of religious broadcasting: By 1987, it had 1,370 religious radio stations and 221 religious television stations, far surpassing that of any other single interest group (aside from corporations)

4. Their mastery of fund-raising skills so essential to sustaining a social movement organization, in terms of grassroots fund-raising (especially through television and direct mail) and contacts with wealthy individuals

5. The New Christian Right's appeal to the growing number of Americans over the age of 65 (3.1 million in 1900 and 29 million in the late 1980s), who are more likely to be religious than a demographically younger population

To this impressive list, Hunter (1991, p. 299) added another: the extensive network of parallel institutions—the schools, colleges, universities, and publishing firms of the New Christian Right. Moreover, despite a number of scandals and difficult times for televangelists, many persist in having widespread influence. A more secularized version of the Christian Right's agenda has a widely heard supporter in the strident voice of syndicated radio and television talk show host Rush Limbaugh.

A number of countertrends call the previously listed predictions into question, however.[6]

1. When Ronald Reagan left the presidency, the initial fervor of the New Christian Right evaporated quickly; Reagan's charismatic authority held together a rather disparate movement with serious internal cleavages.

2. Those people most likely to support the movement—a disproportionate number of the working class and minorities—are traditionally drawn to the more liberal Democratic Party so that few issues unite the movement beyond school prayer and abortion.

3. The institutional resources and power behind the modernist camp—notably the knowledge industry itself, so central to the process of constructing public opinion and with a relatively secular ethos—are probably stronger than those supporting the orthodox camp.

4. Because of its general orientation and modes of operation, the ethos of the modern state does not support an orthodox stance.

5. The ethos of the country's major cultural centers—Washington, D.C., New York, Boston, Chicago, Los Angeles, and San Francisco—tends to be progressive or modernist.

6. Finally, the style of contemporary policy debates assumes the autonomy of rationality, thereby excluding the appeal to traditional authority of the orthodox message. The orthodox camp tends to lose simply by accepting this ground rule for participation in debate. (see Hunter, 1991, p. 306)

The culture wars are far from over in the United States, and bickering occurs within each camp as well. One of the most difficult problems in political culture is the lack of tolerance on both Left and Right. For example, even some of the traditional conservatives were displeased with the New Christian Right's efforts to label those who disagreed with them as immoral people. Senator Barry Goldwater, a longtime spokesperson for the conservative community, declared in 1981 that he was "frankly sick and tired of the political preachers across this country telling me as a citizen that if I want to be a moral person, I must believe in A, B, C, and D. Just who do they think they are?" (Nelson, 1981).

Because of the pluralistic character of the U.S. population, pressures to have a broad, flexible collective moral code have permeated American culture from the country's beginning. Those forces favoring less rigid common morals have always collided with others attempting to install a hegemonic culture and to enforce a particular brand of Christianity on everyone. They have never been fully successful, however, and part of the reason for the

New Christian Right's emergence in the 1970s and 1980s was their perception that they were losing the battle for the country's hearts and minds,
especially as various minority groups began to assert their own subcultures
in the wake of the cultural ferment of the 1960s.

Christian traditionalist and evangelical movements in the United States
provide an important anchor for identity in a sea of change and a means of
expression for people dissatisfied with the direction in which the world is
moving and grieving over the loss of the world as they knew it. Christian
traditionalists also present a serious challenge to American multiculturalism
and religious pluralism because of their strong opposition to religious tolerance and their certainty of the truth on certain moral issues against which
other groups hold competing beliefs just as strongly.

Islamic Traditionalism: Antimodern and Anticolonial

Many believers in the Islamic world have picked up their own banner of
orthodoxy in a manner similar to that employed by the New Christian Right
in the United States. In some ways, these two groups fight the same enemy—
the Western establishment, the modernists who attempt to establish cultural
hegemony in the contemporary world based on secular scientific thought that
undermines the ethos and worldview of the religious community. The Islamic
traditionalists emerged from a very different historical context, however.

In the past century, issues of modernism and colonialism became deeply
intertwined in the Islamic community. On the one hand, life in the 20th century raised the question of how to respond to the globalization of culture, the
interaction among the world's religious traditions, and the challenges to all
religious dogmas presented by science. On the other hand, the humiliation of
colonial subjugation that so many Muslims endured at the hands of Westerners
fed fuel to the fire of conflict. In the polarized climate of charges and counter-
charges, it became virtually impossible, in many parts of the Muslim world, to
be faithful to Islam and also tolerant of other religious faiths and scientific
inquiry—much like the Catholic Church aligned itself against science and
democracy in the West during the 18th and 19th centuries.

By the end of the 19th century, nearly all of the Muslim world had been
conquered by European colonialism: Britain controlled the Gulf area, Egypt,
portions of Africa, India, and Malaysia; France took over North Africa and
much of West Africa and the Middle East; the Soviet Union incorporated
major Muslim areas of Central Asia after the 1917 revolution.[7] During this
colonial period, Jamal al din Afghani (1830–1897) and Muhammad Abduh
(1849–1905) attempted to elevate the dignity of Islamic thought and encouraged self-determination among Muslims in the Middle East. Afghani argued

that Islam was not inherently opposed to modern science but that European domination should not be tolerated; he encouraged the formation of a pan-Islamic federation of states. A number of independent Muslim nation-states were created, although at the beginning they tended to follow, rather than reject out of hand, trends initiated by the European colonials. Ataturk (Mustafa Kemal) founded Turkey, a secular state replacing the disintegrating Ottoman Empire, in 1923. Ataturk essentially disestablished Islam in Turkey, limiting the scope of the Shari'a to personal matters, and adopted many European practices in government and culture.

Iran underwent a similar process, beginning in 1921 with Reza Khan, who proclaimed himself shah (Persian for emperor) and set his country on a process of modernization, despite resistance from the religious scholars, the mullahs, who had always played a significant role in governing the country. Hasan al-Banna founded the Muslim Brotherhood in Egypt, and the movement spread to Syria, Palestine, the Sudan, and elsewhere in the Arab world. In South Asia, scholars such as Sayyid Ahmad Khan, Sayyid Amir Ali, and Muhammad Iqbal participated in an Islamic intellectual revival at the turn of the 20th century, including the founding of the Aligarh Muslim University in India. Muhammad Ali Jinnah pressed vigorously for the formation of a separate Muslim state when India was pushing for its independence from the British Empire, and succeeded in getting it, despite the strenuous objections of Gandhi and others.

Sayyid Abdul Ala Maududi, a Muslim intellectual and religious leader in India and Pakistan in the early 1940s, spoke for many Muslims when he decried the moral decadence and corruption of the West. "Islam and Western civilizations are poles apart in their objects as well as in their principles of social organization," he insisted (Maududi, 1979, p. 23). Moreover, he claimed that Islam was self-sufficient and provided a viable alternative to both Western and socialist ways of life. The Muslim Brotherhood (Jama'at-i Islami), founded in 1941, was vital under Maududi's leadership and laid the basis for an educational campaign, influencing many Muslims outside of Pakistan, including those in Europe and North America (see Cragg, 1965).

Two years later, in 1949, Indonesia gained independence from the Dutch after a struggle led by Sukarno that was motivated by nationalism and Islam. Although the independent republic succumbed to military rule, Indonesia's Muslims remain a vital part of the country, which has the largest population of Muslims in the world. North and West African Muslims also gained independence during the mid-20th century, producing a number of Islamic states. Considerable conflict ensued, however, between orthodox and modernist Muslims within these new nation-states, as well as with external powers. Dreams of establishing a unified Islamic Brotherhood of nations has not materialized, and the difficulties of doing so are exemplified in the 8-year bloody war between Iraq and Iran, in which each side claimed to have God on its side.

The traditional alliance between Islam and the state, an alliance also enjoyed by Western religions until very recently, has exacerbated the tensions between those wishing to make a transition to modernity and forces wanting to revive a more traditional Islamic society. The emergence of Islamic traditionalism—which some call Islamic "fundamentalism"—represents the orthodox camp's effort to reassert itself in the wake of Western colonialism and in the face of rapid changes in many Islamic states. Those internal conflicts have been exacerbated by continued struggles with the non-Muslim world as well, especially since the founding of the state of Israel and the intrusion of the United States after the disintegration of European control over the region. Many of the orthodox thus perceive a dual enemy attacking the sanctity of their traditions: the Western outsiders, especially the United States and Israel, and the modernizing insiders, who reduce the power of religious leaders and the Shari'a, the rule of Islamic law, in regulating society's affairs as they press for a modern, secular state.

These developments in the Muslim world have led to tremendous misunderstandings in the West along with considerable fear fueled both by misleading stereotypes in the media (see Said, 1978) and popular culture and by the visibility of life-threatening terrorist groups, sometimes operating in the name of Allah. Now, for the first time, these groups have launched significant attacks on targets within the United States, heretofore relatively unscathed by the Islamic struggle. The bombing of the World Trade Center in New York in the spring of 1993 and the attacks of September 11, 2001, as well as the subsequent arrests of suspected terrorists planning to bomb other key U.S. targets, has left many in the country suspicious and fearful. On the other hand, the U.S. government, sometimes in league with the other major Western powers, has inflicted much violence on Islamic populations as well, having destroyed the infrastructure of Iraq during the 1991 Persian Gulf War and overthrown the governments in Iraq and Afghanistan in the wake of September 11. The so-called "war on terror" has resulted in rounding up large numbers of Islamic suspects and attacking other targets in predominantly Muslim states. The publication of cartoons of the Prophet Muhammad in Denmark in 2006 became a symbol of perceived insults that provoked rage and protests in many places around the world.

When the battles escalate, religious traditions are often called into play to justify the political stances of the combatants. This practice is particularly obvious in the Islamic world, in which religious rhetoric is a more central part of the political culture than in other societies. Thus, the orthodox attack their enemies—both internal and external—with an intensity that only religious framing can justify.

Islamic Reformers Enrage the Traditionalists

It is one thing to have infidel ideas imposed by foreigners. It may be even more outrageous to have such ideas championed by insiders, who are, in the long run—like all heretics—more dangerous than outsiders because they undermine the faith.

The case of King Amanullah of Afghanistan (who reigned from 1919 to 1929) is instructive.[8] He declared Islam to be the official religion of Afghanistan in 1923, provided the country with a written constitution, and endeared himself to many with his jihad against the British. In 1927, he toured India, the Middle East, and Europe, and his wife Soraya appeared unveiled at receptions in Europe, causing a scandal when her photographs circulated in Kabul. Despite resistance, Amanullah pressed forward with his efforts to reshape his Islamic society. He championed women's rights, outlawing polygamy among civil servants and permitting women to discard their veils. In October 1928, the queen led 100 women in appearing unveiled at an official function in Kabul, outraging the religious establishment. The act was not simply a matter of women's rights, according to some; it rent the entire fabric of the sacred canopy. As one cleric declared, "When reforms come in, Islam goes out" (Hiro, 1989, p. 234). Instead of retreating, the king escalated the conflict. In 1929, he required all Afghans in Kabul to wear Western dress, including European hats. When clerics pronounced this practice blasphemous, the king forbade students to enroll in the famous Deoband seminary. When the Hazrat of Shor Bazaar collected signatures of protest, he was arrested. At this point, rioting broke out in Kabul, and an insurrection ousted King Amanullah from power, despite his last-minute concessions.

The significance of this scenario lies in the role of religion and religious leaders in the debate over the ethos of Afghan society and the extent to which people were willing to allow Western influence to affect the norms and values of their culture. It was not the monarchy itself but the king's alliance with the West and his subversion of Islamic tradition that precipitated both the popular revolt and the organized resistance of the *ulama*. Afghan traditionalism waxed and waned throughout the rest of the century, but it was almost always used as a tool to resist both modern culture and Western intervention in that society.

As the case of Amanullah suggests, one test for ethos-related conflicts between orthodox and modernist camps in the contemporary world is the role and status of women. Some highly visible and influential Islamic women insist on their equality, and many equally visible and influential Islamic men resist their efforts ferociously. The Pakistani Muslim leader Sayyid Maududi, for example, was innovative in trying to develop Islam as an alternative to Western and socialist systems, but on the issue of women he maintained a strict orthodox position. Maududi advocated strict sexual segregation and the necessity of the veil to close the "main gate"—that is, the face: "Nothing can be more unreasonable than to close all the minor ways to indecency but to fling the main gate wide open" (Maududi, 1979, pp. 197–198).

One reason for the intensity of these conflicts now is that the social organization of the family often represents for Muslims the last bastion of traditional Islam. During the colonial period, Muslims were forced by the colonial powers to secularize their legal system, taking power away from the Shari'a, the rule of Islamic law. The family was exempted from many of these developments, probably because it was so important to traditionalists and did not substantially affect the economic interests of the colonizers. By leaving the family under the aegis of the Shari'a, Islamic elites and the general population could be more easily persuaded to cooperate with the colonial government.

Revivalism Around the World

The Islamic world is not the only place where colonialism and modernism are challenged by efforts to revive indigenous religious practices. The prototype is Mahatma Gandhi's revival of Hinduism in India as part of his campaign for independence from Britain. In his famous Hind Swaraj (Gandhi, 1908/1939), Gandhi denounced the corrupt influence of Western, especially British, influence on Indian culture, calling for a return to Hinduism. Gandhi successfully used the stories and rituals of several religions, but especially those of Hinduism, as a vehicle for his development of the Indian freedom movement.

The first major element of the Indian movement's campaign of noncooperation with British rule to protest the colonization of their country was initiated with a day of prayer and fasting. "Do you mean a strike?" he was asked. "No, a day of prayer and fasting," Gandhi replied. The entire country came to a standstill as people prayed and fasted. The British might try to punish people for engaging in a general strike, but how could they suppress a religious celebration? Subsequently, Gandhi used daily prayer meetings and his status as a religious holy man to press his message of Indian home rule and to mobilize the social movement that opposed British rule. Despite the fact that it was presented in an orthodox fashion, Gandhi's interpretation of

Hinduism was novel. His charismatic authority was so great as leader of the freedom movement, however, that no one was able to challenge him successfully on religious grounds.

Other colonial countries followed suit, especially after World War II, often emulating Gandhi's tactics and adapting them to their own situation. A 1947 Nigerian editorial, for example, bemoaned the invasion of Christianity and advocated a revival of traditional religion as a means of resistance.

> The native dweller in Nigeria had a religion before the advent of Christianity. His religion was perfect, and taught him the brotherhood of man and the fatherhood of God. He lived for the other man. His one sole aim was to carry sunshine and happiness into the home of his fellow man. . . . The African has a religion which, unfortunately, is fast giving place to the imported form of worship. His religion takes him closer to the Divine Presence, and enjoins him in true love for his fellow man. Hence the African has always been found a child of nature, docile and unsuspecting. This copyist attitude in all phases of his life has robbed the African of his innate godliness, and it was time our people turned to find God, to worship Him, and to serve Him in the true African way.[9]

In Latin America, Roman Catholicism was used from the beginning of the colonial period to legitimate the imposition of European rule over the indigenous people of that continent. Over the centuries, Catholicism became the core of Latin American popular culture, and people in power there used the faith's symbols and the authority of the church to maintain their hold over the masses. With the emergence of indigenous Christian leaders in the 20th century, however, Christianity began to legitimate movements of resistance against the power structure and was used to mobilize reform movements and even revolutionary activities.

Christian Liberation Theology

A social protest movement challenging the status quo of the modern world—the theology of liberation—represents one of the major religious movements of the 20th century. As with Islamic and Christian traditionalism, as well as the Indian freedom movement, liberation theology frames the desire for freedom from political subjugation in a traditional religious perspective. The birth of this movement among the poor of Latin America signaled a reshaping of traditional Christian symbols in a way that some argue is truer to the spirit and teachings of Jesus and the early Christian church than is the establishment church, which legitimates an oppressive social order. A traditional reliance upon God as a personal savior, or liberator, is taken for granted by people in this movement, but the classical theological questions of modernism—Does God exist? Are the scriptures infallible? Is the pope infallible? Can a Buddhist be saved without becoming a Christian?—mean little to them. They ask instead the following: How can we participate in God's liberating activity in the world?

Christianity From the Bottom Up

Liberation theology grows, first of all, out of Latin American attempts to break out of the historical oppression of first colonialism and then hierarchical systems in which a small wealthy elite is sustained in power by the United States. Penny Lernoux (1982) put it this way:

> From the moment Columbus set foot in the New World cross and sword had been indistinguishable. Priests and conquistadors divided the plunder in people and land—it was a toss-up which was the greedier. And long before Latin America's military regimes installed their torture chambers the Inquisition was at work with whip and rock. By the time of the wars of independence at the beginning of the nineteenth century, the Church was the largest landowner in Latin America. It was also the most conservative political force on the continent.[10] (p. 10)

Other forms of liberation theology have emerged elsewhere in the Third World, especially in Africa, and among feminists, black Christians in the United States, and German philosophical theologians criticizing the middle-class gospel of consumerism. Liberation theology constitutes a new paradigm in Christian theology (Chopp, 1986), as well as a practical liberating activity in the *comunidades de base* ("base communities") of Latin America.

Two events in the 1960s laid the groundwork for the emergence of the liberation theology movement: the Second Vatican Council in Rome, beginning in 1962, and the Medellín Conference (CELAM II) in 1968. Vatican II and Pope John XXIII's attempts to "open the windows" of the church, established a precedent in taking note (Vatican Council II, 1965) that Christians have a special responsibility for "those who are poor or in any way afflicted." This emphasis on the church's responsibility to the poor struck a chord in Latin America, where the church was deeply enmeshed in the lives of the poor. At the General Conference of the Latin American episcopacy in Medellín, Colombia, in 1968, the Roman leaders of the church contended that "the Lord's distinct commandment to 'evangelize the poor' ought to bring us to a distribution of resources and apostolic personnel that effectively gives preference to the poorest and most needy sectors" (Latin American Bishops, 1979, p. 175).

The Medellín Conference gave a green light (Berryman, 1976) to the development of Christian base communities in which small groups of people—often meeting without clergy, because of a shortage of priests and Rome's relaxation of regulations after Vatican II—met throughout Latin America for prayer and Bible study, rediscovering the radical liberating message of the New Testament often lost in contemporary Christianity. This message—and the process by which it emerges from the pages of the Bible in the hands of the peasants—comes through clearly in the four volumes of *The Gospel in Solentiname*, transcripts of Bible studies led by the poet–priest Ernesto Cardenal in Nicaragua. In the discussion following their reading of Luke 4:16–30, in which Jesus announces that he has come "to give the good news to the poor," Cardenal (1982) explained the following:

> The good news is for the poor, and the only ones who can understand it and comment on it are the poor people, not the great theologians. And it's the poor who are called to announce the news, as Jesus announced it. (p. 133)

Similar communities emerged in other parts of the world, including Africa and Asia, although with their own local agendas and languages. In Zimbabwe before independence, people from all across the country would meet early in the morning to pray and sing and ask God for their liberation.

The preeminent figure in articulating the ideas of liberation theology was Gustavo Gutiérrez, whose *Teología de la Liberación* (*A Theology of Liberation*) (1973) served as a manifesto for the movement. Bridging the gap between the church hierarchy and the base communities that flourished in the slums,[11] Gutiérrez's theological reflections were amplified by such figures as Camilo Torres from Colombia (who studied sociology at Louvain, where he was a classmate of Gutiérrez) and Dom Hélder Camara in Brazil.

Ironically, the initial ecclesiastical actions that led to the liberation theology movement grew out of conservative efforts to defend the institutional interests of the Roman Catholic Church in the face of Marxist and other left-wing critics (Adriance, 1992). Moreover, the practical measure of forming base communities in the wake of Vatican II reforms and a shortage of priests cultivated the growth of base communities among the poor, which unexpectedly resulted in a movement that gradually sought some independence from the church hierarchy. Here is Adriance's (1992) examination of developments in Brazil and Chile:

> [they] provide illustrations of the paradox of institutionalization. They show how measures taken by some bishops to restore the church's influence in the context of a secularized, pluralistic society unleashed a potential for social and ecclesial change that may prove to be more radical than the bishops had ever intended. (p. 60)

The Vatican found the liberation theology movement a highly disturbing phenomenon, in part because of its independence not only from the political establishment but also from the ecclesiastical one. The Polish Pope John Paul II was also concerned about its links with Marxism, because in his experience Marxism was more the oppressor than the liberator. When the pope visited Nicaragua in 1983, he shook his finger at Father Cardenal, who was Minister of Culture in the new communist Sandinista government, and scolded him for his political involvement.

Liberation theology had many critics in the church because it was too politicized, too Marxist, and sometimes advocated violence. It also had many supporters at high levels and represented one of the fastest-growing sectors of the church, so the movement was difficult to suppress. As with Islamic and Christian traditionalism, liberation theology provided a vehicle for discontented people to express their protest and try to change the world around them. A similar, and in some ways more successful, movement for change with a religious base occurred in recent nonviolent movements in Asia and Europe, where the church played a key role in organizing resistance against various dictatorships.

A Theology of "Nonpersons"

Although not always articulated by the poor,[12] liberation theology is a perspective of and from the poor or as Gutiérrez (1973) put it, "nonpersons"— that is, the people who are ignored to the extent that they do not even exist for people in power. Erving Goffman (1959) explained the dynamics of this role in social life in the example of the servant, who is a convenient "nonperson" to have around but does not exist for social interaction and is treated as less than human. In the most extreme instances, a "nonperson" in the role of a servant sleeps in the master's bedroom as part of the furniture in case anything is needed during the night. Most of us find ourselves in this role, to a lesser extent, when people of power and privilege simply ignore our presence.

Entire classes of people are nearly invisible to the mainstream, and especially to the elites, of most modern cultures. The invisibility phenomenon enables the rich to cope psychologically with the existence of mass poverty and starvation in their environment. In India, the millions of people living in abject poverty on the street are simply screened out of existence by the middle and upper classes, which go about their daily routines pretending that they do not exist. The number of people who live in such dire poverty that they scarcely live at all is so enormous that we ignore their existence in order to enjoy our own lives. Even the news media, which usually focus on problematic aspects of human life, manage to ignore what is certainly one of the most significant stories about our world every day—that is, that roughly 40,000 children die each day of malnutrition and related causes, and a similar number are permanently damaged each day by the same state.

Since the 1960s, the world has witnessed what Gutiérrez (1973) calls an "eruption of the poor."[13] Religious thinking grows out of the social context in which the thinker lives; liberation theology constitutes a form of reflection on the nature of the sacred from the point of view of those who suffer, from those who constitute the majority of the world's population (see Chopp, 1986). Sociologically, we expect "nonpersons" to view the world differently, of course, from people in other social roles and to have affinities with different religious expressions from those of the rich. Gandhi said that God should appear to the poor in the form of bread. Liberation theology recovers the perspective of the poor from the Judeo-Christian tradition, a rich and deep element that has been conveniently subordinated by the alliance between the church and the Western establishment.

All of these developments in faith communities over the past 200 years or so brings us to a situation in the 21st century in which religious life has been substantially transformed. It is to the religious landscape of the new millennium that I now turn my attention.

7

Religious Movements
for a New Century

A long with the revitalization of ancient traditions in the form of traditionalism and various religious revivals and reformulations of religious traditions, a number of new religious forms have emerged as a response to modernism. These forms include civil religions and syncretic religious movements and new forms of religiosity that draw upon elective affinities between the interests of particular groups of people and the interaction among various religious traditions and secular forces. Religiosity in the new century is a dynamic mixture of the old and the new, of ancient rites and beliefs as well as a plethora of recent religious movements, and the impact of social movements, such as the women's and environmental movements, on religious beliefs and institutions (see, e.g., Robbins & Anthony, 1990). As the pace of globalization picks up, the crosscurrents of the global cultures—and reactions against broad social changes—make spiritual life interesting and diverse.

Civil Religion and Nationalism

One of the most significant forms of quasi-religion in the 20th century has grown out of the social organization of the nation-state, which often replaces traditional religious institutions as a focus of identity and basis of the cultural ethos. This significant element of political culture, especially in modern nation-states, is usually called civil religion, a concept developed by

Jean-Jacques Rousseau (1762/1901) and Alexis de Tocqueville (1862/1945) that is congruent with Emile Durkheim's understanding of the role of religion in collective life. The growth of civil religion has been widespread primarily for two reasons: (1) to provide cultural unity among a set of pluralistic belief systems created by population migration or artificially constructed states (usually established by colonial powers) and (2) as a functional substitute for a religious tradition that was deliberately attacked because of its alliance with the old order (especially in the socialist states).

American society provides a particularly instructive model of civil religion because of the parallels between the multicultural U.S. society and the emerging pluralistic world system. Many of the struggles faced first by Western Europe and then the United States, as they moved into the heterogeneity of the modern world are now encountered in other countries, even when the paths and contents of cultural transformation have been substantially different.

Civil Religion: The American Case

> No matter where we live, we have a promise that can make all the difference—a promise from Jesus to soothe our sorrows, heal our hearts, and drive away our fears. He promised there will never be a dark night that does not end. . . . And by dying for us, Jesus showed how far our love should be ready to go: all the way.
>
> —President Ronald Reagan, February 1984

Civil religion in the United States, though still highly Christian in its tone and basic beliefs, is a general religious orientation that emphasizes belief in a generic God and a specific role for the United States of America in world history. Robert Bellah (1970) is the sociologist who has applied the notion of civil religion most thoroughly to the U.S. context:

> Behind the civil religion at every point lie biblical archetypes: Exodus, Chosen People, Promised Land, New Jerusalem, and Sacrificial Death and Rebirth. But it is also genuinely American and genuinely new. It has its own prophets and its own martyrs, its own sacred events and sacred places, its own solemn rituals and symbols. (p. 186)

Gehrig (1981) concluded that civil religion in the United States is "the religious symbol system which relates the citizen's role and the American society's place in space, time, and history to the conditions of ultimate existence and meaning" (p. 108); it is differentiated structurally from the political and religious communities of the United States and performs "specialized religious functions performed by neither church nor state."

Generic religious images and rituals are frequently evoked to underscore the sacredness of U.S. tradition and culture, practices that reflect the ambivalent and tenuous framing of religion in American political culture: "In God We Trust" is inscribed on the currency, but the referent is deliberately ambiguous. Every presidential inaugural address, except George Washington's second, includes references to God. In the 1950s, the phrase "one nation under God" was added to the pledge of allegiance, which is still recited in most public schools around the country. In the 1980s, Ronald Reagan's unprecedented Christian rhetoric broke traditional rules about the relationship between politics and religion because of his specific references to Jesus. This explicitly Christian presidential rhetoric returned with the presidency of George W. Bush in 2001.

Civic rituals in the United States have a strong religious flavor. The major holidays in the American liturgical calendar unite religion and political culture, if not church and state:

Thanksgiving is at the core of the practice, with its explicit link between the "status legend" (as Weber might call it) of the country's founding and the deity of American civil religion.

The period around "Christmas," now sometimes generically referred to as "the holidays," embracing Christmas and New Year's, and in recent years an adopted holiday of Chanukah and the African American invention of Kwanza.

Martin Luther King Jr.'s birthday, a recent addition for the purpose of inclusivity, celebrates a Christian pastor's contribution to national political life.

Memorial Day links the nation's wars with the deity.

The Fourth of July is explicitly a secular holiday, but God's name is often invoked in official ceremonies.

Hammond (1980) summarized the ideology behind the peculiar alliance between the state and religion in the United States as follows: "(1) There is a God (2) whose will can be known through democratic procedures; therefore (3) democratic America has been God's primary agent in history, and (4) for Americans the nation has been their chief source of identity" (pp. 41–42). This ideology is multifaceted: On the one hand, it has been a useful cultural tool for uniting people from diverse backgrounds into a single body politic; aspects of the civil theology facilitated a critique of slavery in the 19th century and legitimated the civil rights movement in the 20th (see Bellah, 1975). On the other hand, civil religion in the United States has cultivated hegemonic ambitions and a sense of religious superiority that has legitimated some of America's worst episodes of adventurism, from the earlier doctrine of "Manifest Destiny," which was used to justify the genocide of the people living in the New World before the Europeans arrived, to more recent efforts to police the world with the American military.

Debates about Bellah's characterization of American civil religion all reflect the problem of collective religious ideas in a pluralistic society so characteristic of both U.S. culture and the global village. Richard K. Fenn (1972, 1974, 1976) developed a series of critiques, the core of which resurfaced on a broader scale with Stephen Warner's (1993) important proposal for a new paradigm for the study of religion in American society, in which the metaphor of the marketplace replaces that of the sacred canopy. Fenn contended that because cultural integration is impossible in modern societies, civil religion is a useless concept. It is probably more accurate to say that cultural integration is a necessary but extremely difficult process in diverse societies and that the idea of a broad civil religion that lacks any sectarian character but still aids the construction of a collective identity is one important response to the dilemmas of multiculturalism.

Civil Religion in the Global Village

Political elites attempting to achieve national consensus will, of course, find some form of civil religion, or its functional equivalent, desirable to give legitimacy and (at least the appearance of) higher purpose to the political order. The more diverse the citizenry, the more difficult it is to suspend a sacred canopy across the state. Independent India, for example, has had great difficulty developing a secular state that embraces Hindus, Muslims, Sikhs, and Jains because questions of religion and politics are always intertwined. Efforts to use Mahatma Gandhi, the universally acknowledged "father of the country," as a figurehead have sometimes backfired because Gandhi, despite his own universalistic approach to religious belief, alienated large portions of the Muslim and Sikh communities by employing Hindu rhetoric and symbols in his public presentations (although he also included symbols from many faiths, including Muslim, Sikh, and Christian, as well as Hindu).

Like Western Europe following the French Revolution, 20th-century communist countries attempted to replace traditional religious forms with new versions that were not always successfully. Civic rituals in the Soviet Union looked remarkably like Russian Orthodox ones, with some new content. The public processions of the Communist Party and military looked much like ancient religious processions, and the celebrated saints of Christianity were replaced with Saints Marx and Lenin. Meanwhile, opposition movements throughout the Soviet bloc were supported in significant ways by both the ideologies and the institutions of the Christian Church. Religious organizations provided a space for organizing alternative institutions and resistance movements, training workshops in nonviolence, and a general legitimacy within the larger population.

In China, efforts to replace Confucianism, Taoism, and Buddhism with Maoism were only partially successful. Because diverse religious beliefs had lived together in relatively peaceful coexistence in Chinese culture over the centuries, many people simply added Mao to the pantheon, or at least worshipped him in public and their traditional Gods at home. Buddhist religious officials, with a long history of negotiating with Chinese political elites, discussed affinities between Buddhism and Marxism.

The Republic of China (Taiwan) presents an interesting case, because the nationalist leader Chiang Kai-shek was a Christian, as were many of his followers (converted and educated by missionaries). When the government fled to Taiwan from the mainland in 1949, it tried to suppress "superstitious" Chinese folk religions but allowed free worship of Confucianism and Buddhism, which place relatively little emphasis on transcendental theology and have traditionally promoted public order in Chinese culture. Because K'ung-Fu-tzu is regarded more as a philosopher and founder of Chinese social and political thought than as a God, the conflicts do not seem so sharp. Confucian temples are maintained by the state and are the site of major celebrations every year at K'ung-Fu-tzu's birthday, when he is honored—not so much as a God but as the "Great Teacher"—by government officials and the general public.

Efforts to create civil religions within and among national societies are now complicated by intercultural conflicts almost everywhere, even in places where new states have been carved out more or less on the basis of ethnic identities—such as Pakistan and Israel—because of persisting internal diversity and, more important, because no nation lives in isolation and the dilemmas of multiculturalism cannot be avoided.

Religious Syncretism and Alternative Religious Movements

In the new context of multiculturalism, we witness the revival of a venerable phenomenon in the history of the world's religions: religious syncretism— that is, the combination of elements from more than one religious tradition

a new sacred canopy that competes in the cultural marketplace. All
ᴏrld's religions were created over the centuries through processes of
ᴄrcultural conflict, amalgamation, and creative synthesis. The current
.eligious scene presents a dynamic interplay between traditional practices on
the one hand and widespread transplanting of traditions and experiments
with syncretism on the other.

Vast movements of migration and the creation of modern means of communication and transportation have increased the tempo of syncretism in the
late 20th century, now occurring on an unprecedented scale. New versions
of old traditions emerge as alternatives to mainstream culture in contexts
where the tradition has been dominant, as well as in new settings where a
religious movement is transplanted. The potential for conflict between established and alternative religious communities is thereby greatly increased,
posing a significant social problem for the global village.

The Sociology of Alternative Religious Movements

Conflict between religious movements and the religious and social establishment is not a new theme, nor is it a marginal one. It is an age-old story
about the relationship between religious expression and the social order,
between heresy and orthodoxy. Most religious movements—including those
that now constitute the mainstream—begin as small, intense rebel groups at
odds with the religious and political establishments of their own origins.
Early Christians, for example, were in constant conflict with authorities:
Jesus was executed by Rome after complaints by the local religious establishment, and many of his disciples were also put to death. New religious movements (NRMs) tend to attract primarily people who perceive themselves as
marginal or opposed to the dominant society and its value system (Glock &
Bellah, 1976). Consequently, some form of tension between these movements and the larger society they inhabit is inevitable.

This section will examine briefly some alternative religious movements
and traditions. By alternative movements, I mean those religious groups
outside the mainstream popularly called "cults," although sociologists
sometimes refer to them as NRMs to avoid the negative connotations of the
popular term. I will adopt the term *NRM* in this discussion, though reluc-
tantly, because it is something of a misnomer; to quote Ecclesiastes, "There
is nothing new under the sun." Some NRMs are imported versions of
ancient religious perspectives from the other side of the global village; others
are new interpretations of ancient indigenous religions. In this sense, North
American religious life offers a microcosm of the broader processes taking
place on the planet; and the experiments have had mixed results.

New Religions in the United States

The history of religious life in the United States is a story of ongoing struggles among various religious communities to coexist within the same political and geographic area. What is so interesting sociologically about the U.S. case is that it foreshadows many aspects of the process the entire global village is now experiencing. Because of the diversity of religious groups in the New World from the very beginning of American colonial history, a variety of faith communities—some of them with highly incompatible worldviews— have been required to forge a working relationship with each other.

Religious diversity has been normative in U.S. history from the beginning, but the cultural ferment of the 1960s and the waves of new immigrants from various parts of the world in recent decades have precipitated an explosion of NRMs in the United States. The degree of multiculturalism has never been so high in American culture, especially with the introduction of large numbers of Asian immigrants in the late 20th century as a consequence of the relaxation of immigration laws barring them. Although the American religious landscape has always been diverse, this large influx of people from the Far East made a dramatic difference in the number of non-Christian believers living in the United States.

Three related sociological observations I would make about the alternative religious movements of contemporary American life are as follows:

1. Not all religious movements are alike, and they cannot accurately be classed together under the single label "cults" to distinguish them from "legitimate religions."

2. The appearance of large numbers of NRMs around the world in recent decades reflects ongoing globalization processes: The "anticult" movement is often a protest against those changes, sometimes from an ethnocentric point of view.

3. Those characteristics of NRMs that people most detest are as much a consequence of social hostility toward them as of properties inherent in the movements themselves.

Let us examine each of these observations more closely.

Not All "Cults" Are Alike

Alternative religious movements are remarkably diverse in their beliefs and rituals, organization, membership, relationship to the broader society, and virtually any other dimension, with one exception: Most are forged in deliberate contrast to existing mainstream religious groups in the culture or as a subgroup within these groups.

The original meaning of the term *cult* comes from the Latin *cultus*, "care" or "adoration." It also refers to formal religious veneration, worship, or a system of religious beliefs, or its rituals. More recently, the term has taken on a third meaning—as Webster's puts it, "a religion regarded as unorthodox or spurious" ("Cult," n.d.). In its latest sense, the idea of a cult, at least in American popular culture, has absorbed a range of negative images that include brainwashing, fanaticism, mind control, and so on (see Table 7.1).

Table 7.1   Major Religious Movements in the United States

	Movement Name	Roots	Major Figure	Doctrines, Practices
Mormon	Latter-Day Saints	Christian	Joseph Smith	Alternative Christian
Communal/ Countercultural	The Farm			Counterculture
Asian	Buddhism	Buddhist	Buddha, Dalai Lama	Meditation school, parables
	Yoga	Hindu/ Buddhist	Maharishi Mahesh Yoga	Meditation, health
	Transcendental Meditation	Hindu/ Buddhist	A. C. Bhaktivedanta	Meditation, improve clarity
	Krishna Consciousness	Hindu	Maharaj Ji	*Bhagavad Gita*
	Divine Light Mission	Hindu	Sun Myung Moon	New incarnation of God
	Unification Church	Syncretic		New incarnation of God
Feminist	Wicca (Witchcraft)	Syncretic		Feminism, Goddess worship
Neo-Christian	Christian World Liberation Front	Christian		Radical counterculture
	Children of God	Christian		Counterculture Christian
	Campus Crusade	Christian		Evangelical Christian
	Jews for Jesus	Judeo-Christian		Converted Christian

	Movement Name	Roots	Major Figure	Doctrines, Practices
New Thought	Christian Scientist	Christian	Mary Baker Eddy	Healing through positive attitudes
Psychic/New Age/Personal Growth	Scientology	Syncretic	L. Ron Hubbard	Auditing past lives
	EST	Asian/ Syncretic	Werner Erhard	Sensitivity training
	Synanon		Charles Dederich	Substance abuse therapy
Neopagan	Satanism	Pagan	Anton Szandor LaVey	Magic, hedonism
Political	Christian Right	Christian	Jerry Falwell	School prayer, morality
	Sojourners	Christian	Jim Wallis	Evangelical liberation
	Nation of Islam	Muslim	Elijah Muhammed	Counter–white supremacy

Sources: Adapted from Bromley & Shupe (1981); Glock & Bellah (1976); Wuthnow (1976).

Asian and Countercultural Movements

The most significant NRMs in the United States are imported from Asia and derive especially from Buddhist and Hindu traditions. Like other NRMs, the Asian movements were transported through two different routes: (1) immigrants from the region bringing their own native religious beliefs and practices with them and (2) indigenous countercultural movements that emerged, especially in the 1960s. These two distinct groups of practitioners are sometimes in conflict with one another: Whereas the immigrants are busy attempting to assimilate in their new environment, counterculture participants are attempting to jettison the American cultural milieu they reject. Thus, immigrant groups trying to adapt to a new cultural milieu are sometimes embarrassed by practitioners of their religion who are criticizing American culture.

Buddhist groups in the United States, especially the Zen and Pure Land schools, often have Japanese origins, although Tibetan Buddhism has gained a widespread following in American popular culture in recent years, largely because of the popularity of the Dalai Lama and supporters of the Free Tibet movement. Zen Buddhism was a particular favorite of the 1960s counterculture movements because of its spontaneity and its thoroughgoing challenge of Western utilitarian rationality and materialism (see also Hesse, 1971). The Beat generation of the 1950s set the stage for its introduction, with such writers as Jack Kerouac (with his novels such as *On the Road* and *The Dharma Bums*) and Gary Snyder making the precepts of Buddhism familiar to a large American audience. The counterculture of the next decade provided a fertile soil for both Buddhism and Hinduism, demonstrating

> a clear affinity between the pure self-determinist notion of the hippie movement and the law of karma as stated in Buddhism. Despite a present shaped by past actions, freedom of action in each new moment persists, enabling the practitioner to claim complete responsibility for, and control over, the future, religiously and otherwise. Further, the practitioner is enjoined not to look outside of himself to any agency of control such as God. (Prebish, 1978, p. 162)

The association between Asian religions and the American counterculture had mixed consequences for the Buddhist tradition: On the one hand, Asian traditions flourished as an outlet for disgruntled Americans, especially young people, and they received a great deal of attention. On the other hand, versions of Hinduism and Buddhism in the U.S. scene came in a variety of popular forms, often substantially modified from their Asian roots, and focused on a critique of American culture rather than on developing a true religious vision (see Johnson, 1976, p. 48). Yoga classes were given on television, and techniques of Yoga were sold as a form of self-therapy as well as consciousness expansion and relaxation.

Many Indians who were gurus in the yogic tradition—such as Yogi Bhajan of the Happy Holy Organization; Swami Satchidananda, who performed at Woodstock; Swami Vishnudevananda of the International Sivananda Yoga and Vedanta Society; and the Maharishi Mahesh Yoga, with whom the rock group the Beatles associated—came to America with their message. Transcendental Meditation, a particular technique of Yoga promoted by Maharishi Mahesh Yoga, became very popular; by 1975, half a million people had taken his courses.

The International Society for Krishna Consciousness (ISKCON) was founded in 1966 by A. C. Bhaktivedanta, who had been an active Krishna devotee in India before going to New York in 1965. The movement flourished in the 1970s, claiming 10,000 members (half of them in the United States) and 200 centers and communities worldwide (Rochford, 1985).

Bhaktivedanta believed that the world was near the end of the materialistic age of Kali-Yuga, the last cycle of a four-cycle millennium. If the populace could be aroused, a new age of peace, love, and unity would be discovered. With an interesting twist of irony, Western counterculture and modern mass culture technology together created a widespread dissemination of ancient Hindu and Buddhist ideas. When the Beatle George Harrison praised Hare Krishna in his songs, millions of American teenagers had their first positive encounter with Hinduism.

Satguru Maharaj Ji, a young Indian with some 5 million followers (including at least 80,000 Americans), was believed to be an incarnation of God (see Messer, 1976). He established the Divine Light Mission, which maintains households (*ashrams*) throughout the world for devotees who work full time for the movement. When she studied the group, Jeanne Messer (1976) found that the guru was often identified within the movement as Christ, despite the Hindu form of his religious rhetoric and rituals: Initiates were often told that "Guru Maharaj Ji is Christ, that Christ has been on the earth many times as Jesus, as Buddha, as Mohammed, as Krishna, or that Christ has always been on the earth" (p. 58).

One of the most popular—and widely studied—movements with Asian roots is the Unification Church (see Bromley, Shupe, & Ventimiglia, 1979), whose members are nicknamed the "Moonies" because their leader is a Korean minister named the Reverend Sun Myung Moon. Reverend Moon's followers consider him a manifestation of God and incorporate into their practice a number of Korean traditions, from a virulent anticommunism to arranged marriages. Followers of the church's ascetic practices are active in many parts of the world. A number of profitable businesses supporting their activities have become the subject of continued investigation by the U.S. Internal Revenue Service, which may be a form of official harassment.

Another successful group drawing broadly on Asian religions is Scientology, which claims 5 to 6 million members. This movement, founded by L. Ron Hubbard, combines Buddhist, Hindu, Taoist, and Christian ideas along with principles of psychoanalysis and business administration. It has been the subject of much controversy and the object of numerous attacks by U.S. officials and mainstream cultural elites. Scientology features a brand of psychotherapeutic technique called "auditing" that adherents claim helps one to free oneself not only of the burden of painful memories of early childhood, as in psychoanalysis, but also of those stored away from past lives. It has established centers around major universities and stages a Christmas scene with a Santa Claus on Hollywood Boulevard in Los Angeles, just down the street from the Kodak Theatre.

Goddess Worship

Whereas Asian religions have been used as a framework for much countercultural resistance in American culture, a number of feminist movements have turned to so-called pagan religions and witchcraft as a way of embodying their discontent with the patriarchal structure and content of the dominant religions.

What Mary Jo Neitz (1987) called the Goddess movement encompasses a variety of countercultural groups that, although they have "no unifying organization, written scriptures, or dogma, no defining ritual practice" (p. 355), belong to two intersecting currents: neopaganism and feminism. The neopagan movement revived witchcraft in the countercultural ferment of the 1960s, drawing upon pre-Christian Celtic folk religions from ancient Europe. The term *witch* is used deliberately, in spite of (or because of) its negative connotations, because participants in the Goddess movement are protesting the sexism of Christian culture.

Witches organize in covens, small grassroots women's circles that focus on bonding and empowering one another as women (although some neopagan circles include men as well). When witches gather, they draw syncretically upon ancient rituals of religions from what Eisler (1988) called the "partnership" societies of ancient Europe and new rituals that include one reported by Neitz (1987) in which women sit in a circle and give their names and the names of their matrilineal ancestors as far back as they can go. The participants quickly realize that the patrilineal naming practices have caused them to lose even the names of their foremothers. Goddess movement participants have experimented with the use of dance and music to tap the energy of the religious spirit.

Goddess worship is an interesting phenomenon that many North Americans find threatening because it represents a deliberate rebellion against the Christian cultural mainstream. Long a source of denigration by Western patriarchal culture, witchcraft calls up images of devil worship and black magic in the minds of many. For feminist witches, however, the Goddess is "a symbol for the empowerment of women" (Neitz, 1990, p. 353) and provides a vehicle for creating an alternative social space within the patriarchal culture. Some witches have emphasized the more generally accepted positive characteristics of the Goddess image, however, interpreting her as "an archetypal figure based in the early human experience of nurturance from a mother" (Neitz, 1990, p. 356). In that sense, the witches share much in the tenor of their religious movement with other Goddess movements around the world, including Chinese and Indian folk religions and, in a less obvious way, with the cult of the Virgin Mary that has permeated a broad spectrum of the Christian Church.

Because of its deliberately rebellious spirit and negative image in the mainstream culture, Goddess movement participants have often faced considerable opposition. In a recent series of events in Jonesboro, Arkansas, Terry and Amanda Riley were forced to close their shop, the Magick Moon, which sold books on witchcraft, incense, wands, and cauldrons. Area merchants banded together to prevent the two from finding another location. When asked by a *Newsweek* reporter about protecting their religious freedom, one local resident replied that he naturally supported religious freedom but said the Rileys did not have a religion (Shapiro, 1993).

According to Carol Christ (1987), a feminist theologian involved in the movement, the contemporary Goddess movement in North America "is the acknowledgement of the legitimacy of female power as a beneficent and independent power" (p. 121).

Neo-Christian and Charismatic Movements

Some rebels against contemporary American culture turn not to ancient pagan rituals or Eastern religions but to aspects of the diverse Christian tradition not normally emphasized by the religious mainstream. In the 1960s, a number of such movements, such as the Christian World Liberation Front, an outgrowth of the "Jesus Movement," or "Jesus Freaks," as they were sometimes called, emerged as part of the youth counterculture. Their leftist political orientation appealed to campus radicals in Berkeley, California, but these movements were opposed to Marxism and were somewhat introverted, so did not attempt aggressively to recruit new members.

Members of a related movement called themselves the "Submarine Church," after the Beatles song and movie *The Yellow Submarine*, because they considered themselves a Christian community that went underwater, not underground; they emphasized the idea of a radical faith community that challenged the establishment and championed the poor much like the early church. When the United Methodist Church held its quadrennial conference in St. Louis, Missouri, in 1972, the Submarine Church was there as well. At the major local church downtown on the day the conference convened, as a number of international dignitaries were gathered for worship, members of the Submarine Church unfurled a banner from the balcony with a bright yellow submarine and a cross on it. One of the members stood up from the floor just as the worship service began to make a "testimony" of concern about the poor people living in the neighborhood of the church. As he was arrested and escorted from the sanctuary, a number of people in the congregation stood up to voice their protest against the arrest; they, too, were taken off to jail. (Two Methodist bishops later visited the police station and negotiated their release, and all charges were eventually dropped by the church.)

Other new Christian groups had a more conservative message but also represented a counterculture movement within the church. One of these was Campus Crusade, a popular evangelical movement that appealed to a different audience from the other groups mentioned thus far but differed from mainline Christian campus ministries in its aggressive recruiting methods and sectarian organizations.[1] A fast-growing and vital alternative Christian group is the charismatic movement, which swept the United States and many other parts of the world. This movement protests the formality and "dryness" of contemporary establishment Christianity and emphasizes so-called "gifts of the Spirit" (see Neitz, 1987). Although originating in Protestant churches, the charismatic movement became very popular in the American Catholic church following the Second Vatican Council. In 1966, faculty members at Duquesne University in Pittsburgh, Pennsylvania, began meeting regularly to pray for these "gifts," which included "speaking in tongues," or glossolalia, vocalizations that are linguistically unintelligible despite recognizable phonetic features (Lane, 1976) and seem to express intense religious experiences. The Catholic Pentecostal movement, or Catholic Charismatic Renewal as it was later called, spread quickly; at its first annual convention in 1967, 90 persons came; 5 years later, 10,000 attended and the following year, more than 20,000.

The charismatic movement has touched a deep chord in contemporary popular culture, "a concern that the world is falling apart," as Mary Jo Neitz (1987) put it in her ethnographic study of a Catholic charismatic group: "Yet the problems on which their actions focus are overwhelmingly personal, with concern for protecting oneself and one's family" (p. 232). The intense community of the movement and the idea of a powerful personal relationship with Jesus have provided an important source of support for thousands of people living tenuously in the postmodern world.

Who Participates?

Glock and Bellah's (1976) classic study of NRMs in the San Francisco area gives an informative portrait of participants and groups in the major center of cultural ferment at a significant time, the early 1970s. Robert Wuthnow's (1976) contribution to that study identifies three broad categories of movements: (1) countercultural movements that provide an alternative to mainstream American culture (e.g., Zen Buddhism, Yoga, Transcendental Meditation, Hare Krishna), (2) personal growth movements that combine religious practices with psychology and an emphasis on personal development

(EST, Synanon), and (3) neo-Christian movements that are sectarian in nature and draw upon the Christian tradition to create an alternative, usually sectarian, Christian community (such as Campus Crusade, Jews for Jesus, Children of God, and the Christian World Liberation Front).

The countercultural movements tend to overselect young people, with slightly higher than average educational levels. Not surprisingly, they are likely to reinforce values and lifestyles that contrast sharply with convention. Participants in all groups (but especially the countercultural) are more likely to be single and employed part time or looking for work and more geographically unsettled than the general population. Counterculture movements appear to be stronger than their small numbers would indicate because they attract better-educated people and consequently have greater influence. The movements in general may continue to find a more responsive chord among the young, the educated, and those interested in more general forms of cultural and societal transformation.

"Networks of Faith"

Sociologists usually emphasize what Rodney Stark and William Bainbridge (1980) called **networks of faith** in explaining why people join religious groups, whether more conventional, mainstream groups, or the unconventional groups under discussion here. Two slightly different emphases emerge in the literature: (1) on the appeal of the ideology of various groups and the needs of those who join them and (2) on the importance of interpersonal relations. The latter position contends that membership spreads through social networks and that people adjust their religious beliefs to conform to those of people who are important to them.[2] Stark and Bainbridge (1980, p. 1392) provided evidence from studies of three groups (a doomsday group, an Ananda commune, and the Mormon church) for the importance of social networks in recruitment to both types of NRMs. They concluded that a sense of deprivation (e.g., social isolation, people with a grievance of some sort) and ideological compatibility "limit the pool of persons available for recruitment" (p. 1392), but since many people are deprived and ideologically predisposed to cult membership but do not join, a number of situational variables must be explored to explain why some join and others do not.

The number of new religious groups in the United States jumped dramatically between 1950 and 1970: That is the period in which the United States became fully integrated into the world economy; large numbers of immigrants came to the country; tourist travel flourished; and television brought the far corners of the globe into the nation's homes. Many Americans became disenchanted with their own cultural milieu and religious traditions. The same impulse that led to the hippie movement, the New Left, and the civil rights and antiwar movements of the 1960s influenced many to search for alternative religious communities in which to pursue their spiritual quests.

The institutional and technological infrastructure facilitating the spread of NRMs, including cable television and video technologies, developed dramatically in the 1970s and 1980s. ISKCON, for example, developed its own network of print and video newsmagazines, and ISKCON TV began producing instructional videos with lectures from Swami Bhaktivedanta and reports on ISKCON activities around the globe. Distant religious practices no longer appeared so foreign, and as young people rebelled against their own culture and their parents' values and religious traditions, large numbers of Western youth sought spiritual and social support in these alternative communities. In some instances, religious traditions from other parts of the world provided a significant alternative to the conventional (mostly Christian) traditions of their families.

Finally, Stark and Bainbridge (1980, p. 1393) concluded that potential recruits must receive *direct rewards* from their participation if they are likely to join a movement. Although the affection of the movement network is often an important factor—especially for people who suffer from isolation and low self-esteem—successful groups also include other rewards. The Hare Krishnas and Moonies often provide food, clothing, and shelter as well as a meaningful occupation with potential for advancement within the organization. The Mormons have a 13-step recruiting plan that showers potential new members with tangible rewards.

Contemporary empirical research on reasons for joining NRMs does not advance us theoretically much beyond Max Weber's concept of elective affinities between ideas and interests, although it does spell out the nature of the relationship in more detail. People tend to join a NRM primarily for social reasons, adopting first the ethos of a group and gradually accepting its worldview as well. Most people, of course, choose the religious tradition of their families and other immediately relevant social groups. In the global village, however, switching may become more common. The role of elective affinities in the conversion process can be seen quite clearly within some social groups, which, as they undergo dramatic transformation, construct an ethos and worldview they feel a stronger affinity for than those traditionally available to them in their social situation.

In the late 20th century, the number of NRMs in the United States has skyrocketed, in part because of unprecedented new immigration of ethnic groups from parts of the world where Christianity did not predominate (such as Asia) but also because of the cultural upheaval of the 1960s, in which—as we have just seen—many, especially young people, turned to alternative forms of religious expression as part of the counterculture movement. In the first decade of the 20th century, only 11 new religious groups had formed in the United States. In the 1960s, 105 new communities were founded; in the 1970s, it was 177 (see Hinnells, 1984). The only time the trend was reversed during the century was during World War II, which fostered a climate that discouraged cultural experimentation. When the global economy emerged in the postwar world, social networks expanded and cultural patterns were widely disrupted and reformed. The most obvious reason for the birth of NRMs anywhere in the world is that people live increasingly in a global context. People living in a heterogeneous culture generally have more freedom of choice and more religious options in the religious marketplace.

Cult and Anticult

Many Americans find alternative religious communities repulsive and threatening, and the negative label *cult* is popularly used to refer to all of them. Some of the reasons for this criticism are quite valid: Many horrible things are done in the name of these NRMs, just as much evil has been committed in the name of every one of the more established religious communities. People also object to NRMs because they are different, intense, authoritarian, and often cause family members and friends to break with kin and friendship networks, much to the dismay of those left behind. Finally, a great deal of misinformation is spread about alternative religious groups by those who are critical of them.

Ironically, the anticult movement that emerged in the wake of the proliferation of religious movements in the 1960s has often exacerbated the very characteristics that trouble many people about the NRMs in the first place. Under attack by the outside, participants in alternative religious communities become more alienated from the external world and consequently more amenable to authoritarian structures and leadership.

Brainwashing and Deprogramming

As NRMs attracted widespread attention in the 1970s, the anticult movement became increasingly aggressive in its opposition. A major source of the movement was the hostility and fear from family members of converts who felt abandoned, sometimes under stressful conditions. A network of anticult movement organizations mounted a campaign against the NRMs and clashed with the elites of these groups, creating what Bromley and Shupe (1981) called a "social scare. " This means the following:

(1) a sociocultural climate characterized by heightened tension as a result of (2) intense conflict between two (or more) social groups in which (3) the more powerful group mobilizes control claims by (4) denigrating the moral status of the less powerful groups through (5) construction of a subversion mythology. (Bromley, 1988, p. 186)

The key element in the "subversion mythology" constructed about NRMs was the idea of brainwashing—that individual recruits had been robbed of their free will through mind control techniques.[3] The anticult movement focused on Reverend Moon's Unification Church because of its visibility, aggressive recruitment of young adults, direct challenge of the authority of traditional churches, and provision of a single target on which to concentrate its efforts most effectively (Bromley, 1988, pp. 188–189).

The most remarkable part of the campaign was the creation of what anti-cult activist Ted Patrick (1976) called the **deprogramming** process, intended to help "victims" snap out of the hypnotic trance into which cult members were supposedly put by the brainwashing process. In the process, NRM members were kidnapped, taken to a secluded location (such as a hotel room), and subjected to various techniques, including the following:

[eliciting] guilt for rejecting family members and educational plans, expressing love and concern about the dubious future the individual had charted, refuting the group's doctrines, revealing esoteric beliefs and practices that were not known to the individual, challenging the motives and sincerity of the group's leaders, providing testimonials by former members that they had been brain-washed but failed to recognize their own psychological captivity, and threaten-ing that the individual would be released only on the condition that membership in the group was renounced. (Patrick, 1976, p. 194)

Although deprogramming incidents involved varying degrees of coercive-ness, these anticult tactics raise many questions about the reality of religious freedom in the United States. Bromley (1988) contended, "The practice of forcibly separating individuals from religious groups for the purpose of inducing them to renounce their membership is unprecedented in American religious history" (p. 203). The extreme measures of deprogramming were given justification by the mythology of brainwashing and supported by much of the general public. The news media in the United States further legitimated deprogramming and other attacks on the movements by disseminating "atrocity tales" about NRMs (Bromley et al., 1979). By evoking outrage at the alleged acts of religious groups, newspaper accounts facilitated the "social construction of evil" about these religious movements.

In their study of 190 newspaper articles about former members of the Unification Church between 1974 and 1977, Bromley and colleagues (1979) found that all but two contained at least one atrocity story and were primarily hostile toward the church. The most frequently reported atrocities were as follows:

1. Psychological violations of personal freedom and autonomy

2. Economic violations: reports that the church forced members to sell their private property and give it to the church

3. Severing of the parent–child bond—the most sensational of the reports, growing out of the hostility of families who were rejected by members joining NRMs

4. Political and legal atrocities stemming from the fact that the church was founded and run by a foreigner

Although there was an element of truth in many of these stories, the point is that these problems are present in many other organizations and the kind of coverage provided was entirely negative.

That people should be *forced* out of the Unification Church to regain their freedom was an irony lost on many. The movement in and out of most NRMs is relatively high and the average stay within the communities is rather short. Eileen Barker (1988) found "that the majority of members joining the Unification Church seem to leave voluntarily within two years of joining" (p. 167). Long-term affiliation and involvement seems even more tenuous; only 3% of 1,000 people attending a workshop in 1979 were full-time members by the end of 1985. People left the movement for a variety of reasons, and many of the deprogramming efforts were successful,[4] but the cost of extricating them was extremely high to those involved and to the society as a blow to religious freedom.

What to Do About "Cults"

Alternative movements, or NRMs, do present a threat to the established sociocultural order in the United States not only because of what the movements themselves do but also because of the anticult movement's campaign against them. Families are disrupted, and people are cut off from former social networks, sometimes causing pain and anguish; those developments are often symptoms of deeper problems in the relationships and in the broader culture, rather than something caused by the NRM itself. The record of the response to these threats has been mixed. Alternative religious movements have not been accorded the kind of protection guaranteed in the U.S. Constitution. They have been harassed not only by the anticult movement but even by agencies of the state (see Bromley & Shupe, 1981; Wallis, 1976). Unfortunately, such responses are more consistent with historic practices than is a thorough protection of religious freedom.

The tragedy at the Branch Davidian compound in Waco, Texas, in 1993 is an extreme but not entirely unrepresentative case. Federal agents came to arrest the leader of a Christian sect, David Koresh, whose community had stockpiled a vast arsenal of weapons and were awaiting the Battle of Armageddon, which, according to Christian scriptures, would signal the end of the world. Several government agents were killed in the raid; after a long standoff between the FBI and Koresh's followers, remaining members of the religious community perished as their building went up in flames. Confronted by a heavily armed group of "religious fanatics," the federal government responded with violence. The standoff in Waco involved two groups of people who were well armed and willing to fight to the death for higher principles. Although the religious movement involved may have been unjustified in its stockpiling of guns and David Koresh may have misled members of the group by using his charismatic authority to exploit their fears, the question remains as to whether the most appropriate way to respond was in kind—that is, with the same violence and pressure tactics that Koresh was condemned for using. Criticism of the government's response to the Branch Davidian situation was widespread, and an investigation was initiated by the U.S. Attorney General's office. Even though such investigations are often an attempt to justify and cover up the actions taken rather than to expose mistakes made, the norm requiring a clear explanation for such actions was affirmed.

Science-fiction writers have often evoked the theme of what we would do if confronted by an "alien force," by a life-form from another planet with different customs, beliefs, and rituals. Perhaps we can see the answer to that question in how dissident groups are treated in any given society. Kai Erikson's *Wayward Puritans* (1966; Erikson, 1965) concluded that the kind of deviance most likely to be defined and sanctioned in any given society will reflect the important values of that society. The outliers and outlaws provide something of a mirror image of the dominant norms and values, but both criminals and police operate by the same rules and share many of the same values.

Without attempting to prescribe public policy recommendations regarding alternative religious movements, it does seem important for societies to consider the following questions:

1. How should the freedom of all religious movements be protected? This is not an easy, absolute task, and certainly society will draw boundaries around acceptable and unacceptable behavior. Most modern constitutions make religious freedom an official tenet of each nation's political culture (although as we have seen the norm is not always adhered to, for example, in the United States). Consequently, actions taken by the state against any religious group should be done within the boundaries of strict due process and not because of prejudice or theological disagreements. Religious movement members should be accorded the same legal rights as rapists, murderers, and the Ku Klux Klan.

2. How can religious movements be evaluated on the basis of their "fruits"—that is, in terms of the personal and social consequences of their religious beliefs? Disagreements should take place within the confines of the moral debates about what is acceptable and unacceptable behavior, as far as the law is concerned, and questions of theology should be debated theologically. In a heterogeneous society, some people will obviously find some groups more attractive and others more repulsive; is it possible for people to debate those differences freely, rather than trying to suppress those with whom they disagree?

From a sociological point of view, these questions are ultimately the kinds of questions that must be asked about the way in which religious movements will be treated in the global village as we attempt to fashion a global ethos that permits the peaceful coexistence of diverse social groups.

New Forms of Religiosity

One variant of the secularization thesis suggests that the new cultural forms of the modern world are, in fact, a sort of invisible religiosity. Thomas Luckmann (1967) argued that the central cultural themes of "individual 'autonomy,' self-expression, self-realization, the mobility ethos, sexuality and familism," as well as a number of other less important topics, constitute something of an "invisible religion," because they have some claim to a "sacred" status in modern culture, but are not explicitly organized religious traditions (p. 113).

Durkheim's study of aboriginal Australian religion in his formative book, *The Elementary Forms of the Religious Life* (1915/1965), was part of his lifelong search for functional equivalents of the mobilizing power of religion to revitalize the social order and provide a basis for social solidarity. His sociological predecessor Auguste Comte capped his career by elaborating a secular "religion of humanity," basically Catholicism without the supernatural, and Durkheim appeared to be taking somewhat the same path, although with a more sophisticated method. He concluded that the phenomena being worshipped in religious ritual were actually social forces themselves and that ritual could be harnessed for the common good if science were used to provide a moral basis for the collective life.

Durkheim was not alone in his quest for functional surrogates for religion; many other moderns who were "spiritually musical" (that is, were spiritually inclined and talented) but disillusioned with traditional religious forms have sought to replace them with a new worldview and its corresponding ethos. Others have replaced some of the social and psychological functions in a less deliberate manner, through other social activities as voluntarism, dance, music, sports, or nationalism. Even on a relatively mundane level, collective life is now often expressed through secular rituals that take on a quasi-religious character, especially for those most dedicated.

The line between religious and secular rituals is very thin, as both lie along the same continuum of social forms that run from the most sacred to the most profane. In modern and postmodern societies, secular rituals from political campaigns and television ads to rock concerts and Olympic Games fulfill many of the functions identified by Durkheim as religious, promoting social solidarity and facilitating the process of collective identity construction.

Cultic practices often emerge around charismatic cultural icons like popular rock singers or political figures. Many of the groups surrounding particular rock groups, such as the Grateful Dead, take on many characteristics of a religious movement. In its most extreme forms, fans become devotees and exhibit behavior that borders on worship. During the height of their popularity, perhaps the most famous group of the latter 20th century, the Beatles, were thought to have such miraculous powers that people were brought to their dressing rooms for healing. The singers themselves tried to downplay these practices, and John Lennon even wrote a song protesting such attitudes and providing a long list of things he did not believe in, including God and the Beatles, concluding that he believed only in himself.

Women's Movements

"I'd like to be a priest when I grow up," she said.

"You can't," I said, "you're female."

"So what?" she said, in the tone she defies her grandmother but not her mother in.

"Girls can't be priests," I said. "Our Lord said so."

"Where?" she insisted.

I told her He didn't say it in so many words, but He chose no women to be apostles, and priests are successors to the apostles. That means they would have to be like the apostles.

"But the apostles were Jewish, and you're not Jewish," she said.

"What's that got to do with it?" I asked her.

"So, you're not like them, and you're a priest." She glowed with successful argument.

I thought of all the foolish, mediocre men who were permitted ordination because of the accident of their sex. And I thought of this child, obviously superior to all others of her age in beauty, grace, and wisdom. I told her to pray that the Church would change its mind by the time she grew up.

"You pray, too," she said.

I said I would, but it must be a secret between us. And so each morning, at my mass, I pray for the ordination of women.

—Mary Gordon, *The Company of Women* (1980)

Nowhere do interreligious encounters occur more intimately than in the separate voices of men and women as they encounter the sacred not in different parts of the globe but within their own homes. One of the most profound of modern movements—that affects all levels of society as well as all societies—is the demand for equality by women (see Sharma & Young, 1999; Shih, 2010). This movement, in its various forms, challenges the very roots of the major world religions, each of which has its own liberating traditions but all of which have consistently legitimated patriarchal culture and male domination. The related social problem, of course, lies in the ways in which female identity is shaped from a very early age, so that religious legitimations of inferior status affect the lifelong socialization process and the ways in which women (or any other group of people) will be treated and, consequently, think of themselves.

Every major living religious tradition has a patriarchal tendency, and the Western religions, especially Christianity, have been soundly criticized for their sexism. As with critiques of that tradition during the French Revolution, the basis of the complaint is the use of religious institutions and ideas to impose inequality. Some of the early women's leaders in the United States found aspects of Christianity downright immoral. Elizabeth Cady Stanton, for example, felt this way:

[She] was shocked by the frank misogyny of the original [Bible]. Genesis, for example, read to her like "gross records of primitive races," and the stories of Lot's daughters (who got their father drunk and then seduced him) and of

Tamar (who dressed as a whore to seduce her father-in-law) she found unworthy of comment. As for contemporary Christianity, she wrote: "So long as ministers stand up and tell us that Christ is the head of the church, so is man the head of the woman, how are we to break the chains which have held women down through the ages?" (Ehrenreich, 1981, p. 38)

A different strategy of some contemporary critics, like Mary Daly and Rosemary Radford Ruether, has been to forge a feminist theology that reshapes the traditions in a less sexist manner. Ruether (1981, p. 388) developed a feminist critique of religious studies that begins with the historic exclusion of women from religious leadership roles in the Judeo-Christian tradition. She noted that women were prohibited from studying the scriptures in Judaism, as in the rabbinic dictum "cursed be the man who teaches his daughter Torah." Christianity has had similar practices, noted in the New Testament statement, "I do not permit a woman to teach or to have authority over men. She is to keep silence" (1 Timothy 2:12). Moreover, Ruether (1999) claimed this:

> Christianity from its New Testament beginnings exhibited a deep tension between egalitarian and patriarchal views of women. Feminist scholars of the New Testament and early church are increasingly showing that the polemics against women's leadership found in New Testament and patristic writers can only be explained by recognizing an alternative understanding of the gospel as the "good news" of the dissolution of gender hierarchy in the new humanity in Christ that arises from the baptismal font. (p. 214)

Women were usually permitted only an inferior role in shaping the major traditions, and a number of derisive characterizations of women have been made by the core figures of Western thought: Thomas Aquinas, for example, defined a woman as a "misbegotten male." In the medieval scholastic tradition, efforts to ordain a woman for the priesthood were considered simply impossible—the ordination "would not 'take,' any more than if one were to ordain a monkey or an ox" (Ruether, 1981, p. 390). Ruether (1981) argued the following:

> The male bias of Jewish and Christian theology not only affects the teaching about women's person, nature and role, but also generates a symbolic universe based on the patriarchal hierarchy of male over female. The subordination of woman to man is replicated in the symbolic universe in the imagery of divine-human relations. . . . Thus everywhere the Christian and Jew are surrounded by religious symbols that ratify male domination and female subordination as the normative way of understanding the world and God. This ratification of

male domination runs through every period of the tradition, from Old to New Testament, Talmud, Church Fathers and canon Law, Reformation, Enlightenment and modern theology. It is not a marginal, but an integral part of what has been received as mainstream, normative traditions. (pp. 390–391)

Ruether's response to this situation is not to reject the entire tradition but to reshape it, first by documenting the male bias and tracing its sociological roots (see Daly, 1968/1975; Ruether, 1974). The second step is to discover "an alternative history and tradition that supports the inclusion and personhood of women" (Ruether, 1981, p. 391), either within the Jewish and Christian traditions or elsewhere. A number of well-documented studies show, Ruether contended, that the exclusion of women from leadership roles is not the whole story. Women were probably not so excluded in 1st-century Judaism, and the rabbinic dicta against teaching women Torah is only one side of an argument, albeit the side that won (Ruether, 1981, p. 392).

Similarly, the passage in 1 Timothy just cited is a second-generation reaction, Ruether claimed, against the widespread participation of women in leadership positions in the early church. It is unlikely that anyone would bother to oppose female involvement if it were not happening. Ruether argued that the participation of women in early Christianity was a natural part of its theology, in which "baptism overcomes the sinful divisions among people and makes us one in the Christ: Jew and Greek, male and female, slave and free" (Galatians 3:28; Ruether, 1981, p. 393).

Some support for feminist perspectives on the early church have come from unexpected quarters, as in Robin Scroggs's (1972) contention:

We must make a radical reversal in our interpretation of Paul's stance toward women. Far from being repressive and chauvinistic, Paul is the one clear and strong voice in the New Testament speaking for the freedom and equality of women in the eschatological communities he has helped create. (p. 309)

Some authorities in the early church, Scroggs (1972) contended, "found this freedom too radical and quickly rewrote Paul to make his writings conform to the practices of the establishment church" (p. 309). The charge is a serious one, based on textual criticism concluding that portions of the text were altered. This should be taken in light of different norms of the status of texts in the ancient world, where individual authorship rights were not rigidly defined as they are in modern cultures. Paul lucidly sets forth the equal rights and responsibilities of men and women, grounds this freedom in the liberated humanity of the new creation, and assumes that women will live in this freedom.

Women and the Priesthood

Debates about the role of women in religious institutions often focus on their eligibility for the priesthood. Some branches of Protestant Christianity, especially in the mainline denominations, include women in leadership roles so that there are even some female bishops. In the bastion of male clergy, the Roman Catholic Church has proponents of the ordination of women (see the website http://www.womenpriests.org/index.asp, which claims that 9 out of 10 scholars favor ordination and provides over 1,000 documents relevant to the issue). In several areas of Protestantism and in the Roman Catholic and the Eastern Orthodox churches, however, resistance to female leadership in the church is still very strong and unlikely to change soon, despite shortages of priests.

In the meantime, barriers are falling elsewhere, albeit in isolated instances. In Pune, India, in 1984, for example, a group of Hindu women began chanting hymns and conducting rites previously reserved for male Brahman priests. Shankar Hari Thatte, a 76-year-old Brahman, brought them together for that purpose, contending the following:

> The men priests were cheating people, their lives had degenerated, they were unable to honor the holy books and I felt I should organize the women because there is no specific ban on them performing these rites in the religious texts. (Hazarika, 1984, p. 4)

Although some priests and scholars voiced objections, no one attempted to stop them. Ganesh Shastri Shinde, an 86-year-old priest, contended that what they are doing is "against religious traditions," but added, "we will not interfere with their ways. We will let them go on their path and we will continue on ours" (Hazarika, 1984, p. 4).

A year later, Amy Eilberg was ordained as a Conservative Jewish rabbi after graduating from the Jewish Theological Seminary in New York (Goldman, 1985). Following a vote of the Rabbinical Assembly allowing the ordination, Eilberg was admitted into the ancient priesthood as the first Conservative woman member (Reform Judaism had been ordaining women for the previous decade). Although Rabbi Eilberg fulfilled a lifelong dream and broke another barrier, it remains to be seen how long it will take before women rabbis are accepted as equals. Conservative female rabbis are becoming more common, however, and in December 2005, the *Toronto Star* ran a feature story on "dual-rabbinic spouses" featuring Canada's first married Conservative rabbis, Jennifer and Sean Gorman (Csillag, 2005).

Nonpatriarchal Traditions

Some people have used historical and anthropological studies to explore alternative religious traditions that are not patriarchal, including both ancient Mother Goddess traditions and popular contemporary feminist spirituality groups. In a provocative analysis of archaeological research, Marija Gimbutas (1982) contended that because "the task of sustaining life was the dominating motif in the mythical imagery of Old Europe," that is, in the period before about 4500 BCE, "the goddess who was responsible for the transformation from death to life became the central figure in the pantheon of gods" (p. 236). Male and female divinities were presented side by side, however: "Neither is subordinate to the other; by complementing one another, their power is doubled" (Gimbutas, 1982, p. 237).

Rianne Eisler's (1988) *The Chalice and the Blade* analyzed these findings by positing a partnership society, in which men and women were equal partners, that was overrun by male-dominated warrior societies at the end of the Neolithic Age, during the 3rd to 5th millennia BCE. For about 20,000 years, Eisler argued, most European and Near Eastern societies were based on simple, supportive technologies, matrilineal descent, and common ownership of the means of production. They had a cooperative social organization and a gynocentric culture, with the deity represented in female form. Their Kurgan conquerors imposed a dominator model of social organization with male deities, "a social system in which male dominance, male violence, and a generally hierarchic and authoritarian social structure was the norm" (Eisler, 1988, p. 45).

Although the jury is still out on the exact nature of the evidence and its implications, a number of discoveries are quite remarkable. First, representations of weapons appear only after the Kurgan invasions, with "the earliest known images of Indo-European warrior gods" (Eisler, 1988, p. 49; cf. Gimbutas, 1982, 1989). The weapons apparently represent the power and function of the Gods, with the Goddess gradually appearing as the wife or consort of the male deities. Moreover, the nature of burial sites begins to shift at about the same time, from more egalitarian graves to hierarchical "chieftain graves" with marked differences in size and "funerary gifts"— that is, the contents found in the tomb along with the deceased. For the first time in European graves, an exceptionally tall or large-boned male skeleton will be accompanied by the "skeletons of sacrificed women—the wives, concubines, or slaves of the men who died" (Eisler, 1988, p. 50).

If the obvious interpretations of these findings are true—and it seems plausible—the implications are profound. They are both horrifying and hopeful. On the one hand, a radical cultural revolution occurred in human prehistory thousands of years ago, a revolution in which we are still participating, that values hierarchical, male-dominated culture at the expense of partnership models, and warrior Gods over peaceful, nurturing deities. On the other hand, these archaeological discoveries—like some of the existing alternative religions in the 20th century—also suggest that human social organization can be different from the way it now is and that our worldviews might once again be transformed before our own warrior Gods overtake us.

Female Deities and Quasi-Deities

A number of contemporary religious traditions contain a strong female presence. One of the most prevalent figures worldwide is the Virgin Mary, a particularly potent symbol for women and the poor around the world, although she remains a somewhat suspect figure for most Protestant churches. Pope John Paul II, apparently recognizing her importance, and echoing the long-standing popular interest, actually promoted interest in Mary from the Vatican. The Beatles' popular song "Let It Be" captures the spirit of the religious adoration of Mary, claiming that she comes in times of trouble.

Chinese religious symbols include three major female deity cults: Kuan Yin, Ma Tsu, and the "Eternal Mother" (see Sengren, 1983). Although they do not represent a simple transfer of men's and women's social roles, these deities contrast sharply with the male bureaucratic deities in Chinese folk religion. The female deities provide some alternatives to the hierarchical models of the male deities and reflect the mediation role that mothers often play in Chinese family life. Kuan Yin, Ma Tsu, and the Eternal Mother all intercede in times of danger and are approachable in ways that the bureaucratic deities are not, just as Mary seems closer to humans in need (especially women) than the Christian God, as either a father or a son (see Warner, 1976).

Sengren (1983) suggested that whereas one can approach "ancestors or territorial cult deities only as a representative of a patriline or household, one can approach female deities as an individual" (p. 20). Male deities act as officials, responding to justice and bribes and promises of payment, but female deities are moved more by a worshipper's devotion and dependence. Thus, the female deities provide an alternative model of authority within the religious sphere that reflects differences in authority elsewhere. They facilitate the empowerment of women in religious life in a way that is difficult for the male deities, because they respond to the models with which women are socialized in Chinese society.

In practical matters of worship, the female deities in China may act more as opiate than activists despite their "proven" powers, thus covering over

any symptoms of discontent among women that might emerg
inequality within religious institutions. Men tend to run
temples, even those with female deities, who are often mo
domestic Gods. Moreover, the attributes the Goddesses possess reᵢₗₜₜ
of women in the traditional patriarchal society, even if in a subtly subversive
way. In the sectarian cults of the Eternal Mother, a subversion of the hierarchical structure of celestial bureaucracies reflects the coalitions between
children and mothers against their fathers in extended Chinese families
(Sengren, 1983; Wolf, 1968, 1972).

Religious deities or quasi-deities such as Mary, Kuan Yin, Ma Tsu, and
the Eternal Mother thus reinforce the hierarchical, male-dominated social
order under normal circumstances but also contain the seeds of rebellion—
sometimes as a subtle form of covert sabotage but also potentially as a direct
confrontation. Most important, as Neitz (1990) noted, "The symbols of the
goddess movement in themselves represent cultural change. For many the
rituals are new forms of play. For others the rituals express their deepest
hopes for social transformation as well" (p. 370).

The introduction of Goddess worship into the contemporary religious life
of advanced industrial cultures is not limited to those who have rejected
mainstream religious forms altogether and opted for alternative systems, or
even to the more subtle inclusion that surrounds the adoration of Mary in
Catholicism. Controversy recently erupted in the mainline Protestant
denominations of the United States over the use of a ritual devoted to
Sophia, an ancient feminine image of God in Christianity. At an ecumenical
gathering of women in Minnesota, elements of the ritual were included in a
worship service, provoking a series of attacks on the Women's Division of
the United Methodist Board of Global Ministries, which had financed the
participation of several United Methodist representatives at the conference.

One of the most vital of women's movements is the effort of women in
Islam to overcome the patriarchy of cultures in which Islam has taken root
and flourished. Although Muslim scholarship has traditionally been a male
bastion, increasing numbers of women in various parts of the world are
beginning to take up the Qur'an and the Hadith literature in an effort to
reinterpret the tradition on the basis of careful scholarly examination.

Riffat Hassan (1999), for example, explored the basic questions surrounding the role and status of women in the Qur'an and the tradition,
concluding the following:

> The Qur'an, which is the primary source on which Islam is founded, consis-
> tently affirms women's equality with men and their fundamental right to actual-
> ize the human potential that they possess equally with men. Seen through a

nonpatriarchal lens, the Qur'an shows no sign of discrimination against women. If anything, it exhibits particular solicitude for women, much as it does for other disadvantaged persons. (p. 275)

Hassan (1999) contended, by means of a careful examination of key passages in the scripture and tradition, that the Qur'an's position on key issues surrounding women's rights have been superseded and often contradicted by patriarchal interpretations. She claimed that the passages in the Qur'an about creation do not show that men were created before women, showing instead that the language in the cosmogenies refers to both genders. Similarly, the idea that Eve (Hawah) is somehow responsible for the "fall" of humanity is not possible in the Qur'an because it contains no concept of such a fall. Finally, the purpose of women in the Qur'an is not to be servants of men but to be their partners and protectors, as men are to be theirs.

All of the major living religions of the world are in many ways patriarchal and male dominated. They explain the world in masculine terms and tend to reinforce male-dominated social structures. Underneath the surface of these religions, the feminine face of God persists, kept alive sometimes quietly and sometimes loudly. Gimbutas (1989) wrote, "Now we find the Goddess reemerging from the forests and mountains, bringing us hope for the future, returning us to our most ancient human roots" (p. 321). As women become increasingly prominent in public life in the global village, they may change the face of religious beliefs as profoundly as the interaction among the major world traditions now dominating the religious landscape.

Environmental Movements

Many of the religious traditions—like the Goddess worship of Old Europe—that emphasize a harmonic relationship with the natural environment have been destroyed by industrialization, along with the destruction of the environment that has accompanied so much of modern economic development. The more utilitarian attitude toward nature emphasized in the dominant cosmopolitan religions of the 20th century has been challenged on a number of fronts in recent years. For some practitioners of ancient religious traditions like those of some Native Americans and other indigenous groups around the world, the idea of harmony with the environment and a critique of modern ecological destruction have forced the basis of a protest theology that has a clear affinity with the interests of these groups, which are often exploited by the "advance of civilization."

The environmental movement, like the pro-democracy and women's movements, was born largely out of a religious sensibility but contains an ambivalence toward religious traditions because of the legitimation for environmental devastation those traditions have provided. A new sensitivity to ecological issues has emerged in many religious traditions in recent years, however, in part because of the emphasis on placing ethical values and broader universal causes above short-term profit motives that often fuel environmental destruction.

Attention to environmental issues has fanned the flames of multicultural conflict and protest theologies. Of particular interest are questions surrounding the role of Christianity in creating and solving environmental problems. Especially since the 1990s, a number of Christians have attempted to mobilize attention to the environment as a faith issue, including Ian Bradley (1992) in his *God Is Green: Ecology for Christians* and James Nash's (1992) *Loving Nature: Ecological Integrity and Christian Responsibility*. Others claim that the environmental crisis is symptomatic of deep-seated problems in the Judeo-Christian tradition that must be dramatically changed (see Ruether, 1992). A third group attempts to replace or supplement Christianity with other religious perspectives, whereas a final argument contends that environmentalism is an anti-Christian movement undermining the authority of the church (see Wilkinson, 1992).

At the Earth Summit in Rio de Janeiro, Brazil, in 1992, leaders from many religious traditions gathered to focus their attention on the environment along with official governmental representatives and people from various nongovernmental groups around the world. Loren Wilkinson (1992) reported the following in *Christianity Today*:

> The Christian presence at the forum was swamped by a plethora of feminist, universalist, and monist groups, who argued that a new religious paradigm must replace the old one, which was shaped by patriarchy, capitalism, theism, and Christianity. Many blamed the "old paradigm" for the environment's destruction. (p. 48)

The charge against Christianity is considered inaccurate by the Jesuit Drew Christiansen (1992), who contended "that there is something disingenuous in maligning those who, only a generation ago, were considered insufficiently modern as perpetrators of modernity's capital crime" (p. 449). The root of the problem, he claimed, is not any religious tradition but "social systems built on material accumulation" (Christiansen, 1992, p. 451).

Christian theologian Rosemary Radford Ruether argued, however, that such groups as the "deep ecology" and feminist movements are correct in identifying Western culture, sanctified by Christianity, as the major cause of destructive conflict. Ruether (1992) claimed that "ecofeminism brings together . . . two explanations of ecology and feminism in their full, or deep forms, and explores how male domination of women and domination of nature are interconnected, both in cultural ideology and in social structures" (p. 2). Her approach, though not opposed to looking for insights from non-Christian traditions, is to "sift through the legacy of the Christian and Western cultural heritage to find usable ideas that might nourish a healed relation to each other and to the earth" (Ruether, 1992, p. 2).

The green liberals of the 1990s have been joined by the green evangelicals in the new century, with some evangelicals even combating global warming (Goodstein, 2005). Just as issues of racial injustice became a matter of concern for people throughout the church in recent decades, a broad coalition of Christians are now joining forces—despite disparate theological foundations—to promote a new focus on environmentalism as a Christian duty of stewardship for God's creation. Many in this group have become politically active, as Blaine Harden (2005) reported this:

> Frustration with the Bush administration's environmental polices is bubbling up from mainstream churches and synagogues, as reflected in a statement signed recently by more than 1,000 clergy and congregational leaders in about 35 states.
>
> Called "God's Mandate: Care for Creation," the statement says that "there was no mandate, no majority, or no 'values' message in this past election for the President or the Congress to rollback and oppose programs that care for God's creation." (p. A16)

Contemporary environmental movements often turn to indigenous religious traditions for inspiration (Harvey, 2003; cf. the discussion of Chief Seattle in Chapter 4). Although one should not idealize indigenous cultures, many of them had ritualized protections for the environment and managed to live in peace with it.

Constructing and Reconstructing Religious Life

As we have observed in this chapter, religious life has not died in the modern world, as many scholars expected. Instead, religious beliefs and practices have been reformulated in a variety of ways. The ancient traditions of the mainstream have been revitalized, for example, as traditionalist protest movements on the one hand and as liberation theology on the other. Other forms of religiosity, notably civil religion and nationalism, though they are decried by the traditionalists as examples of the disappearance of the "true religion," nonetheless function much as religious traditions do. Religious syncretism at the margins of society and a new ecumenical spirit among the mainstream religious groups also constitute a crucible in which a new generation of religious traditions may now be forming. The emergence of alternative religious movements, especially in the United States, reflects a religious ferment that stimulates creative theologizing in a way reminiscent of the formative periods of the existing mainstream religions.

In addition to the emergence of some deliberately syncretistic religious traditions, such as the Baha'i faith, a new spirit of dialogue seemed to be widespread as the 20th century came to a close. In 1893, at the World's Columbian Exposition in Chicago, the first Parliament of the World's Religions brought representatives from a wide variety of religious traditions together to explore their similarities and differences. A century later, the second Parliament of the World's Religions convened—again in Chicago—to discuss the possibility of constructing a human ethical consensus out of the world's religious traditions and an institutional basis, a sort of religious United Nations, to encourage ongoing dialogue among religious leaders from a wide variety of traditions.

Religious diversity will no doubt remain the hallmark of the global village well into the next millennium—perhaps as long as human life persists—and the major issue is probably not how to eliminate religious conflict among different traditions and perspectives but how to facilitate constructive and creative, rather than destructive, conflict. That issue will be explored in the final chapter.

8

Religion and Social Conflict

Lester R. Kurtz

Mariam M. Kurtz

We but teach
Bloody instructions, which being taught return
To plague the inventor. This even-handed justice
Commends the ingredients of our poison'd chalice
To our own lips.

—Shakespeare,
Macbeth, Scene Vii

T he combining of religious differences with other forms of struggle often leads to conflict and sometimes violence but also to nonviolent civil resistance of injustice. The twin crises of modernism and multiculturalism intensified the religious dimension of many ethnic, economic, and political battles, providing cosmic justifications for the most violent struggles. Multiculturalism produces complex patterns of conflict both between and within religious traditions that feed off one another and often intensify over time, making the 20th century the most violent yet in human history and launching waves of violence at the beginning of the 21st. Given the destructive capabilities of modern weaponry and the consequent necessity for peaceful coexistence, the potential for religious traditions to promote either chaos or community becomes a crucial factor in the global village. The past century also launched a new chapter in human history, the development of nonviolent civil resistance, starting with Mahatma Gandhi and carried in large part by religious institutions and people of faith to topple dictators and challenge injustice around the world.

In this final chapter, we will examine the nature of conflict and the special character of religious conflict and its special relationship with the worst and best forms of conflict. Conflict itself is not inherently a sociological liability, as Simmel (1908/1971) put it; conflicts can be either constructive or destructive, as can the role of religion. As we have seen already in this exploration of the world's religious traditions, conflict is a major source of cultural innovation and the crucible in which our current traditions were forged. We should not fear conflict itself but battles that rage with escalating violence. Violent conflict has been an integral part of human history for thousands of years and has been intimately associated with religious practice. In the global village, where automatic rifles and weapons of mass destruction have usually replaced stones, the cost of violent conflict has simply become too high. The problem created by the special confluence of modernism, multiculturalism, and the modern technologies of violence is not conflict itself but destructive conflict. In this chapter, we will explore the insights that the sociology of the world's religions provides for our understanding of the relationship between religion and social conflict in the global village (see Kurtz, 2008).

The most immediate issue, of course, after September 11 is terrorism and the unprecedented degree to which recent terrorist movements are religiously based (Rapoport, 2008). As Mark Juergensmeyer (2003) observed in his important *Terror in the Mind of God: The Global Rise of Religious Violence*, major terrorist attacks require an organizational network and a moral presumption to justify harm done to innocent victims, as well as internal conviction and "the stamp of approval from a legitimating ideology or authority one respects" (p. 11). Religious beliefs, networks, and organizations have become crucial to the rise in terrorist activity in recent decades, religious terrorist groups did not even emerge until 1980 when they were only 2 of 65 groups active. Only 12 years later, however, the number had increased nearly sixfold, and by 1995, religious groups accounted for nearly half (26) of the 56 known active terrorist groups.

The world of the 21st century is riddled with conflicts with religious motivations or at least justifications, so we close our discussions with a look at religious conflict and violence, as well as religiously promoted alternatives to violence.

A Theory of Religious Conflict[1]

We begin with conflict as a normal part of human life, not to be eliminated but to be transformed so that it is carried out nonviolently rather than destructively. As Georg Simmel (1908/1971) observed, social conflict is a form of social interaction (**sociation**); the social order is itself fashioned out of attractive and repulsive forces:

> Society . . . in order to attain a determinate shape, needs some quantitative ratio of harmony and disharmony, of association and competition, of favorable and unfavorable tendencies. But these discords are by no means mere sociological liabilities or negative instances. (p. 72)

Religious traditions also take shape from a combination of harmony and disharmony. Religious conflict can be extraordinarily bitter, however, and is often destructive because the parties to the dispute view themselves as "representatives of supraindividual claims, of fighting not for themselves but only for a cause," which, as Simmel (1908/1971) argued the following:

> [It] can give the conflict a radicalism and mercilessness. Because they have no consideration for themselves, they have none for others either; they are convinced that they are entitled to make anybody a victim of the idea for which they sacrifice themselves. (p. 87)

Extremes of thought and behavior are often defined as inappropriate for selfish causes but not for religious purposes; in most traditions, believers should be ready to die, perhaps even to kill for their deity. Ironically, the most intense extremism may be reserved for conflict with those whose values and beliefs are in reality not so different from one's own. That is why, for example, ancient Jewish law permitted bigamy but prohibited simultaneous marriage with two sisters. The principle here, Simmel (1908/1971) noted, is "that antagonism on the basis of a common kinship tie is stronger than among strangers" (p. 90). Thus, the most intense form of religious conflict may be that between heretics and religious authorities—that is, conflict *within* a tradition (see Kurtz, 1986). A similar volatility may emerge in relations between sibling religions, such as Judaism and Islam, with the same roots, many overlapping beliefs and practices, but significant divergences as well. Such controversy combines the intensity of religious conflict with that of a sort of kinship. Simmel (1908/1971) suggested that the strongest examples of hatred are "church relations. Because of dogmatic fixation, the minutest divergence here at once comes to have logical irreconcilability—if there is deviation at all, it is conceptually irrelevant whether it be large or small" (p. 90). Although religious conflicts threaten our coexistence in the global village, it is not the struggles between but those within traditions that may be the most dangerous.

The Politics of Heresy

It is no accident that the concept of heresy comes from the Christian tradition, which, in the absence of ethnic or tribal memberships, has placed so much emphasis on right doctrine. Formally, heresy in the Roman Catholic tradition refers to "a sin of one who, having been baptized and retaining the name of Christian, pertinaciously denies or doubts any of the truths that one is under obligation of divine and Catholic faith to believe" (Buckley, 1967, p. 1069). The idea of heresy as evil surfaced during bitter battles in the early church councils (see Cross, 1925; Hughes, 1961). Looking at heresy as a social construction can be useful in understanding the relation between belief systems and social organization.

Elsewhere (Kurtz, 1986, p. 3ff.), I have suggested the following characteristics of heresy:

1. Heresy is simultaneously near and remote. That is, heretics are within the relevant social group and therefore close enough to be a threat, but distant enough to be considered in error. A heretic is thus a "deviant insider" and is considered a danger to the institution and its leadership. Heresies are internal, but they represent pollution from external sources—an inevitability in the global village, where traditions collide.

2. Heresy is socially constructed in the midst of social conflict. As people fight with one another, their interests become associated either with the heresy or with its refutation. Consequently, the problem of heresy is primarily a problem of authority, that of beliefs and structures.

3. Heresy has social consequences as well as social origins. It is a double-edged sword, sometimes disruptive (the conventional view), but often is used for the creation of group solidarity and for purposes of social control. By labeling a group of people as heretics, elites rally support for their own positions but may inadvertently stimulate a rebellion.

4. Campaigns against heresies have doctrinal consequences as well as social ones. In the heat of escalating conflicts, groups often clarify just what they believe about a particular issue.

5. The process of defining and denouncing heresies is a ritual that helps to relieve social and psychological tensions and deal with institutional and

religious crises. Christian rituals for denouncing heresies began in the first centuries of the church's history as the church began to encounter non-Jewish influences, and people fought for control of the institution and its belief systems. Antiheretical rituals reached their peak with the formation of the Inquisition and the use of the formula cogeintrare, "to force them to join" (see Weber, 1968, p. 480), which justified the use of force against heretics, sometimes giving infidels and heretics a choice between either conversion and submission or death.

The phenomenon of heresy is important to our understanding of religious conflict, which is fueled by differences on issues of great significance to people. The issue of heresy is also critical in the global village for two reasons. First, in the face of external attacks by other religious and nonreligious perspectives, people within a tradition will often turn on insiders who have slightly deviant points of view, especially if they echo aspects of the undesirable outsiders' arguments. Second, there is a sense in which everyone is now an insider—as a citizen of the world—so that insider–outsider distinctions become blurred, a situation ripe for escalating conflict and the social construction of evil, as people attempt to redraw boundaries that once seemed safe. The problem, again, is not conflict per se but its potential for destructive violence.

Religious conflict is frequently associated with violence, either directly or as a way of legitimating violent means for handling conflict that originally had a nonreligious basis. Because of the centrality of the problem of violence in the 21st century, it is important to understand the ways in which violence and the sacred are intertwined.

Religion and the Problem of Violence

The norms against the overt use of violence are so strong in the global village that those who would use force must justify it eloquently and defend it in international institutions such as the United Nations, as well as the media. Religion is one of the most convenient and effective ways of making such a justification, and the link between violence and the sacred is an ancient one that grows out of the roots of human cultures (see Ferguson, 1977; Girard, 1977). Religion has also been the source of powerful rationales for nonviolence.

As suggested elsewhere (Kurtz, 1999, 2006, 2008), every major religious tradition has two contradictory themes regarding the use of violence, leading to a good deal of confusion and contradiction. Most sacred texts and faith communities have a warrior motif, on the one hand: It is a sacred duty to stand up and fight against evil. On the other hand, the scriptures and teachings of the world's religions also have a pacifist motif that prohibits harming others and prescribes a love for enemies that seems to challenge the very foundations of the warrior motif.

We should probably speak of the world's religions in the plural, of Hinduisms and Buddhisms, Christianities and Islams, since there is so much variation within each tradition (perhaps more than between them).[2] That is certainly the case with the morality of force and violence, as one cannot characterize any major faith as either warrior-like or pacifist. We can, however, easily identify warrior and pacifist individuals, sectors, and teachings in each of the world's predominant traditions.

It is helpful in this regard to make a distinction among the imperial, institutional, and spiritual aspects or versions of any tradition. The imperial involves linkages between religious and political elites and the use of a tradition for political aggrandizement. It is more likely, of course, to promote the warrior motif than the pacifist, although liberation movements as well as political establishments may employ a warrior motif as well. Institutional religion is the crystallization of a faith tradition in an institutional form and serving the interests of the institutional elites and their status groups, political allies, and so forth.

Finally, the spiritual aspects are the less material and often the more original elements of a tradition. Although they may certainly be carried by institutional and imperial structures, they are sometimes apolitical and often in direct conflict with institutional and imperial interests.

Despite the warrior motif's widespread existence and frequent dominance, most of the world's scriptures and prophets promote a loving of one's enemies that would seem to contradict the warrior approach of controlling or destroying them. This advice is not only a moral admonition but also a strategic wisdom. As the Buddha put it, "Hatreds never cease through hatred in this world; through love alone they cease. This is an eternal law" (*Dhammapada*, n.d., pp. 3–5).[3] Thus, Jesus' admonition to turn the other cheek when struck is an ethical standard but also a strategy in the face of violent attack. It demonstrates the victim's courage and control of the situation even when facing physical abuse or danger.

Violence was simultaneously condemned and condoned in the 20th century (see Turpin & Kurtz, 1997). Although most people abhor violence, many use it either defensively to respond to violence or offensively for coercive purposes. The same violence committed by "illegitimate" forces is sometimes considered acceptable when carried out by the state. Age-old killing practices—usually in the name of one or another God—now produce destruction on an unprecedented scale because the technological means of destruction have become so effective. Not only are individuals or tribes now at risk from acts of violence but so are entire populations. With the advent of nuclear, chemical, biological, and space-based weapons, the ecosphere itself is endangered.

We will now explore the role of religious traditions in the origins of violence, critiques of violence, and the construction of nonviolent alternatives. Then we will examine the cases of two sharply different religious protest movements that put people of faith in intense social conflict.

The Religious Roots of Violence

The use of religious arguments to justify violence waxes and wanes historically, but few justifications of violence are as universally upheld as those supported by the world's religious traditions. Violence is usually done in the name of some good and often in the name of a God. Discerning whether violence is truly religious, or merely a profane act that its perpetrators try to endow with sacred purpose, is a difficult business, perhaps altogether impossible. It is nonetheless an important question to raise if we are to evaluate any empirical case and one full of policy implications for the 21st century. Religious institutions sometimes sponsor violence directly, but the most significant role religion plays in promoting violence is often an indirect one in which sacred traditions are called on to justify its use.

Because violence is considered legitimate only when done for the "right reasons," it often takes on the character of a religious sacrifice (Girard, 1977, p. 1). When a sacrifice is carried out by "divine command," it is defined as more legitimate than if it is done for fame or profit. Thus, when violence is committed or people engage in warfare, such acts are often framed with religious purpose by invoking God's blessing at the beginning and offering thanks after a victory. A portion of the spoils are given to the deity or its representatives in an effort to sacralize the enterprise, allowing the interests of the victors to be masked by religious ideals.

Every major religious tradition includes its justifications for violence (see Kurtz, 1999, 2005b, 2006, 2011a). Behavior that is prohibited under ordinary circumstances becomes obligatory when it carries a divine sanction (Girard, 1977). Although the killing of other human beings is usually prohibited, in times of warfare (especially holy war) it is required. Primal religions typically frame violence as part of the natural process. The creatures of nature die and kill one another on a regular basis as part of their struggle for survival. Killing may not be taken lightly, however, and may not be done often (especially the killing of other human beings), although what is permitted varies widely from culture to culture.

Why do we find so much religiously ordained violence? It may well be that the faith traditions have been hijacked by people with political agendas. The metaphor is apt because it involves capturing a vehicle and using it to take a course contrary to its original destination. The pacifist Christianity of the early church was hijacked by the so-called Holy Roman Empire after the conversion of Emperor Constantine, on through the medieval Crusades, to 19th- and 20th-century European colonialism.

To prevent social or self sanctions from doing harm to others, individuals engage in what social psychologist Albert Bandura (1999) called "moral disengagement." Modern militaries acknowledge that being ordered to kill others can cause psychological trauma so they train soldiers to overcome what "killologist" Grossman called "the natural resistance to killing" that humans share with other animal species (Grossman & Siddle, 1999; cf. www.killology.com).

The "religions of the book" contain ample precedents for justified killing. According to the Hebrew scriptures, held authoritative by Judaism, Christianity, and Islam, God kills people and instructs others to do so when an injustice has been committed or holy law has been disobeyed. In Asian religions, killing is usually prohibited by the principle of *ahimsa* (nonviolence or nonharmfulness) but is sometimes required by one's *dharma*.

The Western God of War

In the Judeo-Christian tradition, God is praised as a God of War. As the psalmist writes,[4]

> I have pursued mine enemies, and overtaken them; neither did I turn back till they were consumed.
>
> I have smitten them through, so that they are not able to rise; they are fallen under my feet.
>
> For Thou hast girded me with strength unto the battle; Thou hast subdued under me those that rose up against me.

God not only caused acts of violence while liberating the Hebrew slaves from the Egyptian pharaoh, according to the tradition, but also gave the Hebrews victory in conquering the inhabitants of the "Promised Land" to which they fled. In the Hebrew scriptures, violence by or on behalf of God seems to favor the poor; it is condemned when undertaken for personal gain, and offensive violence is allowed only in the holy wars against the tribes of the "Promised Land." The Holocaust perpetrated by the Nazis on the Jewish people in the 20th century, in which perhaps a third to a half of the world's Jewish population was killed, imposed such a toll on Jewish identity that all subsequent discussions of Judaism and violence have to be seen in light of that tragedy.

Misunderstandings of the Qur'an and the concept of the jihad perpetuate a similar attitude. Although the idea of Islamic jihad has been widely distorted, it does offer a clear legitimation for violent acts against those who allegedly disobey Allah. As noted earlier, an accurate translation of the word is "struggle," and the "Greater Jihad" takes place internally, within one's own heart, whereas the "Lesser Jihad" refers to struggle of all kinds in the external world, including war (see Esposito, 2003; Juergensmeyer, 2003). Those who do not "heed God" should "be prepared to face war declared by God and His messenger" (2:280). The Qur'an continues, however, that God, not humans, should decide who deserves punishment, and sets strict limits about what kinds of violence can be done under what conditions and also puts a premium on the preservation of human life.

The early Christian Church, as we will see, fostered strong pacifist beliefs marking a break with ancient Judaism, but the church's position shifted dramatically after about 300 years with the conversion of the emperor Constantine to Christianity in 327 CE (see Bainton, 1960). By the 11th century CE, Christianity had shifted from being a pacifist to a warrior religion. It is not difficult to discern why the shift occurred: early Christianity was a small religious sect devoted to a radical new version of Judaism taught by the rabbi Jesus. Although originally popular among marginal social groups in an isolated province of the ancient Roman Empire, gradually its influence reached the elites of the time until the Emperor himself was converted. Gradually an effective affinity emerged between the religious institution and Rome, an alliance that was to transform the church more than the empire in the centuries to come.

Violence in Eastern Religions

The Asian religions are just as complicated in their attitude toward violence as the Western religions, although support for violence is not quite as well developed doctrinally as in the West, for several reasons. First, both of the major religious forces in the East, Hinduism and Buddhism as well as other traditions such as Jainism and Taoism, contain strong nonviolent doctrines in their core teachings. Second, religious doctrines in the East are not as authoritative as in the West (especially Christianity); they are more diverse and less binding, especially in the political sphere. Because the pantheon has a division of labor, some Gods are more likely to support the use of violence than others, so it is unnecessary to develop a definitive stance on the issue of violence. Such a position could not be enforced anyway even if it were proposed. Third, the charismatic authority of first Mahatma Gandhi and then the Tibetan Dalai Lama have transformed much of modern thinking about violence in Hinduism and Buddhism, highlighting some previously subordinate strands valuing the doctrine of ahimsa, or nonharmfulness.

Finally, and relatedly, the major religions of the East are somewhat less closely linked with the political establishment than Western religions are, although that generalization must be carefully qualified. Many heads of state have at least formally adhered to Buddhism, and the teachings of K'ung-Fu-tzu have formed something of a state religion, or quasi-religion, in China until the 20th century. The fortunes of religious institutions are not so closely linked with those of the state, however, as in the West.

Both Hindu and Buddhist leaders have advocated the use of violence and warfare over the centuries, however, tracing their justification to the sacred texts and traditions in which they operate. The Hindu tradition, despite its current nonviolent emphasis, is replete with wars that are understandably interpreted by many believers as sacred legitimation of the use of force in battle. In the *Bhagavad Gita* (Mascaró, 1962), Krishna explained that "there is no greater good for a warrior than to fight in a righteous war" (p. 51). In this case, fighting (and consequently, killing) is required by the God himself not because of the act's inherent moral value but because it is the warrior Arjuna's duty to fight. The warrior battles as a sacrificial act of duty to the God but also because that is what the universe expects of warriors. Fulfilling one's duty to fight becomes a religious sacrifice—as selfless action (*anasakti*) for a larger purpose than personal gain. One should fight regardless of the outcome, because the focus is on the effort of battle, not whether one wins or loses. One does not take personal responsibility for a loss or credit for a victory nor is one to blame for the devastation caused by war. The deaths of warriors are preordained by fate; they will die if their time has come, so the one who "kills" them is only an instrument of a larger cause. Furthermore, one cannot kill the soul but only the body, so the apparent act of killing is merely an illusion.

The Buddhist tradition is more difficult to cite in defense of war because of the Buddha's emphasis on compassion for all creatures and the admonition against killing at the gate of the Eightfold Path. Despite that clear prohibition, some Buddhists (including monks) have made fierce warriors in parts of Asia, especially China.

It is important to note that religion is often the *content* of a conflict, though not its basis (Kurtz, 2005b). The conflict may be political, class, familial, or even psychological, but religious issues become the subject around which the conflict evolves and religious rhetoric is used to carry it out. When fighting breaks out between Protestants and Catholics in Northern Ireland, or Hindus and Muslims in India, the conflict is usually as much economic as religious, though it may be couched in religious rhetoric. In these cases, the conflict's religious significance lies purely in its justification of violence and the sacralization of the cause for which the parties are fighting.

Religious Critiques of Violence

Religious institutions and their adherents have not only justified violence throughout human history but they have frequently criticized it as well. Religious critiques of violence include the contentions that it is as follows:

1. Against human nature, or the nature of things in the universe

2. Ineffective in the long run because it simply produces a spiral of violence

3. Sinful or unjust, and the deity does not like it

In most cultures, a sort of *Realpolitik* hypothesis reigns, however, in which "common sense" dictates that realists must be hardheaded and willing to engage in violence in order to accomplish anything and especially to defeat evil forces that engage in violence. Ideas such as nonviolence may be acceptable—even admirable—for saints and holy people, this thinking goes, but they should not be binding rules of morality for ordinary individuals.

Religious arguments may run counter to these conventional cognitive frames, especially concerning issues of power and politics (Kurtz, 2005c). Religious perspectives are often quite different, especially when they draw upon a theory of the world that allows for a benign universe that will eventually prevail even if evil seems to be temporarily in control. Often this counter-system point of reference can facilitate a nonviolent approach that rejects the more conventional attitude toward violence. In this framework, what appears to be efficacious is only temporarily so, and one needs a larger perspective in order to see the shortcomings of violent methods.

Messianic eschatologies, found especially in the Western religions of Judaism, Christianity, and Islam, are replete with beliefs about the unanticipated turn of events that will elevate the downtrodden. Similar perspectives are found in Eastern religions in which a *bodhisattva* or a God may intervene, or things will just work themselves out according to the law of karma, which helps to explain why things are not the way they superficially appear to be (e.g., one's rewards may come in the next incarnation).

According to the Tibetan Buddhist leader Tenzin Gyatso (the Dalai Lama), people are, contrary to popular belief, naturally more nonviolent than violent; they are naturally filled with compassion, seek affection, and recoil from violence. The universe itself is naturally inclined toward nonviolence; consequently, the Eightfold Path that guides the Buddhist's life contains a set of Five Precepts, the first of which is not to kill (Thich Nhat Hanh). The outcome of taking life is to have an inferior incarnation. Second, even if one does use violence to solve a problem, it is not going to be a successful resolution in the long run. The Dalai Lama (in Kurtz, 2005a) claims, "Even if you achieve something through force, physical force . . . very often it creates a situation [in which] . . . the other party . . . [is] not happy. . . . Therefore, as soon as another opportunity happens, then they'll take retaliation." According to the Taoist principle of *wu-wei*, violence simply begets violence. As it says in the *Tao te Ching* (Lao-tzu, 1995),

> If you used the Tao as a principle for ruling
>
> You would not dominate the people by military force.
>
> What goes around comes around.
>
> Where the general has camped
>
> Thorns and brambles grow.
>
> In the wake of a great army
>
> Come years of famine.

Finally, some individuals refuse to commit violence simply because they believe that God has told them not to do so. Admittedly, this position is a minority one in most religious traditions, but it is significant nonetheless. Russian novelist Leo Tolstoy (1987), for example, contended from the Christian pacifist tradition the following:

> A Christian, whose doctrine enjoins upon him humility, non-resistance to evil, love to all (even to the most malicious), cannot be a soldier; that is, he cannot join a class of men whose business is to kill their fellow-men. Therefore it is that these Christians have always refused and now refuse military service. (p. 11)

From this point of view, refraining from violence has nothing to do with whether or not the alternatives will "work" in any conventional sense of the term but whether or not they are morally superior.

Religious Contributions to Nonviolence

The irony of the relationship between religion and violence is that the most potent violence *and* the most powerful nonviolence both have religious roots. Both Eastern and Western religions have strong pacifist traditions that lay one part of the foundation for the 20th-century tradition of active nonviolence. Violence often appears to require religious justification because it is negatively valued as undesirable and nonviolence because it is equally negatively valued as ineffective. One religious figure who challenged that argument in both theory and practice was Mahatma K. Gandhi (1869–1948). The remarkable growth of nonviolence as a political strategy in recent decades is Gandhi's legacy to the modern world.

Gandhi's Experiments in Nonviolence

Gandhi's response to the problem of violence centers around his formula that one should refuse to cooperate with unjust systems but also to separate the doer from the deed—that is, to hate evil but respect the person, an idea rooted in his religious understanding of the nature of humanity (see Kurtz, 2006). This differentiation of the adversary from his or her behavior changes the nature of conflict; it depersonalizes (and demilitarizes) the conflict but personalizes the opponent. The attack is not against individuals who will respond in kind but against evil itself.

Gandhi's advocacy of nonviolence combines all three religious arguments against violence. First, he claimed that nonviolence is the law of the universe. Because nonviolence is "in harmony with the nature of existence and reality," it "must in the final analysis therefore be action which 'works' and is 'practical'" (Sharp, 1987, p. 41). If the universe itself is basically good, then all people are good; that is why they should be treated with respect. Whether one is convinced or not that one's adversaries are good people, acting as if they are may well evoke a positive response from them. That was not only Gandhi's theory, but also his practice. It sometimes seemed as if he *forced* people to do good rather than evil, even when they seemed inclined toward the latter, by calling out the best in his opponents rather than humiliating them.

Second, Gandhi demonstrated that nonviolence can work even in the most unlikely situations. Perhaps it was the efficacy of nonviolence, more than the ineffectiveness of violence, that shaped his attitudes. Gandhi (1962) contended that "even the most despotic government cannot stand except for the consent of the governed which . . . is often forcibly procured. . . . [When] the subject ceases to fear the despotic force, the power is gone" (p. 154).

Third, Gandhi believed that nonviolence was morally superior to violence and must be carried out even if one feels as if one is acting alone: "There are moments in your life when you must act even though you cannot carry your best friends with you. The still small voice within you must always be the final arbiter when there is a conflict of duty" (*Young India*, August 4, 1920, in Gandhi, 1962, p. 152).

Finally, Gandhi engaged in "experiments with Truth," as he put it, which included creating ashrams—spiritual communities in which people of varied creeds, castes, and races lived (see Chatterjee, 1983). Following the karmic theory that every action has consequences, Gandhi believed that he was helping to transform the universe by bringing people together across the conflict lines in his culture. Indeed, Gandhi's ashrams in South Africa and India inspired Dave Dellinger's Harlem ashram, which became an incubator for the American civil rights and peace movements (Tracy, 1996).

Because religious history is replete with persecution, most traditions have many inspiring models of individuals or groups of people who stand up against overwhelming powers and win. This is one central element of nonviolent action that is always in tension with the efficacy argument: Some would argue that even if nonviolence does not always work, it is always right.

Transforming the Traditions

Despite the widespread presence of nonviolence in the world's religious traditions, the warrior motif remains dominant; the concept of God as pacifist or nonviolent activist is supported only by a minority strand (Kurtz, 1999, 2008). For religion to become a vital force against violence, the traditions would need to be transformed in their dominant manifestations and institutions, as Gandhi tried to do with some success in the Hindu tradition. Although the process would be opposed in many circles, it is already unfolding in a number of religious traditions. Four major developments in this area are worth mentioning here:

1. Criticizing contemporary warfare by means of traditional criteria

2. Spiritualizing the old war stories to make them parables of nonviolent struggle

3. Mobilizing institutional resources of the traditions to oppose violence and promote nonviolence

4. Secularizing religious traditions in such a way as to diffuse their moral lessons into the broader culture without the entire religious "package"

New Critiques With Old Criteria

The Christian concept of a just war, with its roots in the writings of the 4th century CE Saint Augustine and the medieval theologian Thomas Aquinas, contains a fundamental bias against war but allows it as morally justifiable under certain extreme conditions when certain criteria are met. One of the most significant developments in religious responses to violence has been the effort by Christian leaders, beginning with the U.S. Catholic bishops, to reevaluate just war theory in light of contemporary means of mass destruction, which make traditional justifications for war extremely problematic. It is no accident that this radical critique of violence by an established religious institution emerges from a society that threatens to annihilate the species.

In traditional just war theory, the criteria governing the decision to go to war, rely primarily on the issue of proportionality—that is, is the good to be gained from the battle proportionate to the cost to be incurred? According to the U.S. Catholic bishops (National Conference of Catholic Bishops, 1983), a decision to go to war is possible only when the following conditions are met: (1) just cause, (2) competent authority, (3) comparative justice, (4) right intention, (5) last resort, (6) probability of success, and (7) proportionality. The final two criteria especially have forced many Christian authorities to reevaluate the traditional legitimation of warfare at the turn of the 21st century. It is impossible to envision any way in which a total nuclear war could meet either the criteria of probable success or that of proportionality. Even after the decision is made to go to war, the conduct of war itself must be subject to two principles: (1) proportionality and (2) discrimination. The principle of discrimination prohibits action against innocent civilians (see Johnson, 1981, p. 350; Ramsey, 1961, 1968). In modern wars, it is difficult, if not impossible, to discriminate between retaliation against aggressive parties and actions taken against noncombatants.

The Second Vatican Council (Vatican Council II, 1965) concluded that "The horror and perversity of war is immensely magnified by the addition of scientific weapons," requiring us "to undertake an evaluation of war with an entirely new attitude." The implications of that conclusion are profound and have precipitated a rediscovery of the pacifist elements of Christianity lost since the conversion of Emperor Constantine 1,500 years ago. Even though many Christians continue to use just war arguments, and even the "Crusade" tradition, to justify modern wars, responsible religious leaders throughout the church are giving the matter serious consideration. Other religious traditions have undergone similar efforts to reevaluate, or even transform, their positions on violence and warfare. Such a transformation of religious ideas is neither as unusual nor as formidable as some might think. It is a normal aspect of socioreligious development and has occurred innumerable times in the history of religious thought.

Spiritualizing the War Stories

Religious traditions cannot simply repudiate the stories of violence in their history that are often used to legitimate the use of violence in the name of the deity. To escape the trap of violence, these stories must be reinterpreted. One significant model for the process is Gandhi's recasting of the *Bhagavad Gita*.[5] In the pivotal scene, as we have seen, the warrior Arjuna is facing his kinsmen on the battlefield of Kurukshetra, anguished over the decision about whether or not to fight. If he undertakes the battle, he will most likely kill members of his family and his teacher. Finally, Lord Krishna convinces Arjuna that it is his duty as a warrior to go to war.

This difficult legend would appear to create cognitive dissonance for anyone attempting to live a nonviolent life. Gandhi (1930/1987), who relied heavily on the Gita for his own thinking, turned the story into a metaphor: "The battlefield of Kurukshetra," Gandhi wrote, "only provides the occasion for the dialogue between Arjuna and Krishna. The real Kurukshetra is the human heart. . . . Some battle or other is fought on this battle-field from day to day" (p. 8). The meaning of the Gita is thus not that one should engage in physical violence but that one must engage in the struggle that is associated with one's duty. It is, moreover, a detached action: "It is up to us to do our duty without wasting a single thought on the fruits of our action. . . . Gain or loss, defeat or victory, is not in your power" (Gandhi, 1930/1987, p. 11).

Finally, one must be prepared to sacrifice. The gist of the Gita, Gandhi (1930/1987) contended, "is that life is given us for service and not for enjoyment. We have therefore to impart a sacrificial character to our lives" (p. 20). That is not to say, however, that the religious life itself is simply a spiritual exercise for Gandhi. On the contrary, he contended that "the spiritual law . . . expresses itself only through the ordinary activities of life. It thus affects the economic, the social and the political fields" (quoted in Sharp, 1987, p. 39; cf. Sharp, 1979, 2005). The legend of Arjuna and Krishna, which easily could be taken as providing a religious legitimation for violence, becomes instead the direct opposite: It is a story about the obligation to serve others with sacrifice and without self-gain. Instead of legitimating violence, Gandhi's interpretation of the scripture moves toward a radical nonviolence.

This transformation of tradition, like that accomplished by the U.S. Catholic bishops in their study of just war theory, is a characteristic tactic employed by Gandhi, who constantly reinterpreted traditional concepts and "grafted them onto the modern setting" (Kothari, 1970, p. 54).

Institutional Resources

Whereas religious authorities have often rallied populations behind war, many mainstream religious institutions have recently denounced violence and war. One of the most interesting examples is the Roman Catholic Church's teachings against war and efforts by the late Pope John Paul II to denounce violence, reversing the precedents of his predecessors who launched the Crusades and blessed the troops. Not only did he speak out consistently against violence and war—despite his conservative stance on many social issues—but he also called together religious leaders in St. Francis's town of Assisi to proclaim that true religion could not be used to promote violence or terrorism. The Catholic periodical *America* ("Signs of the Times," 2002) reported that "The Assisi gathering brought together Christians from 16 churches and communities, 30 Muslim clerics from 18 nations, 10 rabbis, and representatives from Buddhism, Tenrikyo, Shintoism, Jainism, Sikhism, Hinduism, Zoroastrianism, Confucianism, and traditional African religions" (p. 4). The pontiff declared to his guests, who joined together in prayers and pledges against violence, "Violence never again. War never again. Terrorism never again. In the name of God, may every religion bring upon the earth justice and peace, forgiveness and life, love" ("Signs of the Times," 2002, p. 4).

Secularizing Religious Nonviolent Civil Resistance

Religious ideas never have a widespread impact unless they enter the cultural mainstream and become diffused throughout both elite and popular cultures. In the contemporary world, that process often requires the decoupling of a specific theme from the broader religious message. That is, in fact, precisely what has happened in recent years with the theme of nonviolent resistance. As the idea of active nonviolence (or "nonviolent direct action" or "nonviolent civil resistance") has diffused throughout the world, it has become increasingly secularized (see Zunes, Kurtz, & Asher, 1999). The religious nonviolence of Gandhi and then Martin Luther King Jr. was systematically developed in many modern political movements and in some popular developments such as conflict resolution and mediation. King's strategy was to translate Gandhi's philosophy and strategies—relatively universalistic but with strong Hindu overtones—into Christian rhetoric. Others, such as Gene Sharp (1973–1974, 1979, 1987), removed most of the religious bases for nonviolent action altogether and much of the moral justification for it to construct a set of pragmatic strategies for engaging in nonviolent struggle.

In Eastern Europe, similarly, various Christian churches, such as the Roman Catholics in Poland and the Lutherans in Germany, provided important leadership, but the idea also captured the imagination of the intellectuals, who translated many of the core ideas of nonviolence into a more secular language (see especially Havel, 1990; cf. Goldfarb, 2006). Although religious imagery has persisted in the peace movements in Europe, North America, and elsewhere, and Gandhi and King have remained the "saints" of the movement, many participants have toned down the explicitly religious rhetoric and now speak in more secular terms about the moral aspects of nonviolence.

Religious institutions often have an impact on a nation's policies because of their access to political elites. Many radical ideas, such as human rights and antiwar critiques, begin as politically marginal and are brought into the mainstream by religious leaders whose authority is acknowledged by elites who otherwise would not entertain them.

Islamic Condemnations of Terrorism

What most Americans know about Islam, of course, is from television coverage of September 11 and other terrorist incidents, although stereotypes about Muslims as violent may have been permanently challenged by the Arab uprising of 2011 and especially scenes repeated for weeks of Gandhian-style nonviolent Muslim demonstrators challenging their governmental structures, especially in Egypt. Still, the specter of terrorism and the war against it continue to challenge the role of Muslims in the emerging global civil society.

One young Muslim American scholar, Reza Aslan (2005), has an interesting theory about the September 11 terrorist attacks, which he developed in his book *No god but God*. He summarized the argument himself in the interview below (Chaudhry, n.d.), which, although somewhat lengthy, provides an important Muslim analysis.

We are now living in the twilight of that era of Arab-Islamic reformation. This is a process that began around the time of the colonialist experience, some 100–150 years ago, when Muslims were, for the first time, forced to respond to not just the realities of the modern world—secularism and modernization, and industrialization—but also the western cultural hegemony that came part and parcel with the colonialist experience.

So naturally, there were two broad reactions to it. One, there were those groups of modernists, reformists, and moderates who eagerly accepted these enlightenment principles that the colonialists were preaching—concepts such as human rights, individualism, constitutionalism and rule of law—and to a far lesser degree, democracy and popular sovereignty. They not only adopted [these principles], but also strove to create an indigenous vision of these principles, and an indigenous Islamic enlightenment.

Then there were those Muslims—who at that time I would say represented the majority of the Muslim population—who responded to colonialism by reacting violently against it, by rejecting that western cultural hegemony, including these wonderful principles, as being a part of colonial oppression. They wanted to respond to modernity by reverting instead to what we would now refer to as the fundamentals of their faith. They wanted to go back to a purely and distinctly Islamic identity.

What we're seeing now is a natural evolution of this reformation that began then, and which, in essence, is coming to a close. I call Sept. 11 a part of that internal clash because that precisely was the reason for the attacks on New York and Washington. Whatever we want to say about bin Laden, the savagery of his followers and their murderous inclinations, they're not stupid people. They recognized quite clearly that this kind of spectacular attack on U.S. soil was going to engender an exaggerated response. That is precisely what they were hoping for as an opportunity to galvanize support for what was a losing cause before Sept. 11.

Perhaps what is most remarkable is that some people continue to ask why Muslims did not denounce the terrorist attacks when in fact they did. In statement after statement, the attacks were condemned by such groups as the American Muslim Alliance, the American Muslim Council, the Council on American-Islamic Relations (CAIR), the Islamic Society of North America, the Islamic Circle of North America, the Shari'a Scholars Association of North America, and many more.

In July 2005, the Fiqh Council of North America issued a **fatwa** (a legal opinion from an Islamic scholar) against terrorism, saying that it "wishes to reaffirm Islam's absolute condemnation of terrorism and religious extremism." The fatwa concluded the following:

In the light of the teachings of the Qur'an and Sunnah we clearly and strongly state:

1. All acts of terrorism targeting civilians are haram (forbidden) in Islam.

2. It is haram for a Muslim to cooperate with any individual or group that is involved in any act of terrorism or violence.

3. It is the civic duty of Muslims to cooperate with law enforcement authorities to protect the lives of all civilians.

> We issue this fatwa following the guidance of our scripture, the Qur'an, and the teachings of our Prophet Muhammad—peace be upon him. We urge all people to resolve all conflicts in just and peaceful manners.

Those who claim that Islam is a religion of violence and promotes terrorism simply do not have their facts straight. Certainly, there are Muslims who do so, but they are not in the mainstream of the tradition and do not represent Islam any more than the Ku Klux Klan represents Christianity.

Indeed, faith traditions are carriers not only of violence but also of nonviolent civil resistance in the Gandhi–King tradition, from the prodemocracy and civil and human rights movements of the 20th century through the Arab uprising of 2011.

Religion and Nonviolent Civil Resistance

Gandhian nonviolent resistance was secularized by American scholar Gene Sharp, who removed the concept from its Hindu moorings but then "baptized" it by Christians in the U.S. civil rights movement and in the Philippines' "People Power" movement, who refashioned it with a Christian frame. Martin Luther King Jr. adapted his strategy of nonviolence from Gandhi for the American context. Since it took root in the black church in the United States (see King, 1986), the concept remained explicitly religious, with a strong Christian flavor. African American slaves, historically forced to convert, used the symbols of Christianity for their own liberation not only to talk about rewards in the world to come to sustain them through the suffering but also to mobilize resistance movements and antislavery activities. Thus, African Americans used Christian symbols and myths to resist the very culture that imposed Christianity on them. When the slaves sang spirituals like "Come and Go With Me to That Land," they were talking not only about a heaven in the afterlife, but also about escaping on the Underground Railroad to the North and their freedom.

King and others in the civil rights movement captured that spirit in the black church to give people the courage to struggle against the racism of the American system. The pulpits of Christian churches, both black and white, and other resources of the institution played a crucial role in making the movement. In the daily struggle of the civil rights movement, participants were sustained by their faith and assisted by their religious tactics. Resistance movements elsewhere also found essential support in the ideas and institutions of religious traditions. In South Africa, for example, where antiapartheid forces were often led by religious leaders, protesters encountered advancing police by kneeling in prayer, thus gaining broad sympathy for their cause.

Nonviolent Protests in Asia and Europe

A series of nonviolent social movements in Asia and Eastern Europe modeled their campaigns for social change after Gandhi's freedom movement and the civil rights movement in the United States, especially Martin Luther King Jr.'s, Christian nonviolence. In the Philippines, where a majority of the population is Catholic, base communities were established throughout the country, and an indigenous form of liberation theology emerged. Moreover, in the 1980s, the church became the site of training for nonviolent resistance against Ferdinand Marcos's dictatorship. An astounding People Power revolution in 1986 overthrew the Marcos regime, demonstrating the power of committed people who believed they had God on their side in the face of brutal right-wing oppression (see Deats, 2001; Kurtz, 2001). Members of the Christian Church, many of them trained in nonviolent tactics by the Fellowship of Reconciliation (a nonsectarian religious pacifist organization), took the leadership of the movement.

After Marcos tried to steal the election from popular opposition leader Corazon Aquino and key members of the military defected from the regime, the Roman Catholic cardinal of Manila called on people to join him in the streets. Hundreds of thousands did so. When the troops were sent to stop the rebellion, the demonstrators confronted the tanks and armored personnel carriers, as vividly described by Father José Blanco,[6]

> with our bodies, our prayer, our Filipino piety with images of Mary and the crucifix. Our faith has made us this kind of a people, both we who were resisting, as well as the soldiers who were ordered to attack. We venture to suggest when the soldiers saw praying, unafraid people, cheerful, offering flowers and cigarettes, willing to come under the tank treads; these effectively tied their hands and changed their wills not to carry out their mission of destruction. They might as well have had no tanks and armored cars, because their human concern for the lives of literally thousands was a stronger brake that kept their armored vehicles at a standstill. . . .
>
> This miracle is the mystery of God's grace powerfully working in the hearts of each one of us. The miracle is God bringing about events both big and small, which no one of us thought about, much less planned.

Although the socioeconomic conditions for rebellion were clearly present in the Philippines at the time of the People Power movement, it was the combination of the religiously motivated nonviolence and the institutional resources of the church that made a successful revolt possible.

Similar, and even more dramatic, change took place in the Soviet bloc a few years later, again largely propelled by the church in a context ripe for change. Although some people explain the collapse of the Soviet empire in terms of a bankrupt economic and political system, religious ideas and institutions played an important, and perhaps formative, role. An elective affinity emerged between the interests of dissident forces and the ideas and strategies of a religiously forged nonviolence. Perhaps the turning point in the pro-democracy movements of Eastern Europe was the visit of Pope John Paul II to Poland in 1979. Millions of Poles met him enthusiastically in the streets and were empowered to support movements of change by the introduction of this radically different frame.

Adam Michnik (1992) explained this about the Worker's Defense Committee (WDC, a crucial organization in the initial stages of the dissident movement):

> [It] had a model in Polish civic life: the Catholic Church. Not all WDC members were Catholics, although the overwhelming majority of Poles are Catholics. Not all of them would admit at the time that the Catholic Church was actually the first to provide definite proof that it was possible to be an independent institution in a totalitarian political environment, and that the Church itself demonstrated the first type of antitotalitarian action. (p. 242)

In 1989, the Berlin Wall fell, symbolic of the division between the two superpower camps. Behind the movements for change was the quiet, persistent work of the church and people of faith whose allegiance was to another system. The institutional resources of the church, more than any other single organization, helped to facilitate the dialogue, first quietly and then openly, that led to the resistance movements of Eastern Europe and the Soviet bloc (see Smithey & Kurtz, 1999). At first, much of the opposition took the form of worship services and discussion groups in church basements. When the movement came out into the open, it was not suppressed, in large part because it had the powerful backing of the institutional church and therefore the broad population. Operating in a spirit of reconciliation and nonviolence, the pro-democracy forces fearlessly continued to press for change. It was, in fact, when the military forces tried to suppress the movement that the general population began to voice its opposition and demand change.

Islam and Nonviolence[7]

Islam, like other religious traditions, contains both the warrior and the pacifist motif, although the Western media until the 2011 Arab uprisings usually has falsely portrayed it as having only the former and serving as the major purveyor of violence in the world today. The purpose of this discussion is to provide a more balanced view of Islam by presenting the suppressed aspects. We do not wish to argue that Islam is a religious tradition without violence any more than one could argue that Christianity promotes only pacifism.

Islam is no exception to the general rule of the ambivalence of religious traditions toward violence. Not only does the Qur'an (60:7) assume that God is ultimately in control of any situation but it also suggests, "It may be that God will ordain love between you and those whom you hold as enemies. For God has power over all things; and God is Oft-forgiving, Most Merciful." Strategically, the Qur'an concurs with the Buddha and Jesus that one should return evil with good: "The good deed and the evil deed are not alike. Repel the evil deed with one which is better, then lo!, he between whom and you there was enmity shall become as though he were a bosom friend" (Qur'an 41:34–35).

Some argue that Islam was hijacked in the 7th century CE to serve as a vehicle for the wars of conquest that stretched from Spain to Central Asia (see Freedman & McClymond, 1999, p. 230). Daniel Martin Varisco (2005) contended that some Muslims extremists are "desperately coating an Islamic veneer over political acts for confessional comfort" (p. 150). Moreover, he claimed the following:

> Bin Laden may cave-dream of a return to seventh-century Muslim unity, but the "history" that impels his rhetoric is surely recent placement of American troops in Saudi Arabia, America's political seduction of self-serving Arab leaders, and continued United States support for Israel's oppression of Palestinians. . . . Bin Laden was not born a nomad and is unimaginable without stinger missles [*sic*] and video sermons. (Varisco, 2005, p. 150)

The terrorist attacks of recent years are neither supported by the Qur'an, from many informed perspectives, nor by the mainstream of Islamic leadership. Here is a statement that was issued July 28, 2005:[8]

> The Fiqh Council of North America wishes to reaffirm Islam's absolute condemnation of terrorism and religious extremism. Islam strictly condemns religious extremism and the use of violence against innocent lives. There is no justification in Islam for extremism or terrorism. Targeting civilians' life and property through suicide bombings or any other method of attack is haram— or forbidden—and those who commit these barbaric acts are criminals, not "martyrs."

The Fiqh Council of North America (2005) statement went on to prescribe the proper role of Muslims with regard to terrorism:

1. All acts of terrorism targeting civilians are haram (forbidden) in Islam.

2. It is haram for a Muslim to cooperate with any individual or group that is involved in any act of terrorism or violence.

3. It is the civic and religious duty of Muslims to cooperate with law enforcement authorities to protect the lives of all civilians.

The terrorism of the 21st century is no more representative of contemporary Islam than the Crusades or colonialism of Christianity (see Esposito, 1983). The emergence of waves of terrorism in the previous century is paralleled by waves of nonviolence that provide an alternative means for people wishing to bring about social change.

The Two Hands of Nonviolence

Nonviolence is ancient and complex but has been more fully developed in the last century, in large part as a response to the burgeoning violence precipitated by revolutions in the means of destruction. Nonviolence is often thought of as having two hands—that is, principled or lifestyle nonviolence on the one hand and strategic nonviolence on the other.

The most dramatic development in the field of nonviolence was the emergence of Gandhian nonviolence and the elaboration of strategic nonviolence worldwide in the 20th century. Gandhi's formulations were nothing less than the creation of a new paradigm of conflict, drawing upon both religious motifs and secular strategies from various traditions.

Gandhi synthesized the warrior and pacifist motifs, creating the central notion of the nonviolent activist, or *satyagrahi*, one who "holds fast to the Truth" (see Kurtz, 1997, 2006). From the warrior motif comes the idea of fighting as a sacred duty and from the pacifist theme the prohibition against harming. The nonviolent activist fights like the warrior but does so without harming his or her adversary.

Modern nonviolence was born on September 11, 1906, when Gandhi gathered people at the Empire Theatre in Johannesburg, South Africa, to launch the first nonviolent campaign against racial injustice the world had ever seen. In the century that followed, Gandhi's refinements of nonviolent strategies and tactics were multiplied and diffused, even to some extent institutionalized. Strategic genius Gene Sharp systematized and secularized nonviolence, and Martin Luther King Jr., baptized it. A myriad of groups, movements, individuals, and institutions promoted and actualized nonviolent struggle so profoundly that it shaped much of the geopolitics of the twentieth century from the independence movements in India, Ghana, Tanzania, and elsewhere, to the antiapartheid movement in South Africa and pro-democracy movements in every corner of the globe.

Strategies such as direct action, civil disobedience, and nonviolent resistance and insurrection are now as much a part of the political landscape as the more visible and destructive military infrastructure. An explosion of research, writing, and nonviolent movements marked the advent of the 21st century just as much as the new terrorism and war on terrorism. Gene Sharp's (2005) *Waging Nonviolent Struggle*[9] summarized the strategies, tactics, and history of modern nonviolence, providing an indispensable introduction.

The other hand of nonviolence is what is sometimes called "principled nonviolence," although it should be seen quite broadly in terms of nonviolent lifestyle and social organization. It is chosen for ethical rather than strategic reasons, on the basis of what Weber called "value rationality" (*wertrationalität*) rather than "instrumental rationality" (*zweckrationalität*). As such, it includes efforts to reduce violence in all spheres of one's life, to treat others with respect and civility and to live a lifestyle that reduces violence to humanity and the natural environment. It also involves the creation of social structures that promote justice and human rights, minimize violence in all spheres of life, and facilitate the actualization of every person's potential by providing them with an opportunity structure for the basic necessities of life (food, clothing, shelter, medical care) as well as the fundamental liberties and foundations for self-actualization (education; freedom of speech, expression, and lifestyle; and so forth).

Given this context, we now turn our attention to the relationship between Islam and nonviolence.

The Qur'anic Paradox

A number of scholars have written about the tension previously noted that we find in the Qur'an and the Hadith of Islam (authoritative stories about the life and teachings of the Prophet Muhammad). On the one hand, people of faith are to fight and, on the other, to protect life, a dilemma akin to the warrior–pacifist motifs outlined earlier.

The context of the founding of Islam is a desert culture of the 7th century CE that is full of violence. The new movement itself is constantly under attack, and the Qur'an and Hadith are filled with stories of struggle and advice about it. The warrior motif appears frequently in the sacred text and the stories of the Prophet and his followers. To struggle, both with one's own life discipline (the Greater Jihad) and with others who are unjust (the Lesser Jihad), is part of the faithful life (Qur'an 2:190–193):

> And fight for the sake of God those who fight you; but do not be brutal or commit aggression, for God does not love brutal aggressors.
>
> And you kill them wherever you catch them, And drive them from where they drove you; For civil war is more violent than execution. . . .
>
> But if they stop, God is most forgiving, most merciful.
>
> And fight them until there is no more strife, and there is the religion of God.
>
> And if they stop, then let there be no hostility, except against wrongdoers.

This short passage itself moves deftly back and forth between the warrior and pacifist motifs, advocating now fighting but then mercy and forgiveness. Even the call to fight is full of ambivalence about doing so and places limits on the motives and means with which one should struggle. It is to be for the self-defense of the community—fight those who fight you—and if the attack ceases, remember that God is forgiving and merciful. The execution of aggressors is allowed but only to prevent civil war, which is even more violent. The troubling "kill them wherever you catch them"—perhaps the most difficult verse in the Qur'an—is not a general command to hunt down enemies and slay them but a strategy of limiting the spread of violence in the Prophet's historic context, in which the threat of civil war was imminent. The Qur'an seems to be advocating the execution of those threatening to start a war and injure the well-being of the entire community.

The entire Qur'an is something of a manual on the importance of justice and how to achieve it—it is for justice that one fights (Qur'an 4:75):

> Why would you not fight in the name of God, and oppressed men, women, and children, who say "Our Lord, get us out of this town, whose people are oppressors.
>
> And provide us a protector from You, and provide us a helper from You."

Again, the talk about fighting is interspersed with an emphasis on mercy and the protection of life. The injunction against killing is not absolute in Islam, although it is quite strong. As the Fiqh Council of North America (2005) put it, "The Qur'an, Islam's revealed text, states: 'Whoever kills a person [unjustly] . . . it is as though he has killed all mankind'" (Qur'an 5:32). Moreover, the Qur'an also places a high premium on the sanctity of life. Following the Hebrew scriptures, it declares, "And if anyone saved a life, it would be as if he saved the lives of all the people."

So what are we to make of this puzzle? Are those who wage terrorist campaigns justified in acting in the name of Islam if they can provide an account for their killing on behalf of justice because it is God's will? And if so, what is one to do with the emphasis on the sanctity of life? One of the most helpful explanations of this dilemma is that provided by Thai scholar Chaiwat Satha-Anand (Qader Muheideen) (1993), who concluded the following:

> Yet there is a paradox: if Islam values the sanctity of life, how can Muslims fight "tumult and oppression" to the end? Unless Muslims forsake the methods of violence, they cannot follow the seemingly contradictory injunctions. It is evident that fighting against injustice cannot be avoided. But the use of violence in such fighting can be eschewed. (p. 16)

The solution to the paradox in the Qur'an may be the same that Gandhi found in his quest to reconcile the need to struggle for justice and Indian independence with the moral teachings against harmfulness he had received as a child. The Gandhian solution of fighting without harming would also appear to resolve the apparent dilemma posed by the Qur'an. It is precisely that conclusion that came to Abdul Ghaffar Khan—also known as Badshah Khan and the "Frontier Gandhi"—as he studied the Qur'an in prison during the Indian Freedom struggle:

> As a young boy, I had had violence tendencies; the hot blood of the Pathans was in my veins. But in jail I had nothing to do except read the Qur'an. I read about the Prophet Muhammad in Mecca, about his patience, his suffering, his dedication. I had read it all before, as a child, but now I read it in the light of what I was hearing all around me about Gandhiji's struggle against the British Raj. (quoted in Harris, 1998, p. 102)

Indeed, the relationship between Islam and Gandhian nonviolence is much closer than most would think, and it is to that we now turn our attention.

Nonviolent Struggle in the Islamic Tradition

> *The Prophet (peace be upon him) said: The best fighting (jihad) in the path of Allah is (to speak) a word of justice to an oppressive ruler.*
>
> —Narrated by AbuSa'id al-Khudri[10]

The very birth of modern nonviolence took place when Gandhi was working as a lawyer for Muslim Indian traders in South Africa (Gandhi, 1928). Subjected to the grave indignities of the racist social order there,

Gandhi insisted that the system be challenged and his Muslim companions and employers encouraged, supported, and in some ways guided him.

When the South African government passed a law requiring all Indians to register and carry a pass, the groundwork was laid for the birth of Satyagraha, Gandhi's term for a nonviolent struggle, from the Sanskrit, "holding fast to the truth," sometimes translated as "truth force." On September 19, 1906, Gandhi addressed the Hamadiya Islamic Society to denounce the act, urging those present not to register, and pledging to risk prison before doing so himself. Two days later, at a mass meeting on September 11, the campaign was launched, and Gandhi and his followers burned their registration certificates on the grounds of the Hamidia Mosque in Johannesburg.

Two of the most important figures in the Indian independence movement were Muslims: Maulana Azad, who was part of the inner circle of the Congress Party, and Abdul Ghaffar Khan (also known as Badshah Khan), the inspiration of a recently founded Badshah Khan Peace Initiative.

Although he was not always an advocate of nonviolence, Khan became convinced that it was the imperative of the Qur'an and he mobilized a nonviolent army of 100,000 Pathans to challenge the British with nonviolent resolve. According to Eknath Easwaran (1983), he was among a group of Pathans— known for their fierce, violent culture—who met Gandhi when he arrived in the Northwest Provinces as the champion of the Indian freedom movement:

> A skeptical crowd of . . . Pathans with their guns slung over their shoulders gathered to watch the little figure in his loincloth get up before them. "Are you afraid?" he asked them gently. "Why else would you be carrying guns?" They just stared at him, stunned. No one had ever dared to speak to them like this before. "I have no fear," Gandhi went on; "that is why I am unarmed. This is what ahimsa means." Abdul Ghaffar Khan threw away his gun, and the Pathans, following his leadership, became some of the most courageous followers of Gandhi's way of love. (p. 84)

Badshah Khan's "red shirts," the Khudai Khidmatgars (servants of God), were legendary for their religious and social commitment, taking a holistic approach to the effort to obtain an independent society free of colonialism. They took a vow as they joined to "serve humanity in the name of God," refrain from violence and revenge, live a simple life, refrain from evil, "practice good manners and good behavior," and "devote at least two hours a day to social work" (Easwaran, 1984, pp. 111–112).

Khan and his nonviolent activists believed that the ultimate model for nonviolence was not Gandhi but the Prophet Muhammad himself. As Haji Sarfaraz Nazim put it, "It is wrong to assume that Gandhi was the first to set foot on a non-violence campaign in order to attain *swaraj* [self-rule]. About 1,300 years ago, the Prophet of Arabia had recourse to non-violence" (Banerjee, 2000, p. 149).

Muslim involvement in the development of modern nonviolence did not end with the Indian freedom movement. A number of significant nonviolence movements had key Muslim involvement, from the pro-democracy movement in Thailand (Satha-Anand, 1993) to the antiapartheid struggle in South Africa (Zunes, 1999b) and the largely nonviolent Palestinian *Intifada* of the late 1980s (Zunes, 1999a).

Satha-Anand's (1993) analysis of nonviolent action in Pattani, Thailand, concluded with "eight theses on Muslim nonviolent action" that he presented as "a challenge for Muslims and others who seek to reaffirm the original vision of Islam":

1. For Islam, the problem of violence is an integral part of the Islamic moral sphere.

2. Violence, if any, used by Muslims must be governed by rules prescribed in the Qur'an and Hadith.

3. If violence used cannot discriminate between the combatants and noncombatants, then it is unacceptable in Islam.

4. Modern technology of destruction renders discrimination virtually impossible at present.

5. In the modern world, Muslims cannot use violence.

6. Islam teaches Muslims to fight for justice with the understanding that human lives—as all parts of God's creation—are purposive and sacred.

7. In order to be true to Islam, Muslims must utilize nonviolent action as a new mode of struggle.

8. Islam itself is a fertile soil for nonviolence because of its potential for disobedience, strong discipline, sharing and social responsibility, perseverance and self-sacrifice, and the belief in the unity of the Muslim community and the oneness of . . . [humanity]. (pp. 21–22)

Perhaps one consequence of the new challenges presented by the need to condemn terrorism but still fight against injustices against the Muslim community and others will be a renaissance of Islamic thought about violence along the lines of that advocated by Professor Satha-Anand.

In the winter and spring of 2011, Islamic nonviolent resistance captured the world's attention as a series of uprisings shook the Arab world, combined Islamic principles with Gandhian-style resistance, bringing down dictatorships in Tunisia and Egypt and putting them on alert throughout the Middle

East (see Al Jazeera, 2011; Stolberg, 2011). The iconic image of the Muslim as terrorist on mainstream media was supplanted with tens of thousands of Egyptians praying in Tahrir Square in Cairo, then calling for Hosni Mubarak to step down after decades of authoritarian rule (Kurtz, 2011b).

Nonviolent Lifestyle and Social Organization in Islam

> Be they Muslims, Jews, Christians, or Sabians, Those who believe in God and the Last Day And who do good Have their reward with their Lord.
>
> They have nothing to fear, and they will not sorrow.
>
> Worship nothing but God; Be good to your parents and relatives And to the orphan and the poor Speak nicely to people, Be in constant prayer And give charity.
>
> —Qur'an 2:62, 83

Finally, we will explore the other side of nonviolence and its relationship to Islam—that is, the cultivation of a culture of peace and a just social organization that is nonviolent. Although the Islamic tradition, like all the others, is sometimes used to promote repressive political and social structures and legitimate all kinds of despicable behavior, the overall teachings of the Qur'an and the example of the Prophet Muhammad's life promote mercy, justice, egalitarianism, nonviolent economics, and civility and trustworthiness in daily interactions.

Mercy as the Stamp of Creation

Umar Faruq Abd-Allah (2005) said it is accurate to describe Islam as a religion of peace, but even more so of mercy. He observed the following:

> Islamic revelation designates the Prophet Muhammad as "the prophet of mercy," and Islam's scriptural sources stress that mercy—above other divine attributions—is God's hallmark in creation and constitutes his primary relation to the world from its inception through eternity, in this world and the next. Islam enjoins its followers to be merciful to themselves, to others, and the whole of creation, teaching a karmalike law of universal reciprocity by which God shows mercy to the merciful and withholds it from those who hold it back from others. (p. 1)

Muslims are thus expected to follow the Prophet's example in promoting a "doctrine of universal, all-embracing mercy" that is to be applied not only to other Muslims but to believers and unbelievers and even "the animate and inanimate: birds and animals, even plants and trees" (Abd-Allah, 2005, pp. 4–5). In the end, Abd-Allah (2005) claimed, "The imperative to be merciful—to bring benefit to the world and avert harm—must underlie a Muslim's understanding of reality and attitude toward society" (p. 6). As Mukulika Banerjee (2000, p. 147) noted, the passages in the Qur'an some-times used to justify revenge ("an eye for an eye") are followed by verses often omitted:

> But if anyone
>
> Remits the retaliation
>
> By way of charity, it is
>
> An act of atonement for himself. (Cleary, 2004, 5:45)

Thus, although revenge may be honorable, if kept within bounds ("equal for equal"), "forgiveness is still more worthy in the eyes of Allah" (Banerjee, 2000, p. 147).

Islam and Justice

Another concept at the core of Islamic teachings is the idea of justice, which becomes immediately apparent when one does even the most superfi-cial reading of the Qur'an. Violence and conflict are primarily a result of the lack of justice. Islamic teachings give priority to a notion of justice that one should apply to one's daily life to maintain peace among people. Humans have a responsibility to respect and care for one another, from the family to neighbors to the entire human community.

Islamic egalitarianism is also a Qur'anic principle that is widely mani-fest in Islam ritual and even architecture. The most striking example is the way in which prayers are carried out in the mosque, where all participants stand shoulder to shoulder regardless of class, race, or social status. There is no elevated pulpit from which the imam delivers the word of God down to the people.

One issue of justice rightly raised by many concerns gender inequality. It is necessary, first, to acknowledge the context within which Islamic norms were developed—not only was there apparently female infanticide but there was also little attention to legal rights for women. The Prophet insisted on a number of rights for women that were somewhat revolutionary for the culture of his time and involved women in his movement. Indeed, if it had not been for the encouragement of his wife Khadijah when he first claimed to have heard the voice of God in a cave where he was meditating, he might not have had the courage to speak and to lead.

In one Hadith, we learn that when Fatima, the Prophet's daughter, came to a meeting where he was, he would make room for her to sit beside him. Some even suggest that before he died, the Prophet passed the light of his prophecy to Fatima. This is hardly the practice of *purdah*, or of second-class citizenship for women, which may be a cultural interpretation that contradicts many aspects of early Islam.

Similarly, the Qur'an prohibits racism. Indeed, it states, "The diversity of human languages and complexions is a sign that God exists." Moreover, it states the following:

> To God belongs everyone
>
> In the heavens and the earth
>
> All are obedient to God. (Cleary, 2004, 30:20–27)

A classic example of this recognition of racial equality in Islam appears in the story of Malcolm X. He is originally drawn to the racial teachings of Elijah Muhammad in the Nation of Islam but is transformed when he goes on a pilgrimage to Mecca and sees people of all colors and nationalities standing equally before God in that Holy City (Haley & Malcolm X, 1996).

The Islamic emphasis on justice is also reflected in the nonviolent economics of Islam, beginning with the idea of the *zakat*, the annual donations to the poor that all Muslims are supposed to make. The sharing of resources is a religious obligation and the fourth of the traditional Five Pillars of Islam. Indeed, the crucial ritual of fasting during the holy month of Ramadan is not only a way of advancing oneself spiritually but also of reminding participants of the plight of the poor and the need to establish justice. The purpose of zakat is to reduce the gap between those who have and those who have not, to promote equality among people in human societies. If you have food to eat, you should first look at your neighbor and see if they have something to eat as well. If they do not, you have to provide it.

Finally, Islam traditionally prohibits the charging of interest, which can lead to exploitation and the concentration of wealth. Material wealth is for the benefit of the welfare of humanity, and charging of interest leads to the formation of classes and class conflicts.

Civility and Respect

The model for all Islamic behavior is the Prophet Muhammad, who was reportedly gentle and good-natured, and who was called *Amin* (trustworthy), even by his enemies. Abd-Allah (2005) noted the following, according to the tradition:

> Muhammad jested with children, showed a kindly humor toward adults, and even gave his followers friendly nicknames. He visited the sick, inquired after the welfare of neighbors, friends, followers, and even those who disbelieved in him. He was a warm egalitarian and shared everything with those around him, including their poverty. He was always willing to forgive, rarely chastising those who disobeyed him. . . . The Prophet accepted people at their word and forgave them easily. He harbored no desire for vengeance and rejected the pagan custom of blood feuds and revenge. There was nothing mindless or fanatic about his piety. He was never intransigent or bent on war. (p. 3)

Indeed, the personal character of the Prophet and the way he treated people is not only legendary but it is also part of the explanation for the phenomenal success of the early Islamic movement. As Razi Ahmad (1993) observed, not only was he "respected by the Meccans for his compassion, honesty, purity of character, gentleness, and truthfulness," but he also "neither cursed anyone nor said ill words about anyone" (p. 36), despite the ruthless attacks over the years on him, his followers, and his family. The Qur'an prohibits backbiting and gossip—it says that we are to "speak nicely to people" (Qur'an 2:83) and to show self-restraint (Qur'an 48:26). Muslims are to provide others with a sense of security and well-being, in other words, to create a nonviolent environment for humanity, beginning with the people around them. When you encounter people, you should say Assalamu Alaikum, peace be upon you, to which the other responds, Walaikum Salaam. A wide range of ethical teachings within Islam promotes respect for others.

Harris (1998) suggested that the principle of no-compulsion is what "the rest of the world calls non-violence"—it is "old territory in spiritual terms, but an intellectual frontier" (p. 111). The spiritual nonviolence found in the teachings of the Sufis and other Muslims stands in sharp contrast to the Islam of Fox and CNN. It is a position of power but not domination—of strength but not of violence.

Chaos or Community?

We must realize that all traditions are ambivalent and that it is therefore necessary to be critical about all of them so as to be able to decide which tradition to maintain and which not.

—Jürgen Habermas (1994)

The increasing unity and diversity of the global village has had a profound impact on religious life on the planet. On the one hand, each of the world's faiths has been increasingly forced to take account of the multitude of others. On the other hand, specific religions have become internally diversified as they absorb a broad range of indigenous cultures and as they rediscover diverse elements within their own history. The encounters of American young people with Eastern religions, for example, has helped to stimulate a rediscovery of the mystical traditions of Christianity and Judaism, just as Eastern encounters with Christianity may have revitalized some of the ancient ethical teachings in Hinduism. The richness of individual traditions has been enhanced by encounters among the various faiths.

The Sociological Imagination

Solutions to the cultural dilemmas of the global village will require a great deal of what C. Wright Mills (1959) called the "sociological imagination." The use of that creative imagination in examining contemporary religious life leads, we think, to three fundamental conclusions.

First, religion is intimately linked to social life. Any analysis of religious life has to attend both to social structures and cleavages and to the ways in which religious systems are engaged in sustaining hegemony or rebellion. Indeed, any analysis of the current global human situation must take the significance of religious beliefs, practices, and institutions into account. Religious traditions are vehicles of protest for many currents of social change in the early 21st century, from the anti-Western revolt of the Islamic world to nonviolent and liberation theology struggles of various social movements. Religious institutions—in their pluralistic and contentious variety—constitute the one significant force that possesses some relative autonomy from the dominating centers of international capital and political structures.

Second, the existing religious traditions, having formed out of a long-term process of diffusion across cultural boundaries, are already multicultural, incorporating a variety of indigenous beliefs and practices in the midst of considerable conflict both within and between traditions. No pure monolithic cultures exist. If humanity's future is anything like its past, the challenges of science and diverse religious traditions will not simply go away; rather, they will be essential ingredients in new socially constructed forms.

Third, models do exist for a combination of both unity and diversity in social life that could result in a shared ethos in the global village that does not destroy the rich fabric of human religious life or force anyone who does not wish to do so to participate in religious practice.

Religious traditions are broad cultural abstractions that link worldviews with daily life through rituals and institutions that symbolize and sustain certain values and norms. The link between society and culture is a dynamic one, characterized by elective affinities between ideas of a particular cultural orientation and the interests of particular social groups. Consequently, cultural and social change go hand in hand: When societies change their form of organization, so do cultures; the cultural styles of a previous era will prefigure the way in which societies change, and the kind of change that occurs will shape the worldviews and ethos of the next era.

In recent centuries, a new form of social organization, global in scope, has emerged. Picking up pace in the latter half of the 20th century, it will profoundly affect the nature of human life for some time to come. One consequence of that change is that religious traditions—which play a key role in both precipitating and resisting the transformations—face a crisis of multiculturalism that calls the taken-for-grantedness of every major religious tradition into question.

Multicultural Religious Themes of Tolerance

Two key images of the Egyptian Revolution broadcast around the world were of Egyptian Christians forming a protective circle around Muslims at prayer on Tahrir Square in Cairo and an Egyptian flag stretched across a street between a mosque and a church. The collaboration of Muslims and Christians in the Egyptian prodemocracy uprising could be a symbol of hope for future interfaith coalitions on behalf of justice and peace.

Religious traditions, we have argued, grow out of social life, change over time, and have considerable internal diversity when diffused across a variety of indigenous cultures. As they transform from local to cosmopolitan traditions, they incorporate the diversity of a wide range of beliefs and practices, worldviews and norms yet somehow maintain their integrity. Often when they are in the very process of transformation, the tradition's guardians of orthodoxy claim that it never has changed and never will change. The most serious challenges to religious toleration and diversity usually come from exclusivist truth claims in general and especially from monotheistic traditions because of their inherent tendencies toward intolerance. Yet history does demonstrate that monotheistic religions have found ways of building tolerance into their belief systems, and monotheistic political elites have incorporated it into their public policies. That it has historical precedence is important and encouraging.

The struggle to live together will force people of faith to search their traditions for those points of contact or to fight their neighbors, perhaps to the death. Each of the major traditions in the global village, however, can make a contribution to a multicultural ethos. Hindus suggest that many paths lead to the same summit and that each person and social group must find their own way, guided by their own traditions. Buddhists subscribe to the same "multiple-paths" premise but suggest that the world is so full of suffering that one should try to escape it; the way out of the world, for the Buddhist, is to treat its creatures with compassion for their shared sufferings, thus embracing the world they are rejecting. Judaism begins with a particularism that favors one ethnic group but declares that the group's very purpose is to be a light to others, to lead all of humanity to an ethical lifestyle that sustains a world of justice and peace. Muslims demonstrate an intensity of commitment that could lead to destructive conflict but turns to justice and tolerance whenever they are allowed to practice their worship in peace. Christianity has been spread by world-conquering colonial and neo-colonial powers, but in the violence of the 20th century, Christians redis-covered the nonviolent strength of their founder, who insists that all people are children of God and that the test of one's relationship with God is whether one loves one's enemies and brings good news to the poor.

The fact that these sacred traditions are each formed out of a specific and different social context creates ambivalence. On one hand, faith communi-ties can facilitate the process of ordering a common life since all traditions change and diversify as they encounter other perspectives and all traditions are now sharing more life experiences in common. On the other hand, how-ever, the process is fraught with danger, precisely because of the close link between religious and social life. First, some people's identity and status are threatened by the changes and they will resist, perhaps violently. Second, change may come more slowly, and the resistance may run more deeply than the overall community can tolerate, especially if some people decide that they would rather die than live with pluralism.

Recent developments in the religious world are encouraging, notably the revival of the Parliament of the World's Religions that occurred first in Chicago in 1893 and then again a century later in the same city, followed by subsequent parliaments in Cape Town, South Africa (1999) and Barcelona, Spain (2004). A fresh wind of interfaith dialogue has swept across the globe in recent decades, as well as empirical studies and essays about how to creatively negotiate differences among the world's faith traditions and the benefits of their cross-fertilization (see, e.g., Eck, 1993; Kinnamon & Cope, 1997; Küng, 1991; Martin, 2001; Swidler, 1990). Some scholars, like Mohammed Abu-Nimer (2001), combine interfaith dialogue with empir-ical research to enhance the quality of intercultural communication and interfaith encounters.

So much has happened in recent decades along these lines that we sometimes lose track of the progress made and have high expectations. When I
was appointed to facilitate the discussion of an interfaith group at a
UNESCO seminar in Granada, Spain, in 1998, I found the discussion frus-
trating and the lack of communication appalling. After the session, I ran into
my dear friend S. Jeyapragasam, a renowned Gandhian scholar from South
India, who noticed my frustration. I explained my sour experience in the
discussion, and he wisely responded, "Lester, they're not shooting at each
other!" and I was able to reframe the event in a more positive light. Indeed,
despite its disappointments, the scholars and religious leaders gathered there
were all enriched and enlightened by our time together.

First, if we are to share the planet successfully without redividing it into
a set of isolated fiefdoms, the increased unity and diversity of the world's
cultures and religious traditions will probably intensify. That will inevitably
lead, it seems, to secularized political orders at the broadest level—barring
some religious miracle by which the current religious traditions are quickly
transformed and consensus develops among the world's religious leaders and
at the grassroots. Within that secularized polity—or set of polities—the religious freedom of individual traditions must also be protected if we are to
share the planet peacefully.

Second, we must demilitarize our conflicts and learn how to conduct
them creatively. One of the most hopeful developments for the global village
has been the development, in the 1970s and 1980s, of a set of conflict resolution techniques that have been elaborated by conflict experts and diffused
widely in a number of settings. Our self-conscious development of nonviolent and nondestructive means of conflict and experimentation with them in
various arenas from the school playground to corporations, from interethnic
conflict to international affairs, is still rudimentary but nonetheless promising. Interreligious conflicts, especially when they run across social fault lines,
can be particularly deadly, as we have seen, so they must be attended to in a
deliberate and creative fashion. Secular forms of conflict resolution might be
supplemented with techniques couched in ecumenical religious rhetoric so
that it appeals to each faith community within the cognitive frames of their
beliefs and draws legitimacy from their respective traditions.

Finally, each of the major religions has ethical standards that promote a
basic sense of justice and compassion that could transform the current situation if they were applied to daily life and the construction of human institutions. A set of minimal cultural norms, a global ethos that protects us from
one another, must be codified in secular media. The human rights tradition,
for example, emerging on a global level in recent decades, declares that every
human is entitled to a set of basic rights and due process; this tradition has

been widely accepted in recent decades and is sustained by secular rituals and institutions such as the United Nations. It is unlikely, moreover, that the injustices of the current global economic order, in which half of the world's population lives on the verge of starvation and lacks minimal levels of shelter and clothing, can sustain a peaceful world. Religious traditions must legitimate a global ethos before it can be widely accepted and (more importantly) practiced.

Stories of Conflict and Change

Many stories embedded in religious traditions recall the traditions of courageous men and women who responded to times of social crisis and intercultural conflict by forging new religious systems. Moses came from a group of people who had become enslaved and were under attack from the establishment. In an effort to maintain control over the Hebrew slaves, the Pharaoh ordered the slaughter of their newborn babies. Moses only survived because of the bravery of the Hebrew midwives who defied the order. As a young man, he became enraged when one of his compatriots was beaten; Moses killed the Egyptian overseer and fled for his life only to encounter his God.

Mary, a young Jewish woman living under the shadow of the Roman Empire in a time of social turmoil, proclaimed that God blessed the poor and sent the rich away empty and gave birth to Jesus, who challenged the religious and political establishment of his time. Siddhartha, raised as a prince, turned his back on his heritage of kingship and wealth and sought a different path. The wealthy Arab merchant Muhammad denounced the emptiness and materialism of his day and set his sights on a more meaningful struggle to live an ethical life.

These stories and many more from ancient times to the present suggest that people can rise above fear or apathy and meet the challenge of transformative times. Despite the centuries, if not millennia, of religiously based conflict, the various religious traditions share some essential norms. They posit the significance of collective life and of ethical values; they encourage people to treat others at least with respect, better yet, with compassion or love, and they inspire people to reach beyond their profane everyday lives and to strive for something higher.

The religious path is fraught with danger, however; those who encounter the sacred can be consumed by its fire. Most religions warn through ancient legends that the power of the Gods can destroy those who do not approach them correctly or call upon them for the right purposes. Most malicious deeds, moreover, are usually done in the name of good; humanity's religious traditions are used to destroy as well as to create, for greed as well as altruism. It is a territory conquered by loving action rather than nonresponsiveness, which the great 13th-century Sufi poet Rumi (1995) identified as the great problem:

> The son of Mary, Jesus, hurries up a slope
>   As though a wild animal were chasing him.
>
> Someone following him asks, "Where are you going?
>   No one is after you." Jesus keeps on,
> Saying nothing, across two more fields. "Are you
>   The one who says words over a dead person,
> So that he wakes up?" *I am.* "Did you not make
>   The clay birds fly?" *Yes.* "Who then
> Could possibly cause you to run like this?"
>   Jesus slows his pace.
>
> *I say the Great Name over the deaf and the blind,*
>   *They are healed. Over a stony mountainside,*
> *And it tears its mantle down to the navel.*
>   *Over non-existence, it comes into existence.*
>
> *But when I speak lovingly for hours, for days,*
>   *With those who take human warmth*
> *And mock it, when I say the Name to them, nothing*
>   *Happens.*
>
> *They remain rock, or turn to sand,*
>   *Where no plants can grow. Other diseases are ways*
>
> *For Mercy to enter, but this non-responding*
>   *Breeds violence and coldness toward God.*
>
> *I am fleeing from that.* (p. 204)

According to Rumi's Jesus, we must act. We must, however, also avoid harmfulness. Perhaps religious conflict in the 21st century could follow patterns laid down by Rumi and Gandhi.

Max Weber contended that the Puritans and their specific notion of the "calling" inadvertently laid the groundwork for the modern socioeconomic order of capitalism. It may well be that the ideas of active nonviolence, growing out of a fertile confrontation between Eastern and Western religious traditions, may be preparing the way for a new order that mitigates the spiral of violence and that a new spirit of interfaith dialogue will prepare the way for a multicultural ethos that will respect everyone's beliefs while cultivating our common life. Given the current sophisticated state of violence and the persistent inequalities on the planet, such an order would come none too soon. For, as Martin Luther King Jr. contended, the choice is now between nonviolence and nonexistence: We must learn to live together as brothers and sisters, or we shall die together as fools.

Notes

Chapter 1

1. I am grateful to Sheldon Ekland-Olson for his insights on this matter.

2. Robert Wuthnow (1987) outlined four major approaches in this emerging field in his important work, *Meaning and Moral Order: Explorations in Cultural Analysis*. At the risk of oversimplifying his rich discussion, I have briefly summarized his arguments in the following discussion.

3. Most people have only one death, at least per lifetime, out of the countless events of the life course and will seldom if ever have religious visions. The frequency of sexual intercourse varies significantly, but people have sex much less often than they engage in other activities they consider less significant.

Chapter 2

1. Weber typically surrounded this typology with a series of caveats, admitting that it "may be sketchy" and seeming to imply that its main purpose is to show "how complicated the structures and how many-sided the conditions of a concrete economic ethic usually are" (1922–1923/1946, p. 267).

2. This idea is commonly held but the subject of considerable controversy.

3. Throughout this work, I will use the terms BCE (before the Common Era) and CE (Common Era) to refer to time periods traditionally called BC (before Christ) and AD (anno Domini, "in the year of our Lord"). This compromise position acknowledges the widespread adoption of the so-called "Christian calendar" while moving away from its ethnocentrism.

4. From the *Vammika-sutta* of the Pali text *Majjhima-Nikaya*, quoted in Gard (1962, p. 120) and in Buddha (1954, p. 180).

5. Because a threefold confession of faith is a central ritual in Buddhism, the formula is then repeated twice—quoted in Gard (1962, pp. 53–54).

6. Quoted in Ch'en (1964/1972, p. 474), who excerpted these quotations from Tokiwa, *Bukkyo to jukyo dokyo, 529–531*.

Chapter 3

1. Although this online essay is on a site designed essentially to convert Muslims to Christianity, which would make it immediately suspect as an objective intellectual source, I find the arguments Ibrahim puts forth interesting and intellectually compelling.

2. Quotations from the Qur'an are from Thomas Cleary's (2004) translation.

Chapter 4

1. This spelling, with a capital *P* and a capital *W*, is standard among most tribes.

2. For a critical analysis of the problems with the text of the speech, see Low (1995), who contended that the original publisher probably heard the speech but in a language he did not understand and that various versions published later are "embellished with stereotypes and romanticized tropes of nature, more accurately reflects the projected hegemonic mythology rather than an accurate record of historical event " (p. 418).

Chapter 5

1. I am grateful to Poorno Pragna for his insights into the Vedic traditions.

2. Apparently the fifth precept, the rule against drinking intoxicants, is not considered serious enough for expulsion. Additional rules require either a formal meeting of the order to consider the violation or a confession of guilt. A wide range of practices evolved, however, as the *Sangha* was organized throughout Asia.

3. I have taken the Ten Commandments from the Authorized King James Version in order to use the traditional "Thou shalt" formulation lost in the new, but more accurate, translations.

4. Aaron, who is left in charge, has a wonderful account for Moses, which seems to be an effort to avoid responsibility for the events: They handed him the gold, and "I threw it into the fire, and there came out this calf" (Exodus 32:24). What is remarkable here and elsewhere in the Hebrew scriptures is how details reflecting badly on the heroes are retained in the accounts.

5. I am indebted to Christopher Ellison's insights on this topic and rely heavily on Ellison and Bartkowski (1997) for the discussion that follows.

6. I am grateful to the students in my spring 2011 Sociology of Religion class at George Mason University for an excellent and informative discussion about this topic that helped shape some of the issues and how I framed them.

7. In this section, I am indebted not only to the authors cited but also to Howard Miller and Douglas Laycock, former colleagues at the University of Texas.

8. This is suggested by Yvonne Haddad (1988), in "Islam and the Transformation of Society," a paper delivered at the University of Texas at Austin.

Chapter 6

1. The discussion that follows draws heavily from my discussion of these issues in *The Politics of Heresy: The Modernist Crisis in Roman Catholicism* (Kurtz, 1986).

2. I am grateful to Teresa Sullivan for this way of articulating the problem.

3. The term *culture wars* comes from the *kulturkampf* of the Enlightenment period of 18th- and 19th-century Europe and is used effectively by Hunter (1991) in his analysis of conflicts in contemporary political culture in the United States (cf. Kurtz,1994).

4. See Echo Fields's (1991) discussion of Christian traditionalism ("fundamentalism") along these lines, as well as more theological explanations, but with a similar tone, in Harvey Cox (1984).

5. See Guth's (1983) review of the movement's history and strategies.

6. The first two observations are my own and from Crippen (1988). The others are from Hunter (1991, p. 299ff.).

7. For a brief summary of these developments, see Nanji (1988). For a more detailed discussion of the intellectual issues, see Marty and Appleby's (1993) three-volume edited collection evaluating fundamentalism.

8. The following account is taken from Hiro (1989, pp. 232–235).

9. From "Revive Native Religion" (1947), quoted in Assimeng (1978).

10. This view is not universal, although there is considerable truth in it. Renato Poblete (1970), for example, claims that there was little struggle in the transition from the indigenous religions to Christianity.

11. It is telling that when Gutiérrez was delivering a lecture series in Austin, Texas, as a world-famous visiting theologian, he quipped while struggling with the microphone, "I am not exactly a modern man."

12. The conventional wisdom is that liberation theology emerged from the grassroots poor of Latin America, an assumption challenged by Madeleine Adriance (1986). Certainly its key spokespersons, like Gutiérrez, are college-educated elites, but many come from poor families and others live and work with the poor, listening to and articulating their perspectives on the Christian tradition.

13. The following discussion relies heavily upon Gutiérrez's (1973) important *A Theology of Liberation: History, Politics, and Salvation*, as well as a lecture series he delivered at Austin Presbyterian Theological Seminary from October 24 to 26, 1983. See also Berryman (1984); Boff and Boff (1986); Chopp (1986); Ferm (1986); Gutiérrez (1977, 1983); Lernoux (1982); Novak (1986); Segundo (1976, 1985).

Chapter 7

1. Sectarian organizations are religiously zealous and tend to have somewhat rigid doctrines and minimal tolerance for alternative theologies.

2. On the importance of religious belief in decisions to join, see Clark (1937); Glock and Stark (1965); Smelser (1963); and Wilson (1959). The emphasis on

interpersonal bonds and networks can be found in Bainbridge (1978); Lofland (1977); Lofland and Stark (1965); and Snow, Zurcher, and Ekland-Olson (1980).

3. The concept comes from the purported "mind control" techniques used on American prisoners held in Korea and China during the Korean War. These prisoners were said to have become sympathetic to their captors' belief systems (see Anthony, 1990, pp. 299–300, who considered the brainwashing argument a hoax, used first as an anticommunist propaganda tool during the Korean War and again in the 1970s and 1980s against new religious movements [NRMs]).

4. Barker (1988, p. 175) and Bromley (1988, p. 204) both estimated that about two thirds of the coercive deprogrammings of Unification Church members were successful.

Chapter 8

1. Portions of this section were previously published as Kurtz (2005b), "From Heresies to Holy Wars: Toward a Theory of Religious Conflict" in *Ahimsa Nonviolence*, which is available from http://works.bepress.com/lester_kurtz/31, and in S. Jeyapragasam (1993), *Communalism: The Crisis in India and the Way Out*; they are both used by permission.

2. I am grateful for this insight to comparative religions guru, the late Ninian Smart.

3. This quotation is from the *Dhammapada* (n.d.), retrieved from http://www .unification.net/ws/.

4. Psalm 18:39–40 (retrieved from http://www.mechon-mamre.org/e/et/et2618 .htm)

5. Gandhi used the Gita as a spiritual guide for his daily life and translated it into Gujarati, publishing it on March 2, 1930, the day he marched to Dandi from Sabarmati. While in Yeravda prison, Gandhi received a complaint from a member of his ashram that the Gita was very difficult to understand. Consequently, the Mahatma (a title popularly bestowed on Gandhi meaning "Great Soul") wrote a series of letters (one for each chapter of the Gita) giving his interpretations (see V. G. Desai's translation into English; Gandhi [1930/1987]).

6. This is an unpublished statement from Father José Blanco (personal correspondence, 1995).

7. An earlier version of this section was published as "Solving the Qur'anic Paradox" (Kurtz & Kurtz, 2005), which is available from http://works.bepress.com/ lester_kurtz/10/.

8. A fatwa is a legal decision made by an established religious authority in Islam. The text of this fatwa and a list of endorsing organizations were retrieved from http://www.cair.com/AmericanMuslims/AntiTerrorism/FatwaAgainstTerrorism.aspx

9. If interested, please see my review (Kurtz, 2005b), which is available from http://works.bepress.com/lester_kurtz/31/.

10. This is from a traditional Sunnah (source about the life of the Prophet) (Abu-Dawud, n.d.).

References

100 celebrated chinese women: Queen Mother of the West. (n.d.). Singapore: Asiapac Books. Retrieved from http://web.archive.org/web/20011117024511/www.span .com.au/100women/2.html

Abd-Allah, U. F. (2004). Islam and the cultural imperative. *A Nawawi Foundation Paper*. Retrieved from http://www.nawawi.org/downloads/article3.pdf

Abd-Allah, U. F. (2005). Mercy: The stamp of creation. *A Nawawi Foundation Paper*. Retrieved from http://www.nawawi.org/downloads/article1.pdf

Abhedânanda, S. (1902). How to be a Yogi. *Internet Sacred Text Archive*. Retrieved from http://www.sacred-texts.com/hin/hby/index.htm

Abu-Dawud, S. (n.d.). *Battles (Kitab Al-Malahim)* (Book 37, Number 4330). Retrieved from http://www.usc.edu/dept/MSA/fundamentals/hadithsunnah/abudawud/037 .sat.html#037.4330

Abu-Nimer, M. (2001). Conflict resolution, culture, and religion: Toward a training model of interreligious peacebuilding. *Journal of Peace Research*, 38, 685–704.

Adriance, M. (1986). *Opting for the poor: Brazilian Catholicism in transition*. Kansas City, MO: Sheed and Ward.

Adriance, M. (1992). The paradox of institutionalization: The Roman Catholic Church in Chile and Brazil. *Sociological Analysis*, 53, S51–S62.

Aguirre Beltrán, G. (1967). *Regions of refuge*. Norman: University of Oklahoma Press.

Ahmad, K. (1983). Islamic resurgence. In J. Esposito (Ed.), *Voices of resurgent Islam* (pp. 218–229). New York: Oxford University Press.

Ahmad, R. (1993). Islam, nonviolence and global transformation. In G. Paige, C. Satha-Anand (Qader Muheideen), & S. Gilliat (Eds.), *Nonviolence in Islam* (pp. 27–52). Honolulu: University of Hawaii, Center for Global Nonviolence. Retrieved from http://www.globalnonviolence.org/docs/islamnonviolence/chapter4 .pdf

Ahmed, A. (1988). *Discovering Islam: Making sense of Muslim history and society*. London: Routledge.

Al Jazeera. (2011). Egypt: Seeds of change. Retrieved from http://english.aljazeera.net/ programmes/peopleandpower/2011/02/201128145549829916.html

Almond, G. A., Sivan, E., & Appleby, R. S. (1995). Politics, ethnicity, and fundamentalism. In M. E. Marty & R. Scott Appleby (Eds.), *Fundamentalisms comprehended* (pp. 483–504). Chicago: The University of Chicago Press.

Ammerman, N. T. (1987). *Bible believers: Fundamentalists in the modern world.* New Brunswick, NJ: Rutgers University Press.

Anderson, A. (1993, April). African Pentecostalism and the ancestors: Confrontation or compromise? *Missionalia, 21*(1), 26–39. Retrieved from http://artsweb.bham .ac.uk/aanderson/Publications/african_pentecostalism_and_the_a.htm

Anthony, D. (1990). Religious movements and brainwashing litigation: Evaluating key testimony. In T. Robbins & D. Anthony (Eds.), *In Gods we trust* (2nd ed., pp. 295–344). New Brunswick, NJ: Transaction.

Arjomand, S. A. (1995). Unity and diversity in Islamic Fundamentalism. In M. E. Marty & R. S. Appleby (Eds.), *Fundamentalisms comprehended* (pp. 179–198). Chicago: The University of Chicago Press.

Aslan, R. 2005. *No god but God: The origins, evolution, and future of Islam.* New York: Random House.

Assimeng, M. (1978). Crisis, identity, and integration in African religion. In H. Mol (Ed.), *Identity and the sacred* (pp. 97–118). Oxford, UK: Basil Blackwell.

Azim, M. (2005). A fundamental argument" [Letter to the editor]. *The Economist* 376 (July 23):1. Retrieved from http://ehis.ebscohost.com/ehost/detail?sid=74b54838-8aa7-4897-8e60-6c28d991b07d%40sessionmgr15&vid=1&hid=20&bdata=JnNp dGU9ZWhvc3QtbGl2ZQ%3d%3d

Babcock, B. A. (Ed.). (1978). *The reversible world: Symbolic inversion in art and society.* Ithaca, NY: Cornell University Press.

Bainbridge, W. S. (1978). *Satan's power.* Berkeley: University of California Press.

Bainton, R. (1950). *The Reformation of the sixteenth century.* Boston: Beacon.

Bainton, R. (1960). *Christian attitudes toward war and peace.* Nashville, TN: Abingdon.

Bandura, A. (1999). Moral disengagement in the perpetration of inhumanities. *Personality and Social Psychology Review, 3*, 193–209.

Banerjee, M. (2000). *The Pathan unarmed: Opposition & memory in the North West frontier.* Oxford, UK: James Currey.

Barbour, I. (1960). The methods of science and religion. In H. Shapley (Ed.), *Science ponders religion* (pp. 196–215). New York: Appleton-Century-Crofts.

Barker, E. (1988). Defection from the Unification Church: Some statistics and distinctions. In D. Bromley (Ed.), *Falling from the faith* (pp. 166–184). Newbury Park, CA: Sage.

Barnhill, D. L. (2004). Good work: An engaged Buddhist response to the dilemmas of consumerism. *Buddhist-Christian Studies, 24*, 55–63.

Bayle, P. (1697–1706). *Dictionnaire Historique et Critique.* Rotterdam, The Netherlands: Leers.

Beck, P. V., Walters, A. L., & Francisco, N. (2001). The sacred: Ways of knowledge, sources of life. Tsaile, AZ: Diné College (formerly Navajo Community College) Press.

Bellah, R. N. (1964). Religious evolution. *American Sociological Review, 29,* 358–374.

Bellah, R. N. (1970). *Beyond belief: Essays on religion in a post-traditional world.* New York: Harper & Row.

Bellah, R. N. (1975). *Broken covenant: American civil religion in time of trial.* New York: Seabury.

Bendix, R. (1978). *Kings or people: Power and the mandate to rule.* Berkeley: University of California Press.

Berger, P. L. (1969). *The sacred canopy: Elements of a sociological theory of religion.* New York: Doubleday.

Berger, P. L., & Luckmann, T. (1967). *The social construction of reality.* New York: Doubleday.

Berkes, F. (1999). *Sacred ecology.* New York: Routledge.

Berryman, P. (1976). Latin American liberation theology. In S. Torres & J. Eagleson (Eds.), *Theology in the Americas* (pp. 20–83). Maryknoll, NY: Orbis.

Berryman, P. (1984). *The religious roots of rebellion: Christians in Central American revolutions.* Maryknoll, NY: Orbis.

Bloch, M. (2010). Ancestors. In A. Barnard & J. Spencer (Eds.), *The Routledge encyclopedia of social and cultural anthropology* (2nd ed., p. 54). London: Routledge.

Blumer, H. (1954). What is wrong with social theory? *American Sociological Review, 19,* 3–10.

Boff, C., & Boff, L. (1986). *Liberation theology: From confrontation to dialogue.* San Francisco: Harper & Row.

Boswell, J. (1980). *Christianity, social tolerance, and homosexuality: Gay people in Western Europe from the beginning of the Christian era to the fourteenth century.* Chicago: The University of Chicago Press.

Bradley, I. (1992). *God is green: Ecology for Christians.* New York: Doubleday.

Brecht, B. (1966). *Leben des Galilei* (E. Bentley, Ed. & C. Laughton, Trans.). New York: Grove.

Bromley, D. G. (1988). Deprogramming as a mode of exit from new religious movements: The case of the unificationist movement. In D. Bromley (Ed.), *Falling from the faith* (pp. 185–204). Newbury Park, CA: Sage.

Bromley, D. G., & Shupe, A. D. (1981). *Strange Gods: The great American cult scare.* Boston: Beacon.

Bromley, D. G., Shupe, A. D., & Ventimiglia, J. C. (1979). Atrocity tales, the Unification Church, and the social construction of evil. *Journal of Communication, 29,* 42–53.

Bruce, S. (2010). Secularization. In B. S. Turner (Ed.), *The new Blackwell companion to the sociology of religion* (pp. 125–140). Chichester, UK: Wiley-Blackwell.

Buckley, G. A. (1967). Sin of heresy. *New Catholic Encyclopedia* (Vol. 6, p. 1069). New York: McGraw-Hill.

Buckley, T. (2002). *Standing ground: Yurok Indian spirituality, 1850–1990.* Berkeley: University of California Press.

Buddha, G. (1954). *The collection of the middle length sayings (Majjhima-Nikaya), Vol. 1, The first fifty discourses (Mulapannasa)* (I. B. Horner, Trans.). London: Luzac.

Bullough, V. L. (1976). *Sexual variance in society and history*. Chicago: The University of Chicago Press.

Bullough, V. L. (1979). *The frontiers of sex research*. Buffalo, NY: Prometheus Books.

Bullough, V. L. (1995). Sexuality and religion. In L. Diamant & R. D. McAnultyu (Eds.), *The psychology of sexual orientation, behavior and identity: A handbook* (pp. 444–456). Westport, CT: Greenwood Press.

Bush, R. C. (1988). Buddhism. In K. M. Yates (Ed.), *The religious world* (2nd ed., pp. 113–170). New York: Macmillan.

Cadge, W. (2007). Religion, homosexuality, and the social sciences. In J. Siker (Ed.), *Homosexuality and religion: An encyclopedia* (pp. 19–25). Westport, CT: Greenwood Press.

Campbell, J. (with B. Moyers & B. S. Flowers, Eds.). (1988). *The power of myth*. New York: Doubleday.

Campbell, J. (1968). *The hero with a thousand faces*. Princeton, NJ: Princeton University Press.

Caplow, T., Bahr, H. M., & Chadwick, B. A. (1983). *All faithful people: Change and continuity in Middletown's religion*. Minneapolis: University of Minnesota Press.

Cardenal, E. (1982). *The Gospel in Solentiname* (Vol. 1). (D. D. Walsh, Trans.). Maryknoll, NY: Orbis.

Ch'en, K. K. S. (1972). *Buddhism in China: A historical survey*. Princeton, NJ: Princeton University Press. (Original work published 1964)

Charism. (n.d.). *Merriam-Webster Online Dictionary*. Retrieved from http://www .merriam-webster.com/dictionary/charism

Chatterjee, M. (1983). *Gandhi's religious thought*. London: Macmillan.

Chaudhry, L. (n.d.). The future of Islam. *Alternet*. Retrieved from http://www.alternet .org/story/21891/

Chaves, M. (2004). *Congregations in America*. Cambridge, MA: Harvard University Press.

Chief Seattle. (1887). Chief Seattle's treaty oration—1854. *Seattle Sunday Star*, October 29. Retrieved from http://www.synaptic.bc.ca/ejournal/smith.htm

Chopp, R. (1986). *The praxis of suffering: An interpretation of liberation and political theologies*. Maryknoll, NY: Orbis.

Christ, C. P. (1987). *Laughter of Aphrodite: Reflection on a journey to the Goddess*. Cambridge, MA: Harper & Row.

Christiano, K. (1987). *Religious diversity and social change: American cities, 1890–1906*. Cambridge, UK: Cambridge University Press.

Christiansen, D. S. J. (1992, May 23). Christian theology and ecological responsibility. *America*, 448–451.

Clark, E. T. (1937). *The small sects in America*. Nashville, TN: Cokesbury.

Cleary, T. (Trans.). (2004). *The Qur'an. A new translation*. Chicago: Starlatch, LLC.

Cochran, J. K., & Beeghley, L. (1991). The influence of religion on attitudes toward nonmarital sexuality: A preliminary assessment of reference group theory. *Journal for the Scientific Study of Religion, 30*(1), 45–62. Retrieved from http://www.jstor .org/stable/1387148

Collins, R. (1974). Three faces of cruelty: Towards a comparative sociology of violence. *Theory and Society*, *1*, 425–440.

Comte, A. (1853). *The positive philosophy* (2 vols.) (H. Martineau, Trans.). London: John Chapman.

Confucius. (1998). *The Analects* (D. Hinton, Trans.). Washington, DC: Counterpoint. Retrieved from http://ctext.org/analects

Constitution of the United States. (1938). In C. W. Eliot (Ed.), *American historical documents: 1000–1904. Harvard Classics*. (pp. 180–198). New York: P. F. Collier & Son. (Original work published 1787)

Cousins, L. S. (1984). Buddhism. In J. R. Hinnells (Ed.), *A handbook of living religions* (pp. 278–343). Harmondsworth, UK: Penguin.

Cox, H. (1984). *Religion in the secular city: Toward a postmodern theology*. New York: Simon & Schuster.

Cox, J. L. (1998). *Rational ancestors: Scientific rationality and African indigenous religions*. Cardiff, UK: Cardiff Academic Press.

Coy, P. G., & Woehrle, L. M. (2000). *Social conflicts and collective identities*. Lanham, MD: Rowman & Littlefield.

Cragg, K. (1965). *Counsels in contemporary Islam*. Edinburgh, UK: Edinburgh University Press.

Crippen, T. (1988). Old and new Gods in the modern world: Toward a theory of religious transformation. *Social Forces*, *67*, 316–336.

Cross, G. (1925). Heresy (Christian). In J. Hastings (Ed.), *Encyclopedia of religion and ethics* (Vol. 6, pp. 614–622). New York: Scribner.

Csillag, R. (2005, December 31). Two rabbis, one couple. *Toronto Star*, p. M6.

Cult. (n.d.). *Merriam-Webster Online Dictionary*. Retrieved from http://www.merriam-webster.com/dictionary/cult

Daly, M. (1975). *The church and the second sex* (2nd ed.). New York: Harper & Row. (Original work published 1968)

Dansette, A. (1961). *Religious history of modern France* (2 vols.). Freiburg, Germany: Herder.

Darwin, C. (1952). On the origin of species by means of natural selection. In R. M. Hutchins (Ed.), *Great books of the Western world* (Vol. 49, pp. 1–251). Chicago: Encyclopedia Britannica. (Original work published 1859)

Darwin, C. (1952). The descent of man. In R. M. Hutchins (Ed.), *Great books of the Western world* (Vol. 49, pp. 253–600). Chicago: Encyclopedia Britannica. (Original work published 1871)

Davie, G. (2010). Resacralization. In B. S. Turner (Ed.), *The new Blackwell companion to the sociology of religion* (pp. 160–178). Chichester, UK: Wiley-Blackwell.

Deats, R. (2001). The global development of active nonviolence. *The Fellowship of Reconciliation*. Retrieved from http://forusa.org/nonviolence/0900_73deats.html http://www.forusa.org/nonviolence/0900_73deats.html

Dhammapada, n.d. Retrieved from http://www.unification.net/ws/

Douglas, M. (1966). *Purity and danger: An analysis of concepts of pollution and taboo*. London: Routledge & Kegan Paul.

Dunne, J. S. (1965). *The city of the Gods: A study in myth and mortality*. New York: Macmillan.

Durkheim, E. (1933). *The division of labor in society* (G. Simpson, Trans.). New York: Free Press. (Original work published 1893)

Durkheim, E. (1961). *Moral education: A study in the theory and application of the sociology of education* (E. K. Wilson, Ed., Trans., & H. Schnurer, Trans.). New York: Free Press.

Durkheim, E. (1965). *The elementary forms of the religious life* (J. W. Swain, Trans.). New York: Free Press. (Original work published 1915)

Easwaran, E. (1983). *Gandhi the man* (2nd ed.). Wellingborough, UK: Turnstone Press.

Easwaran, E. (1984). *A man to match his mountains: Badshah Khan, nonviolent soldier of Islam*. Petaluma, CA: Nilgiri Press.

Ebaugh, H. R. (1991). The revitalization movement in the Catholic Church: The institutional dilemma of power. *Sociological Analysis, 52*, 1–12.

Eck, D. L. (1993). *Encountering God: A spiritual journey from Bozeman to Banaras*. Boston: Beacon.

Edwards, J. (1741). *Sinners in the hands of an angry God*. Retrieved from http://edwards.yale.edu/archive?path=aHR0cDovL2Vkd2FyZHMueWFsZS5lZHUvY2dpLWJpbi9uZXddwaGlsby9nZXRvYmplY3QucGw/Yy4yMTo0Ny53amVt

Ehrenreich, B. (1981, February/March). U.S. patriots without God on their side. *Mother Jones*, 35–40.

Eisler, R. T. (1988). *The chalice and the blade: Our history, our future*. San Francisco: Harper & Row.

Eliade, M. (1959). *The sacred and the profane: The nature of religion* (W. R. Trask, Trans.). New York: Harcourt.

Eliade, M. (1970). *Yoga: Immortality and freedom* (W. R. Trask, Trans.). Princeton, NJ: Princeton University Press. (Original work published 1958)

Ellingson, S., & Green, M. C. (2002). *Religion and sexuality in cross-cultural perspective*. New York: Routledge.

Ellison, C. G. (1993). Religious involvement and self-perceptions among Black Americans. *Social Forces, 71*, 1027–1055.

Ellison, C. G. (1999). Introduction to symposium: Religion, health, well-being. *Journal for the Scientific Study of Religion, 37*, 692–694.

Ellison, C. G., & Bartkowski, J. (1997). "Religion and the legitimation of violence: The case of Conservative Protestantism and corporal punishment. In J. Turpin & L. Kurtz (Eds.), *The web of violence: From interpersonal to global* (pp. 45–67). Urbana: University of Illinois Press.

Ellison, C. G., & Sherkat, D. E. (1999). Identifying the semi-involuntary institution: A clarification. *Social Forces, 78*, 793–803.

Erdoes, R. (1972). *The Sundance people*. New York: Alfred Knopf.

Ereira, A. (1992). *The elder brothers: A lost South American people and their message about the fate of the earth*. New York: Alfred Knopf.

Erikson, K. (1965). The sociology of deviance. In E. C. McDonagh & J. E. Simpson (Eds.), *Social problems: Persistent challenges* (pp. 457–464). New York: Holt, Rinehart & Winston.

Erikson, K. (1966). *Wayward puritans: A study in the sociology of deviance*. New York: John Wiley.

Ernst, C. W. (2003). *Following Muhammad: Rethinking Islam in the contemporary world*. Chapel Hill: University of North Carolina Press.

Esposito, J. L. (Ed.). (1983). *Voices of resurgent Islam*. New York: Oxford University Press.

Esposito, J. L. (2003). *Unholy war: Terror in the name of Islam*. New York: Oxford University Press.

"Evangelist Uses Computer Exchange." (1984, August 24). *New York Times*, p. 15.

Fenn, R. K. (1972). Toward a new sociology of religion. *Journal for the Scientific Study of Religion, 11*, 16–32.

Fenn, R. K. (1974). Religion and the legitimation of social systems. In A. Eister (Ed.), *Changing perspectives in the scientific study of religion* (pp. 143–161). New York: John Wiley.

Fenn, R. K. (1976). Bellah and the New Orthodoxy. *Sociological Analysis, 37*, 160–166.

Ferguson, J. (1977). *War and peace in the world's religions*. New York: Oxford University Press.

Ferm, D. W. (1986). *Third World liberation theologies: An introductory survey*. Maryknoll, NY: Orbis.

Fernea, E. (1998). *In search of Islamic feminism: One woman's global journey*. New York: Doubleday.

Fields, E. E. (1991). Understanding activist fundamentalism: Capitalist crisis and the "colonization of the lifeworld." *Sociological Analysis, 52*, 175–190.

Finke, R., & Stark, R. (1986). Turning pews into people. Estimating 19th-century church membership. *Journal for the Scientific Study of Religion, 25*, 180–192.

Finke, R., & Stark, R. (1988). Religious economies and sacred canopies: Religious mobilization in American cities, 1906. *American Sociological Review, 53*, 41–49.

Finke, R., & Stark, R. (1992). *The churching of America, 1776–1990: Winners and losers in our religious economy*. New Brunswick, NJ: Rutgers University Press.

Fiqh Council of North America. (2005). *U.S. Muslim Council issues fatwa against terrorism*. Retrieved from http://www.cair-net.org/includes/ Anti-TerrorList.pdf

Flannelly, K. J., Ellison, C. G., & Strock, A. L. (2004). Methodologic issues in research on religion and health. *Southern Medical Journal, 97*, 1231–1241.

Foucault, M. (1988). *The history of sexuality*. New York: Vintage Books.

Fox, J. (2008). *A world survey of religion and the state*. Cambridge, UK: Cambridge University Press.

Frazer, J. G. (1992). *The golden bough: A study in magic and religion* (Vol. 1). New York: Macmillan. (Original work published 1950)

Freedman, D. N., & McClymond, M. J. (1999). In L. R. Kurtz (Ed.), *Encyclopedia of violence, peace, and conflict* (Vol. 3, pp. 229–239). San Diego, CA: Academic Press.

Freud, S. (1950). *Totem and taboo* (J. Strachey, Trans.). New York: Norton. (Original work published 1913)

Freud, S. (1961). *The future of an illusion* (J. Strachey, Trans.). New York: Norton. (Original work published 1898)

Fromm, E. (1961). *May man prevail? An inquiry into the facts and fictions of foreign policy.* Garden City, NY: Doubleday.

Fukuyama, F. (1992). *The end of history and the last man.* New York: Free Press.

Gandhi, M. K. (1939). *Hind Swaraj or Indian home rule* (Rev. ed.). Ahmedabad, India: Navajivan Publishing House. (Original work published 1908)

Gandhi, M. K. (1930/1987). *Discourses on the Gita* (V. G. Desai, Trans.). Ahmedabad, India: Navajivan Publishing House.

Gandhi, M. K. (1928). *Satyagraha in South Africa* (V. G. Desai, Trans.). Ahmedabad, India: Navajivan Publishing House.

Gandhi, M. K. (1962). *The essential Gandhi* (L. Fischer, Ed.). New York: Vintage.

Gard, R. (Ed.). (1962). *Buddhism.* New York: George Braziller.

Geertz, C. (1968). *Islam observed.* Chicago: The University of Chicago Press.

Geertz, C. (1973). *The interpretation of cultures.* New York: Basic Books.

Gehrig, G. (1981). *American civil religion: An assessment. Society for the Scientific Study of Religion.* Monograph Series No. 3. Storrs: University of Connecticut Press.

Gellner, E. (1981). *Muslim society.* Cambridge, UK: Cambridge University Press.

George, L. K., Ellison, C. G., & Larson, D. B. (2002). Explaining the relationships between religious involvement and health. *Psychological Inquiry, 13,* 190–200.

Gier, N. F. (2007). Was Gandhi a tantric? *Gandhi Marg, 28.* Retrieved from http://www.class.uidaho.edu/ngier/gandtantric.htm

Gimbutas, M. (1982). *The Goddesses and Gods of Old Europe, 6500–3500 B.C.: Myths and cult images* (Rev. ed.). Berkeley: University of California Press.

Gimbutas, M. (1989). *The language of the Goddess.* London: Thames and Hudson.

Girard, R. (1977). *Violence and the sacred.* Baltimore: Johns Hopkins University Press.

Glock, C. Y., & Bellah, R. N. (Eds.). (1976). *The new religious consciousness.* Berkeley: University of California Press.

Glock, C. Y., & Hammond, P. E. (1973). *Beyond the classics? Essays in the scientific study of religion.* New York: Harper & Row.

Glock, C. Y., & Stark, R. (1965). *Religion and society in tension.* Chicago: Rand McNally.

Goffman, E. (1959). *The presentation of self in everyday life.* Garden City, NY: Doubleday.

Goldfarb, J. C. (2006). *The politics of small things: The power of the powerless in dark times.* Chicago: The University of Chicago Press.

Goldman, A. (1985, February 17). Dream of being a rabbi is in sight for a woman. *New York Times.* Retrieved from http://www.nytimes.com/1985/02/17/nyregion/10-year-dream-of-being-a-rabbi-coming-true-for-a-woman.html?scp=3&sq=rabbi&st=nyt

Goodstein, L. (2005, March 10). Evangelical leaders swing influence behind effort to combat global warming. *New York Times*, p. A14.

Gopalan, S. (1978). Identity-theory against the backdrop of the Hindu concept of dharma: A socio-philosophical interpretation. In H. Mol (Ed.), *Identity and the sacred* (pp. 119–132). Oxford, UK: Basil Blackwell.

Gordon, M. (1980). *The company of women*. New York: Random House.

Gottlieb, L. (1995). *She who dwells within: A feminist version of a renewed Judaism*. San Francisco: Harper.

Gottwald, N. K. (1979). *The tribes of Yahweh: A sociology of the religion of liberated Israel 1250–1050 B.C.E.* Maryknoll, NY: Orbis.

Grant, R. M. (1972). *A historical introduction to the New Testament*. New York: Simon & Schuster/Touchstone. (Original work published 1963)

Greeley, A. M. (1989). *Religious change in America*. Cambridge, MA: Harvard University Press.

Griswold, W. (1987). A methodological framework for the sociology of culture. *Sociological Methodology, 17*, 1–35.

Griswold, W. (1994). *Cultures and societies in a changing world*. Thousand Oaks, CA: Sage.

Grossman, D., & Siddle, K. (1999). Psychological effects of combat. In L. R. Kurtz (Ed.), *Encyclopedia of violence, peace and conflict* (Vol. 3, pp. 140–149). San Diego, CA: Academic Press.

Guth, J. L. (1983). The New Christian Right. In R. C. Liebman & R. Wuthnow (Eds.), *The New Christian Right* (pp. 31–45). New York: Aldine.

Gutiérrez, G. (1973). *A theology of liberation: History, politics, and salvation.* Maryknoll, NY: Orbis.

Gutiérrez, G. (1977). *Liberation and change*. Atlanta, GA: John Knox Press.

Gutiérrez, G. (1983). *The power of the poor in history* (R. Barr, Trans.). Maryknoll, NY: Orbis.

Habermas, J. (1975). *Legitimation crisis*. Boston: Beacon.

Habermas, J. (1984). *The theory of communicative action* (T. McCarthy, Trans.). Boston: Beacon.

Habermas, J. (1987). *The philosophical discourse of modernity.* Cambridge: MIT Press.

Habermas, J. (1994, March 24). "More humility, fewer illusions"—A talk between Adam Michnik and Jürgen Habermas. *New York Review of Books*, pp. 24–29.

Haddad, Y. Y. (1988, March 2). Islam and the transformation of society. Lecture, University of Texas at Austin.

Haddad, Y. Y., & Esposito, J. L. (Eds.). (1998). *Islam, gender and social change*. Oxford, UK: Oxford University Press.

Hadden, J. K., & Swann, C. E. (1981). *Prime time preachers: The rising power of televangelism*. Reading, MA: Addison-Wesley.

Haley, A., & Malcolm X. (1996). *The autobiography of Malcolm X*. New York: Chelsea House. Retrieved from http://www.netLibrary.com/urlapi.asp?action=summary&v=1&bookid=38631

Hall, T. W., Pilgrim, R. B., & Cavanagh, R. R. (1986). *Religion: An introduction*. San Francisco: Harper & Row.

Halperin, D. (2002). *How to do the history of homosexuality.* Chicago: The University of Chicago Press.

Hammond, P. E. (1980). The conditions for civil religion: A comparison of the United States and Mexico. In R. N. Bellah & P. E. Hammond (Eds.), *Varieties of civil religion* (pp. 40–85). San Francisco: Harper & Row.

Hanh, T. N. (2008). *For a future to be possible: Commentaries on the five wonderful precepts.* Berkeley: Paralax. Retrieved from http://dharma.ncf.ca/introduction/precepts.html

Harden, B. (2005, February 6). "God's mandate": Putting the White House on notice. *Washington Post*, p. A16.

Hargrove, B. (1989). *Sociology of religion: Classical and contemporary approaches* (2nd ed.). Arlington Heights, IL: Harlan Davidson.

Harris, M. (1974). *Cows, pigs, wars, and witches.* New York: Random House.

Harris, R. T. (1998). Nonviolence in Islam: The alternative community tradition. In D. L. Smith-Christopher (Ed.), *Subverting hatred: The challenge of nonviolence in religious traditions* (pp. 95–114). Boston: Boston Research Center for the 21st Century.

Harvey, G. (Ed.). (2000). *Indigenous religions: A companion.* London: Cassell.

Harvey, G. (2002). Sacred places in the construction of indigenous environmentalism. *Ecotheology*, 7, 60–73.

Harvey, G. (2003). Environmentalism in the construction of indigeneity. *Ecotheology*, 8, 206–223.

Hassan, R. (1999). Feminism in Islam. In A. Sharma & K. K. Young (Eds.), *Feminism and world religions* (pp. 248–278). Albany: State University of New York Press.

Havel, V. (1990). Power of the powerless. In W. Brinton & A. Rinzler (Eds.), *Without force or lies: Voices from the revolution of Central Europe in 1989–90* (pp. 43–127). San Francisco: Mercury House.

Hazarika, S. (1984, July 3). An age-old barrier in India topples: Hindu women assume priestly role. *New York Times*, p. 4.

Herbert, E., Baron Herbert of Cherbury. (1937). *De Veritate* (M. H. Carre, Trans.). Bristol, UK: University of Bristol by J. W. Arrowsmith. (Original work published 1624)

Hertzberg, A. (Ed.). (1962). *Judaism. Great Religions of Modern Man Series.* New York: George Braziller.

Heschel, A. (1962). The Sabbath. In A. Hertzberg (Ed.), *Judaism* (pp. 118–119). New York: George Braziller.

Hesse, H. (1971). *Siddhartha* (H. Rosner, Trans.). New York: Bantam.

Hindu Temple of Atlanta. (2011). Items required for funeral ceremony. Retrieved from http://hindutempleofatlanta.org/AboutPoojaMaterials17.aspx

Hinduism does not condemn homosexuality. (2009, July 3). *Rediff News.* Retrieved from http://news.rediff.com/report/2009/jul/03/hinduism-does-not-condemn-homosexuality.htm

Hinnells, J. R. (Ed.). (1984). *A handbook of living religions.* Harmondsworth, UK: Penguin.

Hirai, N. (n.d.). Shinto. *Encyclopædia Britannica Online.* Retrieved from http://search .eb.com/eb/article?tocld=9105864

Hiro, D. (1989). *Holy wars: The rise of Islamic fundamentalism.* New York: Routledge & Kegan Paul.

Hobbes, T. (1651). *Leviathan: Or the matter, form, and power of a commonwealth, ecclesiastical or civil.* Retrieved from http://etext.library .adelaide.edu.au/h/hobbes/ thomas/h681/chapter33.html

Hoffman, B. (2002). Lessons of 9/11. *Rand Corporation Publication C-201.* Retrieved from http://www.rand.org/pubs/testimonies/2005/ CT201.pdf

Howe, R. H. (1979). Max Weber's elective affinities: Sociology within the bounds of pure reason. *American Journal of Sociology, 84,* 366–385.

Hubbard, T. K. (Ed.). (2003). *Homosexuality in Greece and Rome: A sourcebook of basic documents.* Berkeley: University of California Press.

Hughes, P. (1961). *A church in crisis: A history of the General Councils 325–1870.* Garden City, NY: Hanover House.

Hummer, R. A. Ellison, C. G., Rogers, R. G., Moulton, B. E., & Romero, R. R. (2004). Religious involvement and adult mortality in the United States: Review and perspective. *Southern Medical Journal, 97,* 1223–1230.

Hunter, J. D. (1991). *Culture wars: The struggle to define America.* New York: Basic Books.

Hutt, B. (2010). Russian Orthodox Church raises fears over pro-gay Protestant Churches. Retrieved from http://www.christiantoday.com/articles/russian.orthodox .church.raises.fears.over.progay.protestant.churches/26196.htm

Iannaccone, L. R. (1988). A formal model of church and sect. *American Journal of Sociology* 94(Supplement), S241–268.

Iannaccone, L. R. (1990). Religious practice: A human capital approach. *Journal for the Scientific Study of Religion, 29,* 297–314.

Iannaccone, L. R. (1991). The consequences of religious market structure: Adam Smith and the economics of religion. *Rationality and Society, 3,* 156–177.

Iannaccone, L. R. (1992). Sacrifice and stigma: Reducing free-riding in cults, communes, and other collectives. *Journal of Political Economy, 100,* 271–291.

Iannaccone, L. R. (1994). Why strict churches are strong. *American Journal of Sociology, 99,* 1180–1212.

Ibrahim, A. (n.d.). Jesus—Son of God! An explanation for Muslims. *Arabic Bible Outreach Ministry.* Retrieved from http://www.arabicbible .com/islam/son.htm

Indigenous. (n.d.). *Merriam-Webster Online Dictionary.* Retrieved from http://www .merriam-webster.com/dictionary/indigenous

Ismail, S. (2004). Being Muslim: Islam, Islamism and identity politics. *Government & Opposition, 39,* 614–631.

Jagannathan, S. (1984). *Hinduism: An introduction.* Bombay, India: Vakils, Feffer and Simons.

Jalal al-Din Rumi, M. (1995). *The essential Rumi* (C. Barks, Trans. with J. Moyne, A. A. Arberry, & R. Nicholson). San Francisco: Harper. Retrieved from http:// sologak1.blogspot.com/2009/12/what-jesus-runs-away-from-poem-by-rumi.html

James, W. (1960). *The varieties of religious experience.* New York: Random House. (Original work published 1902)

Jefferson, T. (1802). Jefferson's law of separation letter. Retrieved from http://www .usconstitution.net/jeffwall.html

Jeyapragasam, S. (Ed.). (1993). *Communalism: The crisis in India and the way out.* Madurai, India: The Valliamal Institute.

Johnson, G. (1976). The Hare Krishna in San Francisco. In C. Y. Glock & R. N. Bellah (Eds.), *The new religious consciousness* (pp. 31–51). Berkeley: University of California Press.

Johnson, J. T. (1981). *Just war tradition and the restraint of war.* Princeton, NJ: Princeton University Press.

Jones, R. K. (1978). Paradigm shifts and identity theory: Alternation as a form of identity management. In H. Mol (Ed.), *Identity and the sacred* (pp. 59–82). Oxford, UK: Basil Blackwell.

Jordan, D. K. (1985). *Gods, ghosts, and ancestors: Folk religion in a Taiwanese village* (2nd ed.). Taipei, Taiwan: Cave Books.

Joseph, P. (1993). *Peace politics: The U.S. between the old and new world orders.* Philadelphia: Temple University Press.

Juergensmeyer, M. (1993). *The new Cold War: Religious nationalism confronts the state.* Berkeley: University of California Press.

Juergensmeyer, M. (2003). *Terror in the mind of God: The global rise of religious violence* (3rd ed.). Berkeley: University of California Press.

Karcher, S. (1999). Jung, the Tao, and the classic of change. *Journal of Religion and Health, 38,* 287–304.

Kawagley, A. O. (1995). *A Yupiaq worldview: A pathway to ecology and spirit.* Prospect Heights, IL: Waveland Press.

King, Martin Luther, Jr. (1986). *Testament of hope: The essential writings of Martin Luther King, Jr.* (J. M. Washington, Ed.). San Francisco: Harper & Row.

Kinnamon, M., & Cope, B. E. (1997). *The ecumenical movement: An anthology of key texts and voices.* Geneva, Switzerland: WCC Publications/Grand Rapids, MI: William B. Eerdmans.

Kitagawa, J. M., & Reynolds, F. (1976). Theravada Buddhism in the twentieth century. In M. Demoulin & J. C. Maraldo (Eds.), *Buddhism in the modern world* (pp. 43–64). New York: Collier Macmillan.

Klein, F. (1951). *Souvenirs* (Vol. 4). *Americanism: A phantom heresy.* Cranford, NJ: Aquin Book Shop.

Kuefler, M. (2006). *The Boswell thesis: Essays on Christianity, social tolerance, and homosexuality.* Chicago: The University of Chicago Press.

Küng, H. (1991). *Global responsibility: In search of a new world ethic.* London: SCM Press/ New York: Continuum.

Kurtz, L. R. (1979). Freedom and domination: The Garden of Eden and the social order. *Social Forces, 58,* 443–464.

Kurtz, L. R. (1986). *The politics of heresy: The modernist crisis in Roman Catholicism.* Berkeley: University of California Press.

Kurtz, L. R. (1994). Review of *Culture wars: The struggle to define America*, by J. D. Hunter (1991). *American Journal of Sociology, 99*, 1124–1128.

Kurtz, L. R. (1997, Summer). Gandhi's legacies. *Journal of Peace and Gandhian Studies, 2.*

Kurtz, L. R. (1999). War. In R. Wuthnow (Ed.), *Encyclopedia of religion and politics.* Washington, DC: Congressional Quarterly. Retrieved from http://www.cqpress .com/context/articles/epr_War.html

Kurtz, L. R. (2001). Natural-born activist: Hildegard Goss-Mayr. *Peace Review, 13*(3), 457–461.

Kurtz, L. R. (2005a). Dalai Lama Darshan. *Ahimsa Nonviolence, 1,* 439–443.

Kurtz, L. R. (2005b, March/April). From heresies to holy wars: Toward a theory of religious conflict. *Ahimsa Nonviolence, 1,* 143–157.

Kurtz, L. R. (2005c, January/February). Rethinking power: Its sources and our options. *Ahimsa Nonviolence, 1,* 9–15.

Kurtz, L. R. (2006). Gandhi's paradox. *Manushi: An international journal of women and society, 152,* 19–26.

Kurtz, L. R., & Kurtz, M. R. (2005). Solving the Qur'anic paradox. *Ahimsa Nonviolence, 1,* 350–358.

Kurtz, L. R. (Ed.). (2008). *Encyclopedia of violence, peace, and conflict* (2nd ed., pp. 837–851). Amsterdam: Elsevier.

Kurtz, L. R. (2011a). Abdul Gaffar Khan's nonviolent jihad. *Peace Review, 23*(2), 245–251.

Kurtz, L. R. (2011b). From Egypt to Egypt: Three millennia of nonviolent resistance. *Ahimsa Nonviolence, 7.*

Kushner, H. S. (1981). *When bad things happen to good people.* New York: Schocken Books.

La Feber, W. (1993). *Inevitable revolutions: The United States in Central America.* New York: Norton.

LaDuke, W. (1999). *All our relations: Native struggles for land and life.* Cambridge, MA: South End Press.

Lame Deer, J. (Fire), & Erdoes, R. (1972). *Lame Deer, seeker of visions* New York: Simon & Schuster.

Lane, R., Jr. (1976). Catholic charismatic renewal. In C. Y. Glock & R. N. Bellah (Eds.), *The new religious consciousness* (pp. 162–179). Berkeley: University of California Press.

Lao-tzu. (1995). *Tao te Ching* (P. A. Merel, Trans.). Retrieved from http://www.china page.com/gnl.html

Latin American Bishops. (1979). *The Church in the present-day transformation of Latin America in the light of the council: Second General Conference of Latin American Bishops* (3rd ed.). Washington, DC: National Conference of Catholic Bishops.

Latus, M. A. (1983). Ideological PACs and political action. In R. C. Liebman & R. Wuthnow (Eds.), *The New Christian Right* (pp. 75–99). New York: Aldine.

Ledgerwood, J. L., & Un, K. (2003). Global concepts and local meaning: Human rights and Buddhism in Cambodia. *Journal of Human Rights, 2,* 531–549.

Lee, R. R. (1992). Religious practice as social exchange: An explanation of the empirical findings. *Sociological Analysis*, *53*, 1–35.

Leo XIII, P. (1899). Testem Benevolentiae. In *Leonis XIII Pontificus Maxima Acta*. (Vol. 19, pp. 5–20). Rome: Ex Typographia Vaticana.

Lernoux, P. (1982). *Cry of the people*. New York: Penguin.

Lofland, J. (1977). *Doomsday cult* (Enlarged ed.). New York: Irvington.

Lofland, J., & Stark, R. (1965). Becoming a world-saver: A theory of conversion to a deviant perspective. *American Sociological Review*, *30*, 863–875.

Loisy, A. (1976). *The Gospel and the Church* (C. Home, Trans.). Philadelphia: Fortress Press. (Original work published 1903)

Low, D. (1995). Contemporary reinvention of Chief Seattle: Variant texts of Chief Seattle's 1854 speech. *American Indian Quarterly*, *19*(3), 407–421.

Luckmann, T. (1967). *The invisible religion: The problem of religion in modern society*. New York: Macmillan.

Lyng, S. G., & Kurtz, L. R. (1985). Bureaucratic insurgency: The Vatican and the crisis of modernism. *Social Forces*, *63*, 901–922.

MacCannell, D. (1976). *The tourist: A new theory of the leisure class*. New York: Schocken Books.

Macfarlane, A. (1979). *The origins of English individualism: The family, property, and social transition*. New York: Cambridge University Press.

Malinowski, B. (1954). *Magic, science and religion and other essays*. Garden City, NY: Doubleday.

Mandela, N. (1994). *Long walk to freedom: The autobiography of Nelson Mandela*. Boston: Little, Brown.

Mann, G. S. (2006). The Sikh community. In M. Juergensmeyer (Ed.), *Oxford handbook of global religions* (pp. 41–50). Oxford: Oxford University Press.

Martin, B. (2001). *Technology for nonviolent struggle*. London: War Resisters' International. Retrieved from http://www.uow.edu.au/arts/sts/bmartin/pubs/01tnvs

Marty, M. E., & Appleby, R. S. (Eds.). (1993). *Fundamentalism and society: Reclaiming the sciences, the family, and education*. Chicago: The University of Chicago Press.

Marx, K. (1972a). Contribution to the critique of Hegel's philosophy of right. In R. C. Tucker (Ed.), *The Marx Engels reader* (pp. 11–23). New York: Norton. (Original work published 1843)

Marx, K. (1972b). The German ideology: Part I. In R. C. Tucker (Ed.), *The Marx Engels reader* (pp. 146–200). New York: Norton. (Original work published 1932)

Marx, K., & Engels, F. (1975). *The holy family, or critique of critical criticism*. Moscow: Progress. (Original work published 1844)

Mascaró, J. (Trans.). (1962). *Bhagavad Gita*. Harmondsworth, UK: Penguin.

Maududi, S. A. A. (1979). *Purdah and the status of women in Islam*. Lahore, Pakistan: Islamic Publications.

Mazrui, A. (Writer, Narrator). (1986). *The Africans: A triple heritage* (Documentary film). A coproduction of WETA-TV and BBC-TV.

McLuhan, M. (1960). *Explorations in communication* (E. S. Carpenter, Ed.). Boston: Beacon.

McNeill, J. (1976). *The church and the homosexual.* Kansas City, MO: Sheed and Ward.

Melton, J. G. (2003). *Encyclopedia of American religions* (7th ed.). Detroit, MI: Gale. Retrieved from http://www.lib.utexas.edu:2048/login?url=http://www.netLibrary.com/urlapi.asp?action=summary&v=1&bookid=73343

Merleman, R. (1984). *Making something of ourselves: On culture and politics in the United States.* Berkeley: University of California Press.

Mernissi, F. (1975). *Beyond the veil: A feminist interpretation of women's rights in Islam* (Rev. ed.) (M. J. Lakeland, Trans.). Reading, MA: Addison-Wesley.

Merton, T. (Ed.). (1965). Introduction: Gandhi and the one-eyed giant. In *Gandhi on nonviolence* (pp. 1–22). New York: New Directions.

Messer, J. (1976). Guru Mahara Ji and the Divine Light Mission. In C. Y. Glock & R. N. Bellah (Eds.), *The new religious consciousness* (pp. 52–72). Berkeley: University of California Press.

Michnik, A. (1992). The moral and spiritual origins of solidarity. In W. Brinton & A. Rinzler (Eds.), *Without force or lies* (pp. 239–250). San Francisco: Mercury House.

Miles, J. (1995). *God: A biography.* New York: Alfred Knopf.

Mills, C. W. (1959). *The sociological imagination.* London: Oxford University Press.

Mitchell-Green, B. L. (1994). American Indian PowWows in Utah, 1983–1994: A case study in oppositional culture. PhD dissertation, University of Texas at Austin.

Mizrach, S. E. (2008). Thunderbird and Heyoka, the Sacred Clown. *Heyoka Magazine.* Retrieved from http://www.heyokamagazine.com/heyoka_magazine.25.heyokas credclown.htm

Mol, H. (Ed.). (1978). *Identity and religion: International, cross-cultural approaches.* London: Sage.

Murray, A. (2004). Jack the Ripper, the dialectic of Enlightenment and the search for spiritual deliverance in White Chappell, Scarlet Tracings. *Critical Survey, 16,* 15–66.

Muslims in the American public square: Shifting political winds and fallout from 9/11, Afghanistan, and Iraq. (2004). Report from Project MAPS and Zogby International. Retrieved from http://www.projectmaps.com/AMP 2004report.pdf

Nanji, A. (1988). African religions. In K. M. Yates (Ed.), *The religious world* (pp. 32–51). New York: Macmillan.

Nash, J. (1992). *Loving nature: Ecological integrity and Christian responsibility.* Nashville, TN: Abingdon.

National Conference of Catholic Bishops. (1983). *The challenge of peace: God's promise and our response.* Washington, DC: United States Catholic Conference.

Neitz, M. J. (1987). *Charisma and community: A study of religious commitment within the charismatic renewal.* New Brunswick, NJ: Transaction.

Neitz, M. J. (1990). In Goddess we trust. In T. Robbins & D. Anthony (Eds.), *In Gods we trust* (pp. 353–372). New Brunswick, NJ: Transaction.

Neitz, M. J. (2003). Eve, Lilith, and the sociology of religion: Comments on Zald. *Journal for the Scientific Study of Religion, 42,* 5–8.

Nelson, J. (2008). Household altars in contemporary Japan: Rectifying Buddhist "ancestor worship" with home décor and consumer choice. *Japanese Journal of Religious Studies, 35*(2), 305–330.

Nelson, L.-E. (1981, September 19). Goldwater rips new Right's "threat" tactics. *Austin American Statesman,* p. 1.

Newport, F. (2011, May 20). For the first time, majority of Americans support gay marriage. *Gallup.* Retrieved from http://www.gallup.com/poll/147662/first-time-majority-americans-favor-legal-gay-marriage.aspx

Novak, M. (1986). *Will it liberate? Questions about liberation theology.* New York: Paulist Press.

O'Dea, T. F., & O'Dea Aviad, J. (1983). *The sociology of religion.* Englewood Cliffs, NJ: Prentice Hall.

O'Flaherty, W. (1973). *Asceticism and eroticism in the mythology of the Shiva.* London: Oxford University Press.

O'Flaherty, W. D. (Trans.). (1981). *The Rig Veda: An anthology of one hundred eight hymns.* Harmondsworth, UK: Penguin.

Ogburn, W. F. (1922). *Social change: With respect to culture and original nature.* New York: Huebsch.

Ojo, M. A. (2005). Religion and sexuality: Individuality, choice and sexual rights in Nigerian Christianity. *Lagos: Africa Regional Sexuality Resource Centre.* Retrieved from http://www.arsrc.org/downloads/uhsss/ojo.pdf

Olivelle, P. (2005). *Language, texts, and society: Explorations in ancient Indian culture and religion.* Florence, Italy: Florence University Press.

Overmyer, D. L. (2002). Kuan Yin: The development and transformation of a Chinese Goddess. *Journal of Religion, 82,* 418–424.

Paris, A. (1982). *Black Pentecostalism: Southern religion in an urban world.* Amherst: University of Massachusetts Press.

Parrinder, G. (1980). *Sex in the world's religions.* New York: Oxford University Press.

Parsons, T. (1960). *Structure and process in modern societies.* Glencoe, IL: Free Press.

Parsons, T. (1967). Christianity and modern industrial society. In E. A. Tiryakian (Ed.), *Sociological theory, values, and sociocultural change* (pp. 33–70). New York: Harper Torchbooks.

Parsons, T. (1969). *Politics and social structure.* New York: Free Press.

Patrick, T. (with T. Dulack). (1976). *Let our children go!* New York: E. P. Dutton.

Pattanaik, D. (2006). *Shiva to Shankara: Decoding the phallic symbol.* Mumbai, India: Indus Source Books.

Peers, G. (2001). *Subtle bodies: Representing angels in Byzantium.* Los Angeles: University of California Press.

Pew Research Center for the People and the Press. (2011, May 13). Most say homosexuality should be accepted by society. Retrieved from http://pewresearch.org/pubs/1994/poll-support-for-acceptance-of-homosexuality-gay-parenting-marriage

Pius X, P. (1907). *Pascendi Dominici Gregis: On the doctrines of the modernists.* Retrieved from http://www.papalencyclicals.net/Pius10/p10pasce.htm

Pius X, P. (1908). Lamentabili Sane Exitu. In P. Sabatier (Ed.), *Modernism* (pp. 217–230). London: Unwin.

Piven, F. F., & Cloward, R. A. (1971). *Regulating the poor: The functions of public welfare.* New York: Pantheon.

Plock, D. (1987). Methods for the time being. *Sociological Analysis, 47,* 43–51.

Poblete, R. (1970). The Church in Latin America: A historical survey. In H. A. Landsberger (Ed.), *The church and social change in Latin America* (pp. 39–52). Notre Dame, IN: University of Notre Dame Press.

Pohl, K.-H. (2003). Play-thing of the times: Critical review of the reception of Daoism in the West. *Journal of Chinese Philosophy, 30,* 469–486.

Pollner, M. (1989). Divine relations, social relations, and well-being. *Journal of Health and Social Behavior, 30,* 92–104.

Prabhavananda, S. (1963). *The spiritual heritage of India.* Hollywood, CA: Vedanta Press.

Prebish, C. (1978). Reflections on the transmission of Buddhism to America. In J. Needleman & G. Baker (Eds.), *Understanding the new religions* (pp. 153–172). New York: Seabury.

Ramsey, P. (1961). *War and the Christian conscience.* Durham, NC: Duke University Press.

Ramsey, P. (1968). *The just war: Force and political responsibility.* New York: Scribner.

Rapoport, D. C. (2008). Terrorism. In L. R. Kurtz (Ed.), *Encyclopedia of violence, peace, and conflict* (pp. 2087–2104). Amsterdam: Elsevier.

Redfield, R. (1957). *The primitive world and its transformations.* Ithaca, NY: Cornell University Press.

Regnerus, M. D. (2007). *Forbidden fruit: Sex & religion in the lives of American teenagers.* New York: Oxford University Press.

Religion: Year in review 2010. (2011). In *Britannica book of the year, 2011.* Retrieved from http://www.britannica.com/EBchecked/topic/1731588/religion-Year-In-Review-2010

Ritsema, R., & Karcher, S. (Trans.). (1995). *I Ching: The classic Chinese oracle of change.* New York: Barnes & Noble.

Robbins, T., & Anthony, D. (Eds.). (1990). *In Gods we trust: New patterns of religious pluralism in America* (2nd ed.). New Brunswick, NJ: Transaction.

Robertson, R. (1991). *Religion and global order.* New York: Paragon House.

Robertson, R. (1992a). The economization of religion? Reflections on the promise and limitations of the economic approach. *Social Compass, 39,* 147–157.

Robertson, R. (1992b). *Globalization: Social theory and global culture.* Newbury Park, CA: Sage.

Rochford, E. B., Jr. (1985). *Hare Krishna in America.* New Brunswick, NJ: Rutgers University Press.

Roof, W. C. (1978). *Community and commitment: Religious plausibility in a liberal Protestant church*. New York: Elsevier.

Rousseau, J.-J. (1901). *Social contract*. (C. M. Andrews, Ed.). New York: William H. Wise. (Original work published 1762)

Ruether, R. R. (Ed.). (1974). *Religion and sexism: Images of women in the Jewish and Christian traditions*. New York: Simon & Schuster.

Ruether, R. R. (1981). The feminist critique in religious studies. *Soundings, 64,* 388–402.

Ruether, R. R. (1992). *Gaia and God: An ecofeminist theology of Earth healing*. San Francisco: Harper.

Ruether, R. R. (1999). Feminism in world Christianity. In A. Sharma & K. K. Young (Eds.), *Feminism and World Religions* (pp. 214–247). Albany: State University of New York Press.

Ryan, W. (1976). *Blaming the victim* (Rev. ed.). New York: Vintage.

Safi, O. (Ed.). (2003). *Progressive Muslims on justice, gender and pluralism*. Oxford, UK: Oneworld Publications.

Said, E. (1978). *Orientalism*. New York: Pantheon.

Satha-Anand, C. (Qadar Muheideen). (1993). The nonviolence crescent: Eight theses on Muslim nonviolence actions. In G. Paige, C. Satha-Anand (Qader Muheideen), & S. Gilliat (Eds.), *Nonviolence in Islam* (pp. 7–26). Honolulu: University of Hawaii, Center for Global Nonviolence. Retrieved from http://www.globalnon violence.org/docs/islamnonviolence/chapter4.pdf

Satha-Anand, S. (2001). Buddhism on sexuality and enlightenment. In P. B. Jung, M. E. Hunt, & R. Balakrishan (Eds.), *Good sex: Feminist persectives from the world's religions* (pp. 113–124). New Brunswick, NJ: Rutgers University Press.

Savage, S. D. (2004). The formation of American attitudes toward Islam. Unpublished honors thesis, Sociology Department, University of Texas at Austin.

Scott, J. C. (1985). *Weapons of the weak: Everyday forms of peasant resistance*. New Haven, CT: Yale University Press.

Scroggs, R. (1972, March 15). Paul: Chauvinist or liberationist? *Christian Century,* pp. 307–309.

Segundo, J. L. (1976). *Liberation of theology*. Maryknoll, NY: Orbis.

Segundo, J. L. (1985). *Theology and the Church: A response to Cardinal Ratzinger and a warning to the whole Church*. Minneapolis, MN: Winston Press.

Seidel, A. K., & Strickmann, M. (n.d.). Taoism. *Encyclopædia Britannica Online*. Retrieved from http://search.eb.com/eb/article-9105866

Sengren, P. S. (1983). Female gender in Chinese religious symbols: Kuan Yin, Ma Tsu, and the "Eternal Mother." *Signs, 9*(1), 4–25.

Shaikh, S. (2003). Transforming feminisms: Islam, women, and gender justice. In O. Safi (Ed.), *Progressive Muslims on justice, gender and pluralism* (pp. 147–162). Oxford, UK: Oneworld Publications.

Shapiro, L., with Daniel Glick. (1993, August 23). Do you believe in magick? Witching hour: "Fort God" vs. born-again pagans. *Newsweek,* p. 32.

Sharma, A., & Young, K. K. (Eds.). (1999). *Feminism and world religions*. Albany: State University of New York Press.

Sharot, S. (1991). Judaism and the secularization debate. *Sociological Analysis*, 52, 255–275.

Sharp, G. (1973–1974). *The politics of nonviolent action* (Vol. 1), *Power and struggle* (Vol. 2), *Methods of nonviolent action* (Vol. 3), *Dynamics of nonviolent action*. Boston: Porter Sargent.

Sharp, G. (1979). *Gandhi as a political strategist: With essays on ethics and politics*. Boston: Porter Sargent.

Sharp, G. (1987). Nonviolence: Moral principle or political technique? In V. Grover (Ed.), *Gandhi and politics in India* (pp. 29–53). New Delhi, India: Deep and Deep Publications.

Sharp, G. (2005). *Waging nonviolent struggle*. Boston: Porter Sargent.

Sherkat, D. E. (2002). Sexuality and religious commitment in the United States: An empirical examination. *Journal for the Scientific Study of Religion, 41*(2), 313–323.

Sherkat, D. E., & Ellison, C. G. (1999). Recent developments and current controversies in the sociology of religion. *Annual Review of Sociology, 25*, 363–394.

Sherkat, D. E., & Wilson, J. (1995). Preferences, constraints, and choices in religious markets: An examination of religious switching and apostasy. *Social Forces, 73*, 993–1026.

Shih, F.-L. (2010). Women, religions, and feminisms. In B. S. Turner (Ed.), *The new Blackwell companion to the sociology of religion* (pp. 221–244). Chichester, UK: Wiley-Blackwell.

Signs of the times: In Assisi religious leaders call violence, religions incompatible. (2002, February 11). *America*, p. 4.

Shils, E. A. (1981). *Tradition*. Chicago: The University of Chicago Press.

Siker, J. (Ed.). (2007). *Homosexuality and religion: An encyclopedia*. Westport, CT: Greenwood Press.

Simmel, G. (1971). *On individuality and social forms* (D. N. Levine, Ed.). Chicago: The University of Chicago Press. (Original work published 1908)

Simmel, G. (1978). *The philosophy of money* (T. Bottomore & D. Frisby, Trans.). London: Routledge & Kegan Paul. (Original work published 1907)

Sinha, M., & Sinha, B. (1978). Ways of yoga and the mechanisms of sacralization. In H. Mol (Ed.), *Identity and the sacred* (pp. 133–150). Oxford, UK: Basil Blackwell.

Smelser, N. J. (1963). *Theory of collective behavior*. Glencoe, IL: Free Press.

Smith, H. (1965). *The religions of man*. New York: Harper & Row.

Smith, J. (1999). Transnational organizations. In L. R. Kurtz (Ed.), *Encyclopedia of violence, peace, and conflict* (Vol. 3, pp. 591–602). San Diego, CA: Academic Press.

Smith, M. (1952). The common theology of the Ancient Near East. *Journal of Biblical Literature, 71*, 135–147.

Smith, M. (1973). On the differences between the culture of Israel and the major cultures of the Ancient Near East. *Journal of Ancient Near Eastern Studies, 5*, 389–395.

Smithey, L., & Kurtz, L. R. (1999). We have bare hands: Nonviolent social movements in the Soviet bloc. In S. Zunes, L. R. Kurtz, & S. B. Asher (Eds.), *Nonviolent social movements* (pp. 96–124). Cambridge, UK: Blackwell.

Snow, D. A., Zurcher, L. A., & Ekland-Olson, S. (1980). Social networks and social movements: A microstructural approach to recruitment. *American Sociological Review, 45*, 787–801.

Spinoza, B. (1883). *Tractatus Theologico-Politicus* (Originally published Hamburg, Germany: Kunraht). Vol. 1, *The chief works of Benedict de Spinoza* (R. H. M. Elwes, Trans.). London: Bell. (Original work published 1670)

Stark, R. (1999). Secularization, R.I.P. *Sociology of Religion, 60*, 249–273.

Stark, R., & Bainbridge, W. S. (1980). Networks of faith: Interpersonal bonds and recruitment to cults and sects. *American Journal of Sociology, 85*, 1376–1395.

Stark, R., & Bainbridge, W. S. (1985). *The future of religion: Secularization, revival, and cult formation.* Berkeley: University of California Press.

Stolberg, S. G. (2011, February 16). Shy U.S. intellectual created playbook used in a revolution. *New York Times.* Retrieved from http://www.nytimes.com/2011/02/17/world/middleeast/17sharp.html

Sutton, M. A. (2005). Clutching to "Christian" America: Aimee Semple McPherson, the Great Depression, and the origins of Pentecostal political activism. *Journal of Policy History, 17*, 308–338.

Swidler, L. J. (1990). *After the absolute: The dialogical future of religious reflection.* Minneapolis, MN: Fortress Press.

Taylor, R. J., & Chatters, L. M. (1988). Church members as a source of informal social support. *Review of Religious Research, 30*, 114–125.

Thomas, W. I. (1966). *W. I. Thomas on social organization and social personality.* Chicago: The University of Chicago Press.

Tillich, P. (1967). *Systematic theology.* Chicago: The University of Chicago Press.

Tobin, T. J. (2001). The construction of the codex in classic- and postclassic-period Maya civilization. Retrieved from http://www.mathcs.duq.edu/~tobin/maya/

Tocqueville, A. de. (1945). *Democracy in America* (H. Reeve, Trans.). New York: Alfred Knopf. (Original work published 1862)

Tolstoy, L. (1987). *Writings on civil disobedience and nonviolence* (A. Maud, Trans.). Philadelphia: New Society Publishers.

Tönnies, F. (1957). *Community and society* (C. P. Loomis, Trans.). East Lansing: Michigan State University Press. (Original work published 1887)

Tracy, J. (1996). *Direct action: Radical pacifism from the Union Eight to the Chicago Seven.* Chicago: The University of Chicago Press.

Turner, B. S. (2010). Mapping the sociology of religion. In B. S. Turner (Ed.), *The new Blackwell companion to the sociology of religion* (pp. 1–30). Chichester, UK: Wiley-Blackwell.

Turner, V. (1967). *The forest of symbols aspects of Ndembu ritual.* Ithaca, NY: Cornell University Press.

Turpin, J., & Kurtz, L. R. (Eds.). (1997). *The web of violence.* Urbana: University of Illinois Press.

Tylor, E. B. (1871). *Primitive culture.* London: Murray.

UNICEF. (2010). *2010 annual report.* Retrieved from http://www.unicefusa.org/news/publications/annual-report/Unicef_2010_AnnReport_LR.pdf

Varisco, D. M. (2005). *Islam obscured: The rhetoric of anthropological representation*. New York: Palgrave Macmillan.

Vatican Council II. (1965). *Gaudium et Spes: Pastoral constitution on the Church in the modern world*. Retrieved from http://www.vatican.va/ archive/hist_councils/ ii_vatican_council/documents/vat-ii_cons_19651207_ gaudium-et-spes_en.html

Wach, J. (1944). *The sociology of religion*. Chicago: The University of Chicago Press.

Wadud, A. (1999). *Qur'an and woman*. Kuala Lumpur, Malaysia: Fajar Bakti.

Wald, K. D., Owen, D. E., & Hill, S. S. (1989). Habits of the mind? The problem of authority in the new Christian Right. In T. G. Jelen (Ed.), *Religion and political behavior in the United States* (pp. 93–108). New York: Praeger.

Wallace, A. F. C. (1966). *Religion: An anthropological view*. New York: Random House.

Wallerstein, I. W. (1984). *The politics of the world-economy. The states, the movements and the civilizations*. Cambridge, UK: Cambridge University Press.

Wallis, R. (1976). *The road to total freedom*. New York: Columbia University Press.

Warner, M. (1976). *Alone of all her sex*. New York: Alfred Knopf.

Warner, R. S. (1993). Work in progress toward a new paradigm for the sociological study of religion in the United States. *American Journal of Sociology*, *98*, 1044–1093.

Warry, W. (2007). *Ending denial: Understanding Aboriginal issues*. Toronto, Canada: Broadview Press.

Weber, M. (1946). *From Max Weber: Essays in sociology* (H. H. Gerth & C. W. Milles, Trans. & Eds.). New York: Oxford University Press. (Original work published 1922–1923)

Weber, M. (1947). *Gesammelte Aufsätze zur Religionssoziologie*. Tübingen, Germany: Mohr.

Weber, M. (1958). *The Protestant ethic and the spirit of capitalism*. New York: Scribner. (Original work published 1904)

Weber, M. (1963). *The sociology of religion*. (E. Fischoff, Trans.). Boston: Beacon Press.

Weber, M. (1968). *Economy and society* (3 vols.). Berkeley: University of California Press.

Wei-ming, T. (n.d.). Confucianism. *Encyclopædia Britannica Online*. Retrieved from http://search.eb.com.content.lib.utexas.edu:2048/eb/article-9109629>

Weightman, S. (1984). Hinduism. In J. R. Hinnells (Ed.), *A handbook of living religions* (pp. 191–236). Harmondsworth, UK: Penguin.

Welch, H. (1972). *Buddhism under Mao*. Cambridge, MA: Harvard University Press.

White, A. D. (1896–1897). *History of the warfare of science with theology in Christendom* (2 vols.). New York: Appleton. Retrieved from http://www.cscs .umich.edu/~crshalizi/White/creation/final-effort.html

Wilkinson, L. (1992, July 20). Earth Summit: Searching for a spiritual foundation. *Christianity Today*, *36*, 48.

Willaime, J.-P. (2004). The cultural turn in the sociology of religion in France. *Sociology of Religion*, *65*, 373–389.

Williams, J. A. (Ed.). (1962). *Islam. Great Religions of Modern Man Series*. New York: George Braziller.

Williams, M. D. (1974). *Community in a Black Pentecostal Church: An anthropological study*. Pittsburgh, PA: University of Pittsburgh Press.

Wilmer, F. (2008). Indigenous peoples response to conquest. In L. R. Kurtz (Ed.), *Encyclopedia of violence, peace, and conflict* (pp. 987–1000). Amsterdam: Elsevier.

Wilson, A. (Ed.). (1991). Love your enemies. In B. Schuman, United Communities of Spirit (Archivist), *World scripture: A comparative anthology of sacred texts*. Retrieved from http://origin.org/ucs/ws/theme144.cfm

Wilson, B. (1959). An analysis of sect development. *American Sociological Review*, *24*, 3–15.

Wilson, D. B. (1992, Spring). *Mis Misa*: The power within *Akoo-Yet* that protects the world. *News From Native California*. Retrieved from https://eee.uci.edu/clients/tcthorne/Socec15/mismisa.pdf

Wolf, M. (1968). *The house of Lim: A study of a Chinese farm family*. New York: Appleton-Century-Crofts.

Wolf, M. (1972). *Women and family in rural Taiwan*. Stanford, CA: Stanford University Press.

Woodward, K. L., & Gates, D. (1982, August 30). Giving the devil his due. *Newsweek*, pp. 71–74.

Wuthnow, R., Davison Hunter, J., Bergesen, A., & Kurzweil, E. (Eds.). (1984). *Cultural analysis: The work of Peter Berger, Mary Douglas, Michel Foucault, and Jürgen Habermas*. London: Routledge & Kegan Paul.

Wuthnow, R. (1976). *The consciousness reformation*. Berkeley: University of California Press.

Wuthnow, R. (1987). *Meaning and moral order: Explorations in cultural analysis*. Berkeley: University of California Press.

Zedeño, M. N., Austin, D. E., & Stoffle, R. W. (1997). Landmark and landscape: A contextual approach to the management of American Indian resources. *Culture and Agriculture*, *19*, 123–129.

Zeidan, D. (2003). *The resurgence of religion: A comparative study of selected themes in Christian and Islamic Fundamentalist discourses*. Leiden, The Netherlands: Brill.

Zolbrod, P. G. (1984). *Dine' Bahane': The Navajo creation story*. Albuquerque: University of New Mexico Press.

Zunes, S., Kurtz, L. R., & Asher, S. B. (Eds.). (1999). *Nonviolent social movements: A geographical perspective*. Oxford, UK: Blackwell.

Zunes, S. (1999a). The origins of people power in the Philippines. In S. Zunes, L. R. Kurtz, & S. B. Asher (Eds.), *Nonviolent social movements: A geographical perspective* (pp. 129–157). Oxford, UK: Blackwell.

Zunes, S. (1999b). The role of nonviolence in the downfall of apartheid. In S. Zunes, L. R. Kurtz, & S. B. Asher (Eds.), *Nonviolent social movements: A geographical perspective* (pp. 203–230). Oxford, UK: Blackwell.

Glossary/Index

Aaron, 172, 320n4
Abduh, Muhammad, 235
Abhedânanda, Swâmi, 59
Abraham, sacrifice of, 35
Abrahamic tradition, Judaism, Islam, and Christianity tracing their roots to Abraham and Sarah, 106
  collective worship, 36
  Islam, 106, 111
  Judaism, 90, 92–93, 106
Acts of devotion, 34–35
A.D. (Anno Domini), 319n3
Adam and Eve, in the Garden of Eden, 93–95, 175–176, 183, 186
  See also **Genesis**
Adultery, 184, 187, 230–231
Advani, L. K., 192
Afghani, Jamal al din, 235–236
Afghanistan:
  King Amanullah, 238
  U.S. conflict with, 237, 238
Africa:
  indigenous religions, 136, 141, 142–143
  multiculturalism, 241, 242
  politics and religion, 195, 198
  religious affiliations (2010), 46*figure*
  See also specific country
African Americans:
  civil rights movement, 295–296, 298
  slavery, 298
African National Congress, 231
Agnositcs, 46*figure*
Agrarian Reform Law (Guatemala, 1952), 126

Ahimsa, 67, 286, 288
Ahmed, Akbar, 107
Ajatasattu, King, 146
Ali, Sayyid Amir, 236
Alito, Samuel, 231
Altars, China, 140–141
Alternative religious movements.
  See Cults; **New religious movement** (NRM); *specific movement*
Amanullah, King of Afghanistan, 238
America, 295
American Church Review, The, 211
American Indian Religious Freedom Act (United States, 1978), 126
American Indians:
  Chief Seattle, 143–144
  Hopi, 129
  indigeneous religions, 126, 129–135, 142, 143–144
  Lakota, 138, 139–140, 142
  Navajo, 129–131, 137
  Paiute, 137
  Pitt River Tribe, 131–135
  Pow Wow, 137–139, 320n1
  Pueblo, 129
  ritual, 137–140
  Sundance, 137
  sweat lodge, 137
  vision quest, 139–140
  See also Indigenous religions
Americanism, 213
American Muslim Alliance, 297
American Muslim Council, 297
American Psychiatric Association, 189